WileyPLUS

WileyPLUS is a research-based online environment for effective teaching and learning.

WileyPLUS builds students' confidence because it takes the guesswork out of studying by providing students with a clear roadmap:

- **what to do**
- **how to do it**
- **if they did it right**

It offers interactive resources along with a complete digital textbook that help students learn more. With *WileyPLUS*, students take more initiative so you'll have greater impact on their achievement in the classroom and beyond.

WileyPLUS

ALL THE HELP, RESOURCES, AND PERSONAL SUPPORT YOU AND YOUR STUDENTS NEED!

www.wileyplus.com/resources

1st DAY OF CLASS ... AND BEYOND!

2-Minute Tutorials and all of the resources you and your students need to get started

WileyPLUS

Student Partner Program

Student support from an experienced student user

Wiley Faculty Netwo

Collaborate with your colleagues, find a mentor, attend virtual and live events, and view resources
www.WhereFacultyConnect.com

WileyPLUS

Quick Start

Pre-loaded, ready-to-use assignments and presentations created by subject matter experts

Technical Support 24/7
FAQs, online chat, and phone support
www.wileyplus.com/support

Your *WileyPLUS* Account Manager, providing personal training and support

QUANTITATIVE REASONING

QUANTITATIVE REASONING

TOOLS FOR TODAY'S INFORMED CITIZEN

SECOND EDITION

ALICIA SEVILLA

KAY SOMERS

Vice President & Publisher	Laurie Rosatone
Project Editor	Jennifer Brady
Content Editor	Beth Pearson
Marketing Manager	Kimberly Kanakes
Marketing Assistant	Patrick Flatley
Media Assistant	Courtney Welsh
Production Editor	Kenneth Santor
Designer	Maureen Eide
Product Designer	Thomas Kulesa
Editorial Operations Manager	Melissa Edwards
Cover Photo	Mitchell Funk/Photographer's Choice/Getty Images, Inc.

Founded in 1807, John Wiley & Sons, Inc. has been a valued source of knowledge and understanding for more than 200 years, helping people around the world meet their needs and fulfill their aspirations. Our company is built on a foundation of principles that include responsibility to the communities we serve and where we live and work. In 2008, we launched a Corporate Citizenship Initiative, a global effort to address the environmental, social, economic, and ethical challenges we face in our business. Among the issues we are addressing are carbon impact, paper specifications and procurement, ethical conduct within our business and among our vendors, and community and charitable support. For more information, please visit our website: www.wiley.com/go/citizenship.

Printed in the United States of America.

ISBN-13 978-0-470-59271-7
ISBN-13 binder-ready version 978-1-118-40683-0

10 9 8 7 6 5 4 3 2 1

CONTENTS

Section I: Numerical Reasoning

Section II: Logical Reasoning

Section III: Statistical Reasoning

Appendix

Selected answers to Explorations and graphing calculator versions of the Excel Activities are included on the text website at www.wiley.com/college/sevilla.

ANNOTATED CONTENTS

Section I: Numerical Reasoning

This section of the text explores topics related to pictorial representations of data and relationships and using data, numbers, and functions to help solve problems.

Topic 1: Organizing Information Pictorially Using Charts and Graphs   2

Graphs allow us to absorb quantitative information quickly and accurately and have become an important part of daily communication. Today's informed citizen must have not only the ability to read and understand graphs, but also the ability to create them. In this topic, we analyze different types of graphs routinely used to convey information and tell a story. We discuss examples of bar graphs, pie charts, histograms, and stemplots (stem-and-leaf graphs), and which graphs are appropriate to use when working with categorical variables and which are best to represent quantitative variables. Population data, AIDS data, and postal data are used to create and analyze graphs.

Topic 2: Bivariate Data   52

This topic investigates scatterplots of data sets that involve two quantitative variables: an explanatory or independent variable and a response or dependent variable. We discuss the diference between a relation and a function and also explore the importance of representing relationships in multiple ways—using tables of numerical data, symbols, graphs or charts, and words. Finally, we look at directly proportional relationships. We use SAT grades by state, historical data, and sports data to explore these ideas.

Topic 3: Graphs of Functions   82

In this topic, we discuss characteristics of graphs of functions and how to interpret these different characteristics in the context of the function variables. This topic includes the vertical line test for graphs of functions and discussion of increasing and decreasing functions, concavity, and absolute and relative maximum and minimum. In addition, we introduce the concept of average rate of change and the relationship between concavity and variations in the rate of change of a function. We explore these ideas using federal budgets, number of students enrolled in private and public schools, sports salaries, and poverty data.

In this topic, we examine situations where the response (dependent) variable depends on several explanatory (independent) variables. We continue to emphasize different modes of communicating relationships between variables—words, tables, symbols, and graphs. We also investigate how a response variable behaves when all but one independent variable is held constant. Contexts for these investigations include basal metabolic rate, body mass index, wind chill equivalent temperature, and blood alcohol level.

We review the concept of proportional relationships and discuss linear and piecewise linear functions and their equations. Examples include basal metabolic rate, medicine dosage, income tax, and the price of letter postage. We explore the significance of the y-intercept and the slope of a linear function using physical exercise–related data. We also discuss the concept of inversely proportional functions using data on the average speed of winners of the Daytona 500 auto race for several years.

This topic begins with a discussion of how to recognize a linear relationship between two quantitative variables. We explore how and when to fit a least-squares regression line to data that are not exactly linear. Exponential relationships and exponential growth are compared to linear growth, and we look at an example in which growth is even more rapid than exponential growth. We study these ideas using population data from around the world, education data, and financial data such as salaries and federal debt over time.

This topic introduces the common logarithm and the exponential function, base 10, using the decibel scale for sound and the Richter scale for earthquake intensity. We explore the relationship between the graph of a function and the graph of its inverse function and review basic properties of logarithms. We also discuss the use of scientific notation to make estimates when very large or very small quantities are involved; we use financial data and earth science data to explore these ideas.

In this topic, we look at indexes, such as the Consumer Price Index and the Consumer Confidence Index, to investigate trends over time. We also examine the Fog Index, which assesses the reading difficulty of a passage of text, and the Human Development Index. Rating systems that are set up to compare people or places are examined. These indexes are used to help understand trends in minimum wage data, the rise in sporting-event costs, and the readability of newspaper editorials, while rating systems are applied to cities and colleges.

In this topic, we introduce terminology associated with personal financial management and discuss simple and compound interest. We analyze and use formulas for computing interest received on savings and calculate costs and payments associated with various types of loans and annuities. In particular, we consider examples that relate to saving for a down payment on a home and paying credit card debts and student loans.

In this topic, we use Pólya's four-step problem-solving process and analyze basic techniques for problem solving. We identify where used each techniques was in the previous topics. We apply these techniques to revisit the problem of finding a person's body mass index, and to analyze problems in financial and educational contexts.

Section II: Logical Reasoning

This section investigates inductive and deductive reasoning approaches and applies these ideas to decision making, apportionment issues, and problem-solving techniques.

This topic focuses on methods for helping us make decisions in which information related to the decision is known. (These types of decisions are called decisions under certainty.) We discuss criteria that impact various decisions as well as two methods for making decisions: the cutoff screening and weighted sum methods. These methods are applied to decisions such as purchasing a digital camera, deciding which job to accept, or determining which candidate to elect.

This topic starts with a discussion of the difference between deductive and inductive reasoning. We then explore various forms of inductive reasoning—prediction, generalization, causal inference, and analogy—using a variety of examples taken from news articles and other readings. We present a variety of scenarios, including medical and educational studies.

In this topic, we focus on the basic elements of deductive reasoning. We discuss statements and their negations as well as how to formulate and analyze compound statements and their negations. The contrapositive and converse of conditional statements and quantified statements are examined, leading up to deductive arguments. We investigate these ideas using published speeches, news articles, and advertising statements.

In this topic, we discuss different methods of apportionment that the House of Representatives has used or that were once proposed. We investigate two quota methods: Hamilton's method and Lowndes' method. We also discuss the following divisor methods: Jefferson's method, Adam's method, Webster's method, and the Huntington-Hill method, which is the method used currently to determine the number of representatives for each state based on the population numbers determined by the most recent census. We explore all these methods using data from the 2010 census.

Topic 15: More on Problem Solving   422

This topic builds on Topic 10 and discusses additional problem-solving techniques. These techniques are linked to deductive and inductive reasoning. We identify how these techniques were used in previous topics and apply them to solve problems such as a salary negotiation, the best payment option to select after winning the lottery, and designing a year-long community service project.

Section III: Statistical Reasoning

This section introduces basic concepts of probability and statistics and applies them to the study of sampling and surveys and making decisions that involve uncertain data.

Topic 16: Averages and Five-Number Summary   436

In this topic, we investigate several measures of center and spread. We discuss the concepts of mean, median, mode, quartiles, range, and interquartile range. We also calculate the five-number summary and graph the boxplot for data sets. In exploring these concepts we use data sets on number of waste sites by state, state governors' salaries, and calorie content in popular brands of brownies and ice cream bars, among others.

Topic 17: Standard Deviation, z-Score, and Normal Distributions   462

In this topic, we investigate the standard deviation as a measure of variability within a data set. We look at normal curves and compare normal curves with different means and standard deviations. We consider standardized z-scores as a way to compare values obtained from data sets with different units of measure. We apply these ideas by looking at various data sets such as ages of presidents and chief justices, SAT scored, and number of home runs.

Topic 18: Basics of Probability   492

We introduce basic probability concepts—random process, sample space, outcomes and events, relative frequency, probability of an event—as well as basic probability rules. We explore these ideas using coins, dice, and playing cards. We also use data on single vehicle crashes by size of vehicle and people in the U.S. without health insurance by region.

Topic 19: Conditional Probability and Tables   522

We explore how to use a two-way table to represent data in which each individual in the data set is characterized in two different ways. We discuss how to analyze this data and look for relationships using conditional probabilities. We examine how identifying independent events helps us discover additional relationships. These ideas are investigated using Olympic gold medal data, vehicle crash data, and congressional voting records, and analyzing diagnostic tests.

Basic components of observational studies and experiments are investigated in this topic. We examine how to understand the results of a study or experiment by identifying the sample, the population, and the relevant variables. We discuss various sampling methods and explore sampling variability and biases. We also discuss one method to elicit honest responses to sensitive survey questions. These ideas are explored through studies reported in newspapers, periodicals, and on the Internet.

In this topic, we discuss the concept of expected value (or mean or average value) of a probability distribution and how to use it to make decisions that involve uncertain information. We also examine several other approaches to decision making when uncertainty is present. To explore these ideas, we analyze decision-making problems regarding accepting a magazine raffle offer, choosing a health insurance alternative, and deciding on an investment approach.

Appendix

PREFACE

Empowering students to use quantitative information to make responsible financial, environmental, and health-related decisions in their personal and work lives is at the core of *Quantitative Reasoning: Tools for Today's Informed Citizen*. This book's main objective is to help students become better critical thinkers by engaging them as active learners of the quantitative methods of analysis discussed in the text. Through numerous examples, explorations, and activities featuring real data, students develop the skills necessary to

- Identify, analyze, and solve real-world problems that involve quantitative information

- Reason quantitatively and make numerical arguments

- Interpret and communicate the results of quantitative analyses

- Use technology and Internet resources efectively and build skills in working with data

- Develop and improve "numerical intuition" and confidence in the ability to engage in quantitative thinking

Quantitative Reasoning: Tools for Today's Informed Citizen is organized into three sections. Section I, "Numerical Reasoning," provides a foundation for quantitative reasoning and communication. It includes topics related to using numbers, functions, and graphs and an introduction to problem solving. Section II, "Logical Reasoning," addresses different types of reasoning and applications and concludes with a further discussion of problem-solving techniques. Section III, "Statistical Reasoning," includes investigations of descriptive statistics, probability, and sampling. It concludes with a discussion of decision making when uncertainty is involved.

Throughout the text, students use a variety of methods of analysis: inductive and deductive reasoning; tabular, symbolic, verbal, and graphical forms of functions and relations; graphs and pictorial representations of data; interpretations of probabilistic data; surveys and statistical studies. Nearly all of the examples, exercises, and activities in the text use real data or draw on real-life situations to demonstrate the significance of quantitative reasoning in students' daily lives and to illustrate misapplications of mathematics and quantitative reasoning. The use of real data highlights the relevance and practicality of the material. In this way, students gain a better understanding of the concepts. This text shows over and over again how useful and relevant mathematics is for understanding the world we live in and for making informed decisions based on the proper use of quantitative information and reasoning.

Key Features

Each of the three sections of the text consists of **topics** that introduce mathematical concepts and terminology necessary to explore different approaches to solving the problems presented. These topics can be investigated typically in one or two class periods and include the following features.

Objectives are outlined at the beginning of each topic so that students can preview the ideas they will explore.

TOPIC 1
OBJECTIVES

After completing this topic, you will be able to:

- Distinguish between quantitative and categorical variables.

- Draw bar graphs and pie charts, and interpret them in the context of the data they represent.

- Compute percentage of the whole and percent change.

- Interpret stemplots and histograms and group data to create them.

- Decide when each type of graph is appropriate.

Objectives preview the ideas students will explore in each Topic.

Examples are integral to illustrating the concepts and tools introduced and showing how they can be applied to real-life situations using actual data. Worked-out solutions to the examples explain how to solve a problem and what the solution means in the context of the problem.

Examples illustrate concepts and tools introduced.

Example 6.6

In the year 2000, a newspaper article pointed out that India's current population of approximately 1 billion may double to 2 billion in just another 100 years. The article predicted that India's population by 2050 will be 1.6 billion. Assume that the population growth of India is exponential and that the prediction of India's population in 2050 is correct; use this information to determine what the population of India will be at the end of this century, in 2100.

Solution

Using the information given, we assume that the population of India grows from 1.0 billion to 1.6 billion in the 50-year time period from 2000 to 2050. Thus the growth factor is $\frac{\text{population in 2050}}{\text{popuation in 2000}} = \frac{1.6}{1.0} = 1.6$. Therefore, assuming that this growth will continue, (population in 2100) $= 1.6 \times$ (population in 2050) $= 1.6 \times 1.6 = 2.56$ billion. If the growth continues at this rate, India will add more than 1 billion people in the next 100 years.

A **summary** of the topic reviews the concepts that are critical in helping students master the objectives introduced at the beginning of the topic.

Summary

A Summary reviews main concepts for each topic.

In this topic, we studied two methods we can use to help us make decisions that involve "certain" information—that is, information that is known or that we assume to be known. These methods are the cutoff screening method and the weighted sum method. We discussed various criteria that might be important to consider for making different decisions, and we also investigated how to create consistent ratings of possible choices, relative to each criterion. Finally, we looked at the link between the weighted sum method and rankings of various alternatives.

Explorations give students a chance to apply their understanding of the main concepts to additional real-life situations. These explorations allow students to broaden their problem-solving skills and their understanding of the mathematics involved, thus enabling them to see new contexts for the applications discussed.

Explorations

Explorations enhance students' understanding of the main concepts and broaden problem-solving skills.

1. Three students each have $1,000 to invest from their summer jobs. Armen invests his money in an account that earns simple interest at an annual interest rate of 5%. Barok invests his money in an account that earns 4.9% interest per year compounded annually. Carrie invests her money in an account that earns 4.8% interest per year compounded monthly. Find how much money each student has in his or her account after:

   a. 2 years

   b. 10 years

   c. 30 years

2. Here's an old story: A man walks into a New York City bank and asks for a $5,000 loan, offering his Ferrari, worth $250,000 as collateral. He tells the loan officer that he needs the money for two weeks for an important venture. The loan officer, having the car as security and after checking references, gives the man the money he requested, with a signed agreement that he will pay the money plus $45 in interest when he returns in two weeks. The bank officer takes the car keys and the car is parked in the bank's underground lot. The man returns in exactly two weeks, pays the loan and interest, and reclaims his car. The bank officer asked him why he was willing to pay such a high interest rate. His reply: Where else can I safely park my car for two weeks in New York City for only $45?

   a. What annual interest rate did the man pay?

   b. How much would the man need to repay at the end of two months if he borrowed $5,000 with the same rules and same annual interest rate?

Activities engage students as critical thinkers by investigating real-life situations using the concepts learned in greater depth. Each of the activities is designed so that the bulk of the activity can be completed by most students in a typical class period, with some time remaining for a review of background material and wrap-up. Activities can also be assigned as out-of-class work if desired.

Activities complement the material in topics and encourage critical thinking.

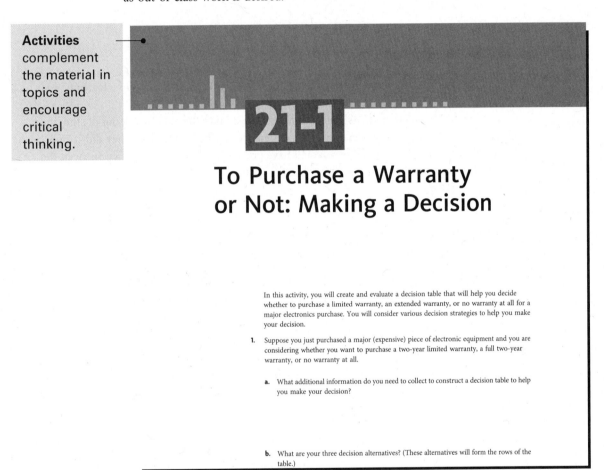

21-1

To Purchase a Warranty or Not: Making a Decision

In this activity, you will create and evaluate a decision table that will help you decide whether to purchase a limited warranty, an extended warranty, or no warranty at all for a major electronics purchase. You will consider various decision strategies to help you make your decision.

1. Suppose you just purchased a major (expensive) piece of electronic equipment and you are considering whether you want to purchase a two-year limited warranty, a full two-year warranty, or no warranty at all.

   a. What additional information do you need to collect to construct a decision table to help you make your decision?

   b. What are your three decision alternatives? (These alternatives will form the rows of the table.)

Technology

Students are encouraged to use technology in a fundamental way to help visualize data and to facilitate calculations. Because most students have access to Microsoft Excel and will use it in their work environment, the activities in the text feature Excel. Data sets to accompany the Excel Activities are included on the text website which can be accessed at **www.wiley.com/college/sevilla.** For instructors who prefer that students use a graphing calculator, activities that use the T1-83 or T1-84 Plus graphing calculator, instead of Excel, are available on the text website. Instructions for using Microsoft Excel and graphing calculator technology are integrated into the activities carefully so that students can concentrate on ideas rather than computational details when investigating problems. An index of Excel Commands by Activity and Graphing Calculator Commands by Activity are also provided in the back of the text and the text website, respectively, as a quick reference tool for locating specific technology functions.

WileyPLUS

WileyPLUS is an innovative, research-based online environment for effective teaching and learning.

WileyPLUS builds students' confidence because it takes the guess-work out of studying by providing students with a clear roadmap: what to do, how to do it, if they did it right. Students will take more initiative so you'll have greater impact on their achievement in the classroom and beyond. For more information, go to www.wileyplus.com.

TO THE STUDENT

Welcome to an experience that will help you better understand mathematical concepts and tools and the quantitative information you need to make informed decisions in our changing world. This book was developed with you in mind—to help you understand and solve problems that are relevant to your family, your community, your workplace, your country, and your world. Because all of the concepts introduced in this book are anchored in real-world situations, you will never need to ask, "What is this good for?"

The key to your success in this course is active involvement when reading and solving problems. You should read and work on examples for a particular topic before the topic is visited in class; by doing so, you will be ready with questions for your instructor and for participation in class discussion. Be an active reader, and read with pencil and paper close by to check computations, record questions, and provide alternative answers. You will find that some of the questions in this book have more than one good (or "right") answer. Just as you become proficient at a sport by practicing, you become skilled at problem solving by solving a lot of problems. You might even discover that problem solving can be just as much fun as tennis or golf or basketball!

The **examples** in this text are essential in demonstrating how mathematical concepts and tools are used in the real world. Read each example and try to come up with your own solution first. Then work through the given solution carefully, following each step in detail. Analyze the graphs and charts and think about different ways to visualize information and problems. Use technology tools, such as Microsoft Excel or a graphing calculator, to help create additional graphs. Ask yourself whether the solution makes sense; that is, ask whether or not the solution is reasonable in the context of the problem. Discuss the examples and your findings with your classmates and your instructor. Think about related questions you might ask to better understand the topic and other situations in which you might be able to apply what you have learned.

Jump in and work on the **explorations** at the end of each topic. The issues and problems covered in the explorations build on those in the topic and include additional, related concepts. In the explorations, you will bring ideas and tools together to investigate new examples and practice your problem-solving and analysis skills. Your explanations will help you better understand the concepts discussed in the topic and make connections with those covered in previous topics.

The **activities** build on the readings, examples, and problems presented in the explorations. The activities generally focus on an interesting application of the material in the topic in more depth. By completing the activities, you will better understand the topic. Because detailed instructions on how to use technology to help create graphs and perform calculations are included in the activities, you will also gain facility in using this technology.

You will gain experience and learn new techniques and different ways of looking at problems as you solve interesting and relevant problems in which quantitative information and techniques play an important role.

Student Support Materials

The text website at www.wiley.com/college/sevilla contains the following resources to assist you in your studies:

- Excel data bank to accompany Excel Activities
- Selected answers to Explorations
- Graphing Calculator Activities
- Graphing Calculator Commands by Activity

TO THE INSTRUCTOR

This textbook grew out of notes written for a course we have been teaching at Moravian College since the year 2000. This course, developed with a grant from the National Science Foundation (Grant No. 9950229), has been very successful with students who liked mathematics before coming to college and feel comfortable with quantitative analysis, as well as with students who declare on the first day of class, "I cannot do math."

In developing this course and writing the text, we were especially conscious of the need to reach all types of students, especially this last group. Our philosophy in developing this course was to help students see the need for what they are learning; thus the emphasis of this text is to teach applications using real data. As instructors, our objective is to facilitate the learning process and support students as they learn, and with this objective in mind, we developed an activity-based course that combines class instruction with in-class student activities. This approach engages the student as an active learner and makes the classroom atmosphere much more enjoyable for both students and instructors.

To prepare students for the class in which they will be investigating a topic, we suggest assigning a section of reading from the book (sometimes a full topic, sometimes a portion of a topic) to be completed along with some **explorations** from the same topic. We use time at the beginning of the class to discuss the assigned reading material and explorations. Our students then spend the majority of the class period working on the carefully constructed **activities** for each topic, which build on and complement the assigned reading. For courses with a shorter class period, activities can also be completed during two class periods or assigned as homework. Some activities require students to work together, and we always encourage students to discuss their progress and results with one another.

Because we want students to feel comfortable with technology, and not threatened by it, we provide carefully written instructions for using Excel, or the graphing calculator. Generally, we introduce only the necessary functions in each activity so as to not overwhelm the students with the technology and cause them to lose track of the main objective of thinking critically about the activity.

We strongly recommend an activity-based approach, but this text can also be used in a more traditional setting. For example, an instructor can give an interactive lecture on a particular topic using some of the examples, explorations, and/or activities to illustrate the concepts. Additional explorations and activities can be assigned to be completed out-of-class.

It is essential that students reach their own conclusions and show their thinking processes. The **Instructor Resources**, available in electronic form, contain complete solutions to explorations and activities, suggestions for teaching each topic, and information on typical student errors and misconceptions. Furthermore, suggestions for designing a course syllabus and sample questions for quizzes and exams are provided in the Instructor Resources.

Because the emphasis is on the student as an active learner and the context of each example is real, students become engaged with the course and the material quickly. Students will see that quantitative reasoning is a necessary skill in order to be an informed and productive citizen.

Instructor Support Materials

The **Instructor Resources**, available on the text website, www.wiley.com/college/sevilla, contain the following resources to assist you in preparing for class quickly and effectively:

- Teaching ideas for each topic and accompanying activities
- Complete solutions to Explorations, Excel Activities, and Graphing Calculator Activities
- Sample course syllabi
- Sample questions for quizzes and exams

ACKNOWLEDGMENTS

The authors and publisher wish to thank the following reviewers for their outstanding and invaluable advice on guiding this text to its successful completion:

First Edition Reviewers:

Ron Barnes, University of Houston, Downtown Campus

Linda E. Barton, Ball State University, Indiana

Pamela B. Cohen, Southern New Hampshire University

Larry Cusick, California State University, Fresno

Caren L. Diefenderfer, Hollins University, Virginia

Carrie Muir, University of Colorado, Boulder

Jessica Polito, Wellesley College, Massachusetts

Juli D'Ann Ratheal, West Texas A&M University

Mariano Rodrigues, Rhode Island College

Allan Rossman, California Polytechnic State University, San Luis Obispo

Terry Tolle, Southwestern Community College, North Carolina

And the following consultants from whose expertise we benefited:

Christian Aviles-Scott, Consultant in Educational Mathematics

Marcy Lunetta, Permissions Consultant

Second Edition Reviewers:

Linda Barton, Ball State University

Larry Cusick, California State University, Fresno

Steven Ledbetter, Austin Community College

Frederick Schineller, Mesa Community College

Kira Hamman, Penn State University, Mont Alto

Heather Gallacher, Cleveland State University

Tanweer Shapla, Eastern Michigan University

Dana Draghicescu, CUNY Hunter College

Glenn Sproul, Johnson State College

Maggie Chenoweth, Marshall University

Raydo Bugayong, MacMurray University

Teresa Graham, Baldwin Wallace College

Christopher Herald, University of Nevada, Reno

Tracy Wang, Curry College

We gratefully acknowledge the contributions of Allan Rossman, an author of Wiley's **Workshop Statistics** series. His vision and preliminary work on a Workshop Quantitative Reasoning project influenced the development of our text. We also thank the National Science Foundation (Grant No. 9950229) for support that helped us create the course and materials from which this textbook grew.

We especially thank our students at Moravian College for their support and feedback as they learned from early versions of the text. Finally, we thank our families for their continuing love and encouragement, their patience, and most of all, for maintaining a sense of humor.

QUANTITATIVE REASONING

Organizing Information Pictorially Using Charts and Graphs

After completing this topic, you will be able to:

- Distinguish between quantitative and categorical variables.

- Draw bar graphs and pie charts, and interpret them in the context of the data they represent.

- Compute percentage of the whole and percent change.

- Interpret stemplots and histograms and group data to create them.

- Decide when each type of graph is appropriate.

Why are pictures so enticing? A picture can tell a story faster than many words. Newspapers, magazines, books, and television news all use charts and graphs to present information and help us understand articles that contain numerical data. Because technology makes it possible for us to make graphs and charts easily, graphic representations are used frequently. In this topic, we discuss four types of graphs: bar graphs, pie charts, histograms, and stemplots (stem-and-leaf graphs).

When trying to understand some phenomena or to make sense of the relationship between two or more factors, pictures help us to see patterns, identify relationships, and describe main ideas. Graphs and charts can show patterns and trends not readily evident in the raw data. We will investigate **variables** and the various ways to represent them pictorially.

A **variable** is a characteristic of an object or a person (sometimes called an **individual** or a **case**) that can change from one object or person to the next. If the variable is assigned a number, then it is a **quantitative variable**. If it is assigned a category, like "male" or "female" in response to a question about gender, then it is called **categorical variable**. In the following example, we identify cases and variables, and decide whether each variable is quantitative or categorical.

Example 1.1

Suppose the individuals in our data set are movies. The following table gives the studio that produced it and the total domestic gross for each of the top 10 highest-grossing films released in 2011.

Rank	Movie Title	Studio	Total Gross (in $ millions)
1	Harry Potter and the Deadly Hallows Part 2	Warner Brothers	$381
2	Transformers: Dark of the Moon	Paramount/ DreamWorks	$352
3	The Twilight Saga: Breaking Dawn Part 1	Summit	$281
4	The Hangover Part II	Warner Brothers	$254
5	Pirates of the Caribbean: On Stranger Tides	Buena Vista	$241
6	Fast Five	Universal	$210
7	Mission Impossible–Ghost Protocol	Paramount	$209
8	Cars 2	Buena Vista	$191
9	Sherlock Holmes: A Game of Shadows	Warner Brothers	$187
10	Thor	Paramount	$181

a. Identify the cases and whether the following variables are quantitative or categorical: rank, studio, and total gross.

b. Why would we want to classify variables as quantitative or categorical?

Solution

a. The cases are the movies. Each case (or movie) has the three characteristics of rank, studio, and total gross associated with it. When data are given in a table format, each row represents a case and each column represents a different variable. The variables of rank and total gross are quantitative variables; the studio is a categorical variable.

b. These classifications can help us determine an appropriate way to present the information graphically.

PERCENTS

A percent represents a fraction out of 100. (The term "percent" means per 100; the "cent" portion of the word has the same origin as the word "century," which means a 100-year period.) We often use percents because they are easier to use than fractions, and they make it possible to compare parts of totals, when the totals have very different sizes. For example, California and Montana both experienced an approximately 10% increase in population from the 2000 census to the 2010 census. However, the population of California grew by more than 3 million people during that decade [approximately 10% of its 2000 population of 33,871,648: $\left(\frac{10}{100}\right) \times 33{,}871{,}648 = 3{,}387{,}648$] whereas the population of Montana only increased by a little over 87,000 people.

We also use percents to communicate information about growth or decline of a quantity as a percentage of its original value. For example, the population of Kendall County, IL, in the 2000 census was 54,544, whereas the 2010 census showed that the population of this county was 114,736. Similarly, the 2000 population of Cook County, IL, was 5,376,741, whereas the 2010 population was 5,194,675.

We evaluate the percentage change in population in each case by computing the following: $\frac{\text{new value} - \text{original value}}{\text{original value}} \times 100\%$. Over the years from the 2000 census to the 2010 census, Kendall County, FL, experienced an increase of $\frac{114736 - 54544}{54544} \times 100\% \approx 110.4\%$. (Note that the increase is larger than 100%, indicating that the population more than doubled.) Cook County, IL, showed a percent change of $\frac{5194675 - 5376741}{5376741} \times 100\% \approx -3.4\%$. The Cook County population decreased over that 10-year period, so we incorporate the negative sign in our description of the percent change and say that the population of Cook County decreased by approximately 3.4%.

Example 1.2

Data from 2001 to 2010 give the number of new privately owned housing units, in thousands, completed each year, as shown in the following table:

Year	Total New Privately Owned Housing Units Completed (in Thousands)
2001	1,570.8
2002	1,648.4
2003	1,678.7
2004	1,841.9
2005	1,931.4

Year	Total New Privately Owned Housing Units Completed (in Thousands)
2006	1,979.4
2007	1,502.8
2008	1,119.7
2009	794.4
2010	651.7

Source: U.S. Census Bureau, www.census.gov.

a. Find the percent change in total new privately owned housing units completed over the years 2001 to 2005.

b. Find the percent change in total new privately owned housing units completed over the years 2005 to 2010.

Solution

a. We look at the change in new housing units over the requested time period, which reflects a growth in this case, and then look at that change as a percent of the original number of housing units completed in 2001. The change is $\frac{1931.4 - 1570.8}{1570.8} \times 100\% \approx 23\%$. The total number of new privately owned housing units increased by approximately 23% over the years 2001 to 2005.

b. A similar computation for the time period 2005 to 2010 shows $\frac{651.7 - 1931.4}{1931.4} \times 100\% \approx -66.3\%$. The total number of new privately owned housing units decreased by approximately 66.3% over the years 2005 to 2010.

■ ■ ■ ■ ■ ■ ■ ■ ■ ■

BAR GRAPHS AND PIE CHARTS

A **bar graph** and a **pie chart** are two ways of representing categorical variables pictorially. (Note that the terms *graph* and *chart* are often used interchangeably.) These tools are also used to represent quantitative variables when the numbers fall into only a few categories. Because many people will look only at the graph in a news article (and not read the whole write-up), bar graphs and pie charts should be labeled so they are easy to understand. On the other hand, the graph should not be too cluttered with words and other symbols that mask the basic point. When examining a set of data, we will sometimes want to look at one variable at a time; other times we will want to study relationships between two or more variables. Next, we look at some examples of bar graphs and pie charts, and think about what story they tell us.

Example 1.3

The following bar graph and pie charts show information about new cases of AIDS during the mid-to-late 1990s and in 2010. Explain what the graph and charts show about the AIDS epidemic.

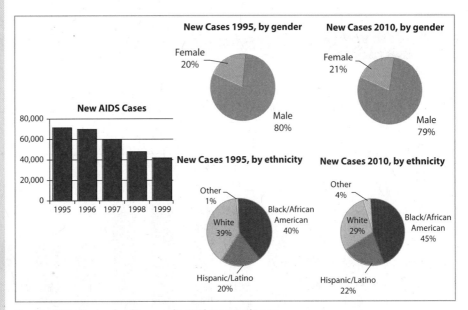

Source: U.S. Center for Disease Control, www.cdc.gov.

Solution

The bar graph shows how the total number of new AIDS cases reported fell fairly steadily over the years from 1995 to 1999, from close to 70,000 to slightly more than 40,000. The pie charts tell us that the percentage of new cases among Hispanics increased slightly from 1995 to 2010, while the percentage of new cases involving white patients decreased, and those involving black patients increased. Similarly, the percentage of new cases among women increased by approximately 1%, while men represent a smaller percentage of the new cases.

When constructing a bar graph or pie chart for a categorical variable or a quantitative variable for which the numbers fall into one of several categories, we first need to decide what the groupings or categories will be. We then determine how many cases fall into each of the categories. If we are creating a bar graph, we can represent the height of the bars either as the total number of cases that fall into each category (as the graph in Example 1.3 does, showing total new AIDS cases reported) or as the proportion or percentage of the total number of cases that fall into that category. Which option we choose (total number of cases, proportion of total, or percentage of total) will determine how the vertical axis of the graph is labeled, but the form of the bar graph and relative heights of the bars will be the same for all three options.

Example 1.4

The following graph, created with data from the United States Postal Service, displays information on the volume of advertising mail and its share of all mail, from years 2002 to 2011.

Source: U.S. Postal Service, www.usps.com.

Explain what the labels on this graph represent; specifically, why is the bar for 2002 labeled 87.2 bil. and 43.0%, while the bar for 2011 is labeled 84.7 bil. and 50.4%?

Solution

The height of the bar for each of the years from 2002 to 2011 gives the number of pieces of "junk mail" for each of those years. The percentage given for each year is the percentage of all pieces of mail handled by the post office that are "junk mail." In 2002, the number of pieces of advertising mail handled by the post office was 87.2 billion, which was 43.0% of the total number of pieces handled that year. In 2011, the number of advertising mail pieces handled by the post office was 84.7 billion. This is 2.5 billion fewer pieces than in 2002, but represents 50.4% of that year's total mail. So the proportion of "junk mail" was larger in 2011 than in 2002. When giving percentages, we need to make sure we know what the percentage represents.

Pie charts are useful when we want to tell a story about what percentage of the whole each category represents. Pie charts given at two different points in time, like those in Example 1.3, show how percentages of the whole have changed.

To construct a pie chart from raw data, we need to find what percentage of the whole each category represents. If we are drawing the pie chart by hand, it is helpful to represent each category not only as a percentage, but also as a portion of the whole circle that will be easy to visualize as, for example, halves, quarters, eighths, or sixteenths. We discuss how to do this in the following example. (Although computer programs can easily be used to draw nice pie charts, it is useful to know how to create a pie chart without the aid of a computer. This way we gain a deeper understanding of such graphs and we can better tell whether or not a chart created on a computer is correct.)

Example 1.5

The total population of the United States was shown in the 2010 census to be 308,745,538. The number of people in each of six age groups is as follows:

Age Range	19 Yrs and Younger	20 to 39 Yrs	40 to 59 Yrs	60 to 79 Yrs	80 to 99 Yrs	100 Yrs and Older
Population	83,267,556	82,829,589	85,562,485	45,849,148	11,183,396	53,364

Source: U.S. Census Bureau, www.census.gov.

a. Make a table that shows the percentage of the population in each age group.

b. Draw a pie chart that represents the estimated population by age group.

Solution

a. There are 83,267,556 people 19 years of age or younger, and the total population is 308,745,538. To find what percentage of the total population 83,267,556 represents, we divide to obtain the ratio $\frac{83,267,556}{308,745,538} \approx 0.270$, and then multiply by 100 to convert to percent: $0.270 \times 100 \approx 27.0$. So, approximately 27.0% of the total population in 2010 is age 19 or younger.

Since the number of people in the age group 20 to 39 years is 82,829,589 and $\frac{82,829,589}{308,745,538} \approx 0.268$, we see that approximately 26.8% of the population in 2010 is between 20 and 39 years old. In the same manner, we compute the percentage of the total population that corresponds to each of the remaining age groups. The following table shows the percentages:

Age Range	19 Yrs and Younger	20 to 39 Yrs	40 to 59 Yrs	60 to 79 Yrs	80 to 99 Yrs	100 Yrs and Older
Percentage	27.0%	26.8%	27.7%	14.9%	3.6%	0.0%

b. A pie chart for these data will be a circle divided into six sectors, each representing one of the age groups. The size of each sector is determined by the size of the population in the age group that sector represents. For example, the sector representing the group between 80 and 99 years of age should be a sector of the circle that covers 3.6% of the area of the whole circle. To visualize the corresponding portion of the circle, we write $3.6\% = \frac{3.6}{100} = 0.036 \approx \frac{1}{32}$. We also write the portions corresponding to the remaining groups as fractions with denominator 32. For the group 19 and younger, we write $27.0\% = 0.270 = \frac{(0.279) \cdot (32)}{32} \approx \frac{9}{32}$.

In the same way, we find that portions corresponding to the remaining groups can be converted, approximately, to fractions as follows:

Age Range	19 Yrs and Younger	20 to 39 Yrs	40 to 59 Yrs	60 to 79 Yrs	80 to 99 Yrs	100 Yrs and Older
Proportion	$\frac{9}{32}$	$\frac{9}{32}$	$\frac{9}{32}$	$\frac{5}{32}$	$\frac{1}{32}$	0

(Note that the number of people age 100 and older is not really 0, but $\frac{(0.00017) \cdot (32)}{32} \approx \frac{0.006}{32}$ is practically 0. In fact, sometimes when we approximate, the sum of the proportions will not equal 1 because of roundoff errors, as happens with these data.)

To draw the pie chart, we divide a full circle into two halves, then cut each half into two equal parts, so each is a quarter of the circle. We then cut each of those in eight equal parts to obtain sectors of size $\frac{1}{32}$ of the whole circle, as shown in the following figure:

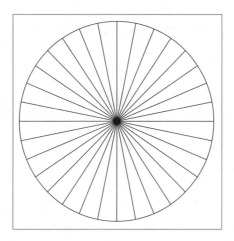

Now we mark how many $\frac{1}{32}$ portions correspond to each population group and label each sector. Here is the pie chart we obtained:

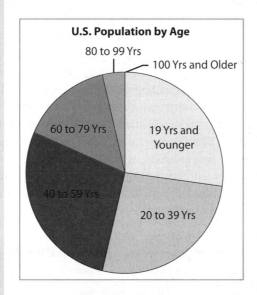

To construct this pie chart, we used the approach of dividing the circle into equal sectors, because it gives us a convenient way to visualize the size of each piece of the pie. Pie charts are most often constructed using calculators or computers to find the size of each sector.

A pie chart is not appropriate to represent all data. A pie chart provides a good representation only when the data represent parts of the same "whole" and consist of a small number of categories or data that can be grouped into a small number of categories.

Example 1.6

Explain why a pie chart would not be an appropriate graphical method to show the variable "audience rating" of top-rated TV specials, from Example 1.1.

Solution

Although the quantitative variable "audience rating" is a percentage, we wouldn't be interested in what proportion of the total percentage is represented by the audience rating of each single show. These percents are not portions of the total audience at the same time.

HISTOGRAMS

Another type of graph that is useful for visualizing the distribution of quantitative variables is the **histogram**; that is, a histogram shows how the data are distributed.

Quantitative variables such as a state's population may take on many different values in a range of, for example, 500,000 to 30 million or more. To make sense of the data, we often group the data into **classes**. If the data are given as raw data, we would first need to determine the groupings or classes and then proceed to count how many data values fall into each class. Steps 1−5 give a procedure for constructing a histogram for a quantitative variable. We illustrate this procedure in the next example.

1. Divide the range of data into classes of equal width. We usually choose between 5 and 20 classes, depending on how many cases we are working with. We also want to specify the classes so each data value falls into exactly one class.

2. Count the number of data values that fall into each class.

3. To draw the histogram, first construct a horizontal axis and mark the scale for the variable being graphed. On this scale, mark the boundaries for each class, using consistent measurements.

4. On the vertical axis, mark a scale for the counts for each class.

5. Draw a bar for each class, with the base of the bar covering the class on the horizontal axis and the height of the bar determined by the number of data values in the class. Bars for adjacent classes will touch one another (unlike the bar graph, where bars are generally separated by a space).

In a histogram, we can see the overall pattern and spread of the data. Because bars are of equal width, the area of each bar is determined by its height, and all the data are fairly represented.

Example 1.7

The following table gives a list of 12 well-known U.S. universities, along with the percent acceptance rate for applicants of the class of 2015 wishing to be admitted to the university in fall 2011.

College or University	Percent Accepted
Harvard University	6
Yale University	7
Princeton University	8
Johns Hopkins University	27
Georgetown University	18
Notre Dame University	24
Duke University	13
Virginia Tech	65
George Washington University	32
Northwestern University	18
American University	41
Cornell University	18

Create a histogram for these data and describe what the graph shows.

Solution

For Step 1, we see that the data values range from 6 to 65, so if we choose classes 10 units wide, starting at 0, we will have seven different classes: 0 to 9, 10 to 19, 20 to 29, 30 to 39, 40 to 49, 50 to 59, and 60 to 69. In Step 2, we count the number of data values that fall in each class. The result is shown in the following table:

Class	Number of Universities
0 to 9	3
10 to 19	4
20 to 29	2
30 to 39	1
40 to 49	1
50 to 59	0
60 to 69	1

In Steps 3 and 4, we set up the axes for our graph. Finally, in Step 5, we create the graph with seven adjacent bars, one for each class, 10 units wide and with a height equal to the number of data values in the class.

This histogram shows that most of the universities in the given list accept between 0 and 29% of the applicants. Only three universities in the list accept more than 30%, one of which accepts more than 60% of its applicants. The graph also shows a gap; none of the universities accepted between 50 and 59% of its applicants.

Note that in Example 1.7, we labeled the first class "0 to 9" and the second class "10 to 19." In this case, because all the data values were integers, we knew that no data value would fall between 9 and 10. In general, to allow for data values that are not necessarily all integers, we label the classes in such a way that every real number is in one of the classes. Instead of "0 to 9" and "10 to 19," these classes would be labeled "0 to 10" and "10 to 20." Because these names do not indicate whether the value 10 is considered in the first or the second class, we need to clarify this either on the histogram itself or in a separate sentence. This process guarantees that each data value falls in one and only one class. Example 1.8 illustrates this.

Example 1.8

The following table contains a list of the states and District of Columbia, along with the percent change in population from the census of 2000 to that of 2010. Construct a histogram for these data.

State	Percent Change	State	Percent Change	State	Percent Change
Alabama	7.5	Alaska	13.3	Arizona	24.6
Arkansas	9.1	California	10.0	Colorado	16.9
Connecticut	4.9	Delaware	14.6	District of Columbia	5.2
Florida	17.6	Georgia	18.3	Hawaii	12.3
Idaho	21.1	Illinois	3.3	Indiana	6.6
Iowa	4.1	Kansas	6.1	Kentucky	7.4
Louisiana	1.4	Maine	4.2	Maryland	9.0
Massachusetts	3.1	Michigan	−0.6	Minnesota	7.8
Mississippi	4.3	Missouri	7.0	Montana	9.7
Nebraska	6.7	Nevada	35.1	New Hampshire	6.5
New Jersey	4.5	New Mexico	13.2	New York	2.1
North Carolina	18.5	North Dakota	4.7	Ohio	1.6
Oklahoma	8.7	Oregon	12.0	Pennsylvania	3.4
Rhode Island	0.4	South Carolina	15.3	South Dakota	7.9
Tennessee	11.5	Texas	20.6	Utah	23.8
Vermont	2.8	Virginia	13.0	Washington	14.1
West Virginia	2.5	Wisconsin	6.0	Wyoming	14.1

Source: U.S. Census Bureau, www.census.gov.

Solution

The first thing we need to do is identify the range of the data. The smallest percent change is −0.6, whereas the largest is 35.1. We divide that range into suitable classes or intervals of equal length. We could use convenient intervals of 5 units. If we start with −5 as the left boundary of the first interval, the first interval would be from −5 to 0; the second interval would be from 0 to 5, and so on. To ensure that each point is in exactly one interval, we include the right endpoint of each interval in the interval and do not include the left endpoint. Thus, the first interval is −5.0 < percent change ≤ 0.0; the second interval is

0.0 < percent change ≤ 5.0; the third interval is 5.0 < percent change ≤ 10.0; the next interval is 10.0 < percent change ≤ 15.0; the next interval is 15.0 < percent change ≤ 20.0; the next interval is 20.0 < percent change ≤ 25.0; the next interval is 25.0 < percent change ≤ 30.0; the next interval is 30.0 < percent change ≤ 35.0; the final interval is 35.0 < percent change ≤ 40.0. Note that we would not need to start our first interval at −5. We could have used −2, or even −1, but −5 was a convenient choice. All the intervals need to be of equal width and each data value must be in one and only one interval. (Choosing our intervals the way we did ensures this, but it is not the only possibility.) Now we tally the number of data values that fall into each interval.

Interval (percent change)	Number of Data Values (states) in Interval
−5.0 < percent change ≤ 0.0	1
0.0 < percent change ≤ 5.0	15
5.0 < percent change ≤ 10.0	16
10.0 < percent change ≤ 15.0	9
15.0 < percent change ≤ 20.0	5
20.0 < percent change ≤ 25.0	4
25.0 < percent change ≤ 30.0	0
30.0 < percent change ≤ 35.0	0
35.0 < percent change ≤ 40.0	1

We sketch the histogram as shown next. We want to look at the general shape of the histogram, its center, and see how spread out the histogram is to get an idea of the general data pattern.

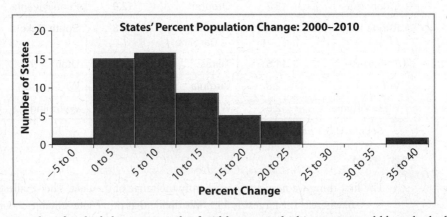

Note that if we had chosen intervals of width 10 or 8, the histogram would have looked a bit different.

STEMPLOTS

Another type of graph, a **stemplot** or **stem-and-leaf graph**, is often used to display a quantitative variable, particularly if the data set is not too large. This type of graph shows not only the general pattern of the data, as a histogram does, but also displays all the individual data values. Here are the steps for constructing a stemplot:

1. Divide each data value into two parts: a **stem** consisting of all but the single rightmost digit, and a **leaf**, consisting of the rightmost digit. For example, if data values represent test scores, the score of 82 would have a stem of 8 and a leaf of 2. If data values represent math SAT scores, the score of 625 would have a stem of 62 and a leaf of 5.

2. Write the stems in order in a vertical column with the smallest at the top. (We must include all possible consecutive stems, even if there are no values in our data set with that particular stem; otherwise, the data are distorted.) We draw a vertical line to the right of the column of stems.

3. Write each leaf in the row to the right of its stem, in increasing order from left to right. Take care to be consistent with the spacing and size of the numbers representing each leaf.

Example 1.9   ■ ■ ■ ■ ■ ■ ■ ■ ■ ■

Create a stemplot for the data on the percent acceptance rate in the following table. (This is the same data used in Example 1.7.) Describe any patterns that emerge.

College or University	Percent Accepted
Harvard University	6
Yale University	7
Princeton University	8
Johns Hopkins University	27
Georgetown University	18
Notre Dame University	24
Duke University	13
Virginia Tech	65
George Washington University	32
Northwestern University	18
American University	41
Cornell University	18

Solution

The data values consist of two digits, so we will use the "tens" digit as the stem and the "units" digit as the leaf. We start by listing the stems from 0 to 6, in a vertical column.

0

1

2

3

4

5

6

We add a vertical line after the column of stems and then add the leaves one at a time. The final step is to order the leaves on each stem and produce an ordered stemplot. We also include a title and a key:

Percent Acceptance Rate: $2|5 = 25\%$

```
0 | 6  7  8          0 | 6  7  8
1 | 8  3  8  8       1 | 3  8  8  8
2 | 7  4             2 | 4  7
3 | 2                3 | 2
4 | 1                4 | 1
5 |                  5 |
6 | 5                6 | 5
```

The stemplot shows that for this small collection of data values, most of the schools have an acceptance rate in the 0 to 27% range. Only three universities accepted 32% or more of their applicants. Also, there is a gap; none of the universities in this group accepted between 41 and 65% of its applicants. The "center" of the data is probably somewhere in the low 20s. From the stemplot, we can see the shape and spread of the data and get a general idea of the center and any gaps; we can also see the actual data values.

For some data, when creating a stemplot, it may be necessary to truncate the data values. See Example 16.3 for an instance of this. In any case, when creating a graph, we always want to consider what story our graph will tell.

Summary

We have seen examples of quantitative and categorical variables and learned how to interpret information given in the four types of graphs.

Bar graphs and pie charts are used to represent categorical data. In a bar graph, each bar represents a category and the height of each bar represents a count or percentage for the category. To make a pie chart, sometimes it is necessary to group the data. Each portion of a pie chart shows what percentage each category is of the whole. To find the corresponding portion of a category in a pie chart, we need to use fractions and compute percents.

Histograms and stemplots (stem-and-leaf graphs) are used to display numerical data. A histogram shows the shape and spread of the data and the frequency of data values in determined classes; that is, how many times data values fall in a particular class. A stemplot shows the shape and spread of the data and gives the actual data values.

Explorations

1. For each of the following variables, indicate whether it is quantitative or categorical. Then, identify the individual (or case) and write a sentence explaining why you might be interested in such a variable.

   a. The number of calories in a cup of breakfast cereal

   b. Salaries of last year's college graduates

   c. Preferred brand of cola

   d. Time it takes college freshmen to read a particular editorial

   e. Race of small business owners in Philadelphia

2. A store advertises its discounts in two parts. Here is the offer: "Take 50% off of the original price, then take another 30% off of that." Suppose a shirt you want to buy originally cost $25.00.

   a. If there is a single discount of 50%, how much would the shirt cost?

   b. If we now apply the second discount of 30% to the discounted price of the shirt from part (a), what is the cost?

   c. With the "double discount" as described, what is the total percentage discount deducted from the original price?

   d. Would that "double discount" percentage be the same for an item with a different original price? Explain your answer.

   e. Why do you think the store uses this two-part advertising method?

3. The total population of the United States was 151,325,798 in 1950; it increased to 281,421,906 in 2000 and to 308,745,538 in 2010.

    a. Find the percent change in population of the United States from 1950 to 2000.

    b. Find the percent change in population of the United States from 1950 to 2010.

    c. Find the percent change in population of the United States from 2000 to 2010.

    d. Explain what your answers to parts (a)–(c) show about the total population growth in the United States.

4. In 1990, the population of Michigan was 9,295,297. During the 10-year period 1990–2000, Michigan's population increased by 6.9% and during the 10-year period 2000–2010, Michigan's population decreased by 0.6%.

    a. Use this information to find Michigan's population in the year 2000.

    b. Find Michigan's population in the year 2010.

    c. Find the percent change in population of Michigan from 1990 to 2010.

5. In the graph shown in Example 1.4, we saw that the number of pieces of "junk mail" in 2002 was 87.2 billion. This represented 43.0% of the total number of mail pieces handled by the post office during that year.

    a. Find the total number of mail pieces handled by the post office during 2002.

    b. Use the information given in Example 1.4 to find the total number of mail pieces handled by the post office.

        i. During the year 2004

        ii. During the year 2006

        iii. During the year 2008

        iv. During the year 2010

6. Data from the United States Postal Service show that in 2007 the post office handled 212.2 billion pieces of mail; in 2008, the volume of mail was 4.5% lower than its 2007 level. The volume declined again in 2009 to 176.7 million pieces.

    a. Find the volume of mail handled by the post office in 2008.

    b. Find the percent change in the volume of mail handled by the post office from 2008 to 2009.

    c. Find the total number of mail pieces handled by the post office in 2011 and find the percent change in volume from 2009 to 2011.

7. In 2003, over 10,000 drivers, whose primary vehicle is equipped with safety belts, were asked the following question: When driving this (car/truck/van), how often do you wear your (lap/shoulder) belt? The responses are summarized in the following table:

Gender	All of the time	Most of the time	Some of the time	Rarely	Never
Male	79%	12%	4%	2%	3%
Female	89%	6%	3%	1%	1%

Source: U.S. National Highway Traffic Safety Administration, www.nhtsa.gov.

   a. Create an appropriate graph for these data.

   b. Explain what your graph shows about safety belt use.

   c. Discuss any difficulties that may have been associated with collecting these data.

8. Write a paragraph to describe what the following bar graphs show. The graphs accompanied the article "Children of the Gun" that appeared in *Scientific American*, June 2000.

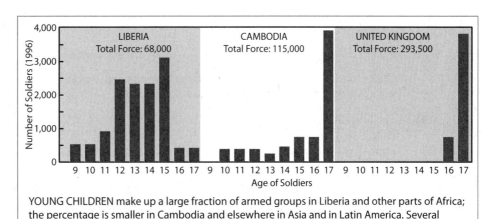

YOUNG CHILDREN make up a large fraction of armed groups in Liberia and other parts of Africa; the percentage is smaller in Cambodia and elsewhere in Asia and in Latin America. Several developed nations recruit 16-year-olds but are now raising the minimum age.

9. Group the data about Liberia's armed forces on the bar graph in Exploration 8 into three age groups. Then use this grouped data to create a pie chart that shows the composition of Liberia's armed forces in the three age groups. Clearly indicate which age groups you are considering and give the percentage of the total armed forces that corresponds to each group.

10. Explain what the pie chart in the solution of Example 1.5 shows about the population of the United States in 2010.

11. The following table containing U.S. Bureau of Labor Statistics gives the number of workers by race and ethnic origin and gender for the three largest ethnic groups of workers in the United States for the years 2000, 2005, and 2009:

Numbers (thousands)	2000	2005	2009
White, non-Hispanic	108,264	111,844	114,996
Men	59,119	61,255	61,630
Women	49,145	50,589	53,366
Black or African American	14,444	14,777	15,025
Men	6,741	6,901	6,817
Women	7,703	7,876	8,208
Hispanic or Latino	14,762	17,785	19,647
Men	8,859	10,872	11,570
Women	5,903	6,913	8,077

Source: U.S. Department of Labor, Bureau of Labor Statistics, www.bls.gov.

Create one or more appropriate graphs that show how the racial and ethnic makeup of the U.S. workforce has changed over this period of time.

12. The following table contains a list of the states and District of Columbia, with the average critical reading SAT test scores for high school seniors (for the academic year ending in 2010) and the percentage of high school seniors who take the test:

State	Average Critical Reading SAT	Percent Taking Test	State	Average Critical Reading SAT	Percent Taking Test
AL	556	7	MO	593	4
AK	518	48	MT	538	24
AZ	519	25	NE	585	4
AR	566	4	NV	496	43
CA	501	50	NH	520	77
CO	568	18	NJ	495	76
CT	509	84	NM	553	11
DE	493	71	NY	484	85
DC	474	76	NC	497	63
FL	496	59	ND	580	4

State	Average Critical Reading SAT	Percent Taking Test	State	Average Critical Reading SAT	Percent Taking Test
GA	488	74	OH	538	21
HI	483	58	OK	569	5
ID	543	19	OR	523	54
IL	585	6	PA	492	71
IN	494	64	RI	494	67
IA	603	3	SC	484	66
KS	590	6	SD	592	3
KY	575	6	TN	576	10
LA	555	7	TX	484	53
ME	468	92	UT	568	6
MD	501	70	VT	519	66
MA	512	86	VA	512	67
MI	585	5	WA	524	54
MN	594	7	WV	515	16
MS	566	3	WI	595	4
			WY	570	5

Source: U.S. Department of Education, National Center for Education Statistics, www.nces.ed.gov.

d. Make a stemplot of the percentage of high school seniors taking the test.

e. Make another stemplot of the average verbal SAT scores of the states.

f. Describe the shape of each of the distributions of the variables "percent taking" and "average verbal SAT score."

g. Look at the states by the region of the United States in which they are located, and make any preliminary observations about the variables "percent taking" and "average verbal SAT score."

13. In the 2005 Electronic Monitoring and Surveillance Survey conducted by the American Management Association, the following information about what employers record and review relative to electronic eavesdropping was obtained. The information was based on responses from 526 companies. Create an appropriate graph to represent these data and explain why you chose the graph you did.

Recorded	Percent of Companies
Telephone conversations	3
Voicemail messages	7
Computer files	30
E-mail messages	38
Video recording of job performance	6
Telephone use (time spent, numbers called)	31
Computer use (time logged on, keystroke counts)	21
Video surveillance for security	32
Website connections	76

14. The following graph shows the number of murders in New York City each year from 1988 to 1999.

Additional data on the number of murders reported to police in New York City for selected years since 1999 appear in the following table:

Year	2001	2003	2006	2007	2008	2009	2010
No. of Murders	649	598	596	494	516	471	536

Source: NYC Police Department statistics, www.nyc.gov.

    a. Explain in detail what the graph shows about the number of murders in New York City during the 1990s.

    b. What was the percent change in the number of murders in New York City from 1990 to 2006?

    c. What was the percent change in the number of murders in New York City from 2006 to 2010?

    d. Construct a new bar graph with the additional data. What story do the graphs tell us about reported murders in New York City over the years 1988 to 2010?

15. Describe the differences and similarities between the histogram created in Example 1.7 using data on acceptance rates in 12 well-known universities and the stemplot in Example 1.9 created using the same data set.

16. Use the Internet to find current data on the AIDS epidemic, and create several graphs like those in Example 1.3 to show how the epidemic is continuing to change.

17. From each student in the class, collect the following data, and then create an appropriate graph or chart for each variable and explain what it shows.

    a. The yearly salary to the nearest hundred of dollars that they estimate they will earn in their first post-college job

    b. Their most preferred leisure-time activity, chosen from among the following: watching a movie; watching television; playing a sport or exercising; reading; going out to eat; playing computer/video games; other

18. Write a summary paragraph explaining when each of the different types of charts or graphs is useful.

19. Give an example of a situation in which speaking of a percent over 100% (such as 125% or 300%) makes sense.

World Motor Vehicle Production: Bar Graphs and Pie Charts

In this activity, you will create several different types of graphs to help understand patterns in world motor vehicle production. You will investigate when each type of graph is appropriate for the data.

1. Consider the following data on leading world producers of passenger cars and commercial vehicles in 2010:

Location	Cars (in thousands)	Commercial Vehicles (in thousands)
Brazil	2,828.3	820.1
China	13,897.1	4,367.6
France	1,922.3	305.4
Germany	5,552.4	353.6
India	2,814.6	722.2
Japan	8,307.4	1,318.6
South Korea	3,866.2	405.7
Spain	1,913.5	474.4
United States	2,731.1	5,030.3
Other	14,431.4	5,547.7

Source: International Organization of Motor Vehicle Manufacturers, www.oica.net.

a. If you were to create a graph or chart of this data set, what would you like it to show?

b. Explain why a bar graph is appropriate for the car production data.

c. Draw axes in the space provided and create a bar graph for the car production data. Be sure to label the axes and show the scales you are using.

d. Explain what your bar graph shows that the original data table did not show and what the original table contains that your graph does not retain.

e. If you create a pie chart for the car production data, what will each "piece" of the pie show?

f. Use the Excel instructions that follow to help you fill in the third column of the next table to show the percentage of the total car production that falls into each row of the table:

Location	Cars (in thousands)	Percentage	Approximate Fraction of the Whole
Brazil	2,828.3		
China	13,897.1		
France	1,922.3		
Germany	5,552.4		
India	2,814.6		
Japan	8,307.4		
South Korea	3,866.2		
Spain	1,913.5		
United States	2,731.1		
Other	14,431.4		
Total			

Instructions to Calculate Percentages

1. Enter the car production data into columns A and B of an Excel worksheet. Use column A for location and B for the number of cars produced, and use row 1 for titles. (A portion of the table follows.) Use the arrows, or press **Enter** after you type the entry in each cell, and fill in the numbers in cells B2 (column B, row 2) through B11.

2. To adjust the column width of column A to fit all the descriptions, use the mouse (or touchpad) to move the cursor to the letter A at the top of column A and click. This will highlight the first column. Then position the cursor at the right border of the column where the cursor changes to a black "plus sign with left and right arrows" and double-click the left mouse button. Repeat with column B, if needed.

	A	B
1	Location	Cars (in thousands)
2	Brazil	2,828.3
3	China	13,897.1
4	France	1,922.3

3. To get the total car production, position the cursor in cell B12. To instruct Excel to perform a computation, you must use the "equals" sign. Type =**sum**(**B2:B11**) and then press **Enter**.

4. To calculate what percentage each value in the table is of the whole, divide each number by the total, and then ask Excel to convert to a percentage. You can tell the computer to do this with the first entry in column B and then "drag down" using the mouse to get the percentages for each row. Put the cursor in cell C2 and type =**B2**/ (the number you got for the total car production in cell B12. Do not write the total with a "comma").

5. To convert this decimal value to a percentage, put the cursor in cell C2, and on the **Home** tab (on the Ribbon), in the **Cells** group, click **Format**. On the drop down menu, choose **Format cells**, and on the **Number** tab, click **Percentage**. In the **Decimal places** box, type **0**, because you just need a general idea of the percentage of the total for each country. Click **OK**.

6. Now to fill in the percentages in the rest of the column, put the cursor in cell C2, and move the cursor to the lower-right corner of the cell, until it changes to a black "plus" sign. Press and hold the left mouse button and drag down to row 8. The rest of the percentages should fill in as you drag down.

g. Use the next Excel instructions to get a rough approximate fraction of the whole represented by each row of the table and insert those values in the final column of the table shown previously. (This will help you create a pie chart.)

Instructions to Convert Percentages into a Fraction

1. Put the cursor in cell D2. Type =**C2** and then press **Enter**. Click cell D2 and go to **Format**, then select **Format Cells** and on the **Number** tab, click **Fraction**. In the **Type** box, choose **As sixteenths (8/16),** and press **OK**. This will give you the numbers as fractions with the denominator 16. You may use other forms, but for drawing the portions of the pie chart, it is easier to use denominators that are powers of 2.

2. Now fill in the rest of the column by using the "drag" function of Excel [see Step (4) in the previous section, "Instructions to Calculate Percentages"].

h. Create a pie chart for the car production data using the last column of the table to help you determine approximately how big each piece of the circle should be. Be sure to label your chart.

i. Use the data given at the beginning of this activity to create a bar graph showing commercial vehicle production in 2010 and explain what the graph shows.

2. For what types of data is a bar graph appropriate and for what types of data is a pie chart appropriate?

Additional Questions

3. The following table shows data on new passenger cars exported to the United States by country of origin in 1999 and 2009:

Country	1999	2009
Japan	1,707,277	1,238,773
Germany	461,061	348,093
Italy	1,697	3,067
United Kingdom	68,394	78,899
Sweden	83,399	27,017
France	186	16,909
South Korea	372,965	476,912
Mexico	639,878	649,740
Canada	2,170,427	1,164,849
Total*	**5,639,616**	**4,276,163**

*Note that the total includes countries that are not listed separately.
Source: The World Almanac and Book of Facts 2011, p. 90.

a. What kinds of information, related to these data, would you want to get from a graph?

b. Would both bar graphs and pie charts be appropriate for presenting these data? Would one be preferable?

c. Create two graphs (bar and/or pie) to present these data, and explain what each graph shows specifically about these data.

4. The following table gives the most popular colors for the 2009 model year for full/intermediate size cars:

Color	Percent
Silver	19
Gray	16
White/white pearl	15
Black/black effect	14
Blue	12
Red	11
Beige/brown	7
Green	4
Gold/yellow	2
Other	1

Source: The World Almanac and Book of Facts 2011, p. 92.

a. Create a pie chart for these data or explain why it is not appropriate to do so. Explain how you determined the size of each "pie piece."

b. Create a bar graph for these data or explain why it is not appropriate to do so.

c. Explain what your chart(s) show.

Summary

In this activity, you learned how to use Excel to find percentages. You gained experience in deciding when a bar graph or a pie chart is appropriate to represent data and in creating bar graphs and pie charts. You also practiced how to "read" and explain information from bar graphs and pie charts.

1-2

Medical Data and Class Data: Graphs with Excel

Microsoft Excel is a powerful computer program that allows you to manipulate data and create graphs. (In Excel, a graph is also called a *chart*.) In this activity, you will use Excel to create bar graphs and pie charts and more importantly, you will use these charts to help understand and interpret the data.

1. Consider the following data on principal reasons given by patients for emergency room visits in 2003:

	A	B
1	Reason	Number of Visits (in thousands)
2	Accidents	3,999
3	Arm/leg injuries	3,758
4	Back symptoms	2,696
5	Breathing problems	4,568
6	Chest pain	5,838
7	Cough/throat symptoms	6,295
8	Earache/ear infection	1,867
9	Fever	5,732
10	Head/neck symptoms	4,641

11	Nausea/dizziness	5,280
12	Nonspecific pain	2,234
13	Skin rashes	1,688
14	Stomach pain	7,583

Source: The World Almanac and Book of Facts 2006, p. 190.

a. What type of graph or chart would help you interpret the data?

b. Explain why a bar graph is appropriate for these data.

c. Use Excel to create a bar graph of these data (instructions follow).

Instructions to Enter Data and Create a Bar Graph

1. Enter the data and titles into the first two columns of an Excel worksheet, just as they appear in the previous table (cells A1 through B14).

2. Click on cell **A1** and drag to cell **B14** to select the data set. Then click the **Insert** tab and select **Column** from **Charts** and select **Clustered column** (the first type in the top row) for **Chart sub-type**.

3. To change the title of your graph, click on the title in the graph and then click again. You can now backspace or delete and retype a new title.

4. To add axis titles to your graph, click inside the border of the graph. You will see the **Chart Layouts** group near the top of your screen. Click the bottom down arrow on the right edge of the **Chart Layouts** group. Click on Layout 9 (the rightmost one in the third row.) In your graph, click on the words "Axis Title" and type appropriate titles. Also click on the Legend and press **Delete** to remove it.

5. Click in a blank area of the graph. The chart now has eight "handles" indicating that it has been "selected." If you select a handle (getting a double-headed arrow) and then press and hold the left mouse button on drag it, you can resize the chart. By clicking on the white interior of the chart and holding down the mouse button, you can drag the chart to another location. Click outside the chart to "deselect" it.

6. To sort data, return to cell A1, and then click and drag to cell B14 to select the complete set of data. Select the **Data** tab and then select **Sort**. Sort by **Number of visits**, sort on **Values** and Order **Smallest** to **Largest**.

---

d. Explain what your bar graph shows that the original data table did not show.

e. Now you'll use Excel to create a pie chart of these data.

Instructions to Create a Pie Chart and Change a Bar Graph to a Pie Chart

1. To create a pie chart, as you did before, highlight the data and labels. Go to the **Insert** tab and select **Pie** and the first chart sub-type, **Pie**, from among the choices.

2. Click on the title to change it if needed.

3. Click inside the graph and then click the bottom down arrow on the right edge of the **Chart Layouts** group. You will see pie chart options that allow you to label the pie pieces and/or show values or percents. (Note that you want to be able to clearly read the chart, so including a legend and data labels will make your chart easy to read.) Click on one of the **Layouts**.

4. To change a bar graph to a pie chart, go back to the graph you created. Select the completed bar graph by clicking inside the border.

5. From the **Type** group at the top left of your screen, select **Change Chart Type**. In the left field, choose **Pie** (for chart type) in the first box and **Pie** (for chart sub-type), the first choice in the bottom row, in the second box. Click **OK**.

6. Now click inside the border of your graph to access the **Chart Layouts** group and choose the layout you want for your pie chart.

f. Is there any information in your pie chart that was not in the bar graph? Explain.

2. Suppose the following data on number of miles from the college to home was collected from a group of 12 students.

Student	Miles	Student	Miles
John	23	Sally	3
Jean	45	Mitch	1250
Harry	11	Taro	322
Bruce	134	Gary	95
Fred	62	Audrey	76
Ann	35	Josi	28

a. Use Excel to create a pie chart for this data set and explain what your pie chart shows. (Think about how you want your pie chart to show the data. You will need to put the data into categories first. You can use columns D and E of the same Excel worksheet.) How does your pie chart help you understand the data?

b. Instruct Excel to convert the pie chart you created to a bar graph.

c. Which of the two charts do you think is more helpful to understand the data in part (a) and why?

Some Other Useful Excel Functions

1. To save your Excel file, click on the **Office Button** in the top left corner of your screen. Then select **Save as** and the first option **Excel Workbook**. (Or you can choose the option **Excel 97−2003 Workbook** if you want to use this file with an earlier version of Excel.) Choose the location in which you wish to save the Excel file. You can now give your file an appropriate name in the text box at the bottom of the dialog and click **Save**. When you want to come back and work on the file again, you can go to the **Office Button**, select **Open**, and retrieve the file from your saved file location. After you have accessed this file and made some modifications, if you want to save it again, just go to the **Office Button** and click **Save**. Excel will save your file in the same location.

2. You may want to go to a new worksheet in your Excel file. Each Excel file document is called a *workbook*, and each workbook has multiple sheets. To go to a new sheet in the same workbook, click **Sheet 2** at the bottom of the Excel page. You will get a new worksheet to use. When you save the file, all the worksheets will be saved in the one file.

3. You will need to get data files from the text website at www.wiley.com/college/sevilla or WileyPLUS. To download a file from this site, double click on it. When you are finished working, you can save your file as described above.

Additional Questions

3. Use the "EA1.2 Class Data Gender.xls" data file from the text website or WileyPLUS.

   a. Create two charts that graphically illustrate this data set. You don't need to use all the data. Select appropriate data for what you want to illustrate with your graphs. (You may want to sort or group the data and create new tables, copy the data into a new portion of your worksheet, and so on.) Experiment with the copy-and-paste capabilities by highlighting your data and selecting Edit from the menu bar.

   b. Explain why you picked the data and the graphs you did.

   c. Explain what your graphs show.

   d. What information about the variables you used is in the table but is not reflected in your graphs?

4. Look at the graphs shown in Example 1.3 of Topic 1.

   a. Enter the data into an Excel worksheet, and re-create the bar graph and one of the pie charts (you may choose which one).

   b. Describe any difficulties you encountered in creating these graphs.

Summary

In this activity, you learned to enter data in an Excel file and created bar and pie charts using Excel. You also learned how to save an Excel file and to work with more than one worksheet within an Excel workbook. You learned that sometimes it is necessary to group the data to create an appropriate pie chart.

SATs and the Super Bowl: Creating and Interpreting Histograms

A histogram is a graph of a frequency distribution and is a useful way to give a pictorial summary of a set of data involving a quantitative variable. In this activity, you will create and interpret histograms.

1. How is a histogram similar to a bar graph and how is it different from a bar graph?

2. The following table gives the percentage of high school seniors who took the SATs in the academic year ending in 2010 for a group of 14 southeastern states and Washington, DC:

State	AL	AR	DE	DC	FL	GA	KY	LA	MD
Percent Taking Test	7	4	71	76	59	74	6	7	70

State	MS	NC	SC	TN	VA	WV			
Percent Taking Test	3	63	66	10	67	16			

a. What is the range of the "percent taking test" data shown in the table?

b. If you want to group the data into approximately six classes, what intervals could you use?

c. Why is it important to group the data?

d. Why do you want to choose intervals of equal width?

e. Fill in the following table and then construct, by hand, a histogram for the percentage of high school seniors in the 15 southeastern locations who took the SATs. Be sure to label the axes of your histogram.

Interval (percent taking SATs)	Frequency (no. of states in each interval)

f. Describe the overall shape of your histogram.

3. There are several ways to create a histogram using Excel. We'll illustrate the first way using the SAT data for all states.

a. Import the "EA1.3.1 SAT data.xls" file from the text website or WileyPLUS into an Excel worksheet.

b. Follow the Excel instructions given next to create a histogram for the SAT data.

Instructions to Use Excel's Insert Tab to Create Histograms

One way to create histograms is to use the **Insert** tab discussed previously. The **Insert** tab is designed to use with data that have been grouped—either categorical data or quantitative data—into classes or intervals. To create a histogram using the **Insert** tab, you will group the data, which is not too difficult if you first sort it.

1. You'll create a histogram of the "percent taking SATs" for all the states, so first sort the data by this column; however, be sure to highlight both columns and then begin to perform the sort. Select the **Data** tab and **Sort**, and then sort by the appropriate column from smallest to largest.

2. Next decide what the "bins" will be. That is, you need to decide what classes (of equal width) to use to group the data. For this data set, it appears that $0 < \% \leq 10; 10 < \% \leq 20; 20 < \% \leq 30$, and so on will be convenient classes. Use your sorted data to count the number of data values that fall into each of these classes, and create a table in columns E and F of your Excel worksheet with the class labels in column E and the frequencies for each class in column F. (Include axis titles.)

3. Use this table of classes and frequencies, and create a bar graph like you did previously.

4. To make this completed bar graph look like a histogram, adjacent bars should touch one another. To change the bar width, right-click on one of the bars of the finished graph to access the **Format data series** dialog box. From **Series Options**, change the **Gap width** from **150%** to **0%**. Then click **Close**.

5. You may add titles and experiment with some of the other features of bar graphs (like adding or taking away axis labels and so on).

c. Describe the overall shape of the SAT histogram for all states. How does this histogram compare with the one you created by hand for the southeastern states? Using the two histograms, what can you say about the percentage of seniors taking SATs in the southeastern states and in all the states?

4. Here is another method to create a histogram with Excel. For this histogram, you'll use a file containing data about the points scored by the winning team in the Super Bowl for the years 1967 to 2011.

    a. Retrieve the "EA1.3.2 Super Bowl.xls" file from the text website or WileyPLUS. You'll make a histogram of the variable "points scored by winner."

Instructions to Use Excel's Analysis ToolPak to Create a Histogram

You can also create a histogram using Excel's **Analysis ToolPak**, which groups the data into classes for you.

    The **ToolPak** may need to be installed; to check, choose the **Data** tab. If the Ribbon shows an **Analysis** group, select **Data Analysis** from the **Analysis** group. If Data Analysis is not shown on the Ribbon, proceed as follows to install the Analysis ToolPak.

    Click on **File** and then click **Options** at the bottom of the menu window that opens. Choose **Add-Ins** on the left box under **Excel Options**, and at the bottom of the **Add-Ins** box, for the **Manage** box, select **Excel Add-Ins**.

    Click **Go**. In the **Add-Ins available** box, select the **Analysis ToolPak** check box, and then click **OK**. (If **Analysis ToolPak** is not listed in the **Add-Ins available** box, click **Browse** to locate it.) If you get a prompt indicating that the Analysis ToolPak is not currently installed on your computer, click **Yes** to install it.

1. From the menu bar, choose **Data** and then **Data analysis** from the **Analysis** group; scroll down to the **Histogram** option and click **OK**.

2. In the dialog box, type the reference for the range of your data, **B1:B46** in the **Input range** area, or click and drag from cell B1 to cell B46. Leave the **Bin range** field blank to allow Excel to select the bins (or groupings) for your data. Check the **Labels** box because B1 (where there is a label for the column) has been included in the **Input range**. Click to select **Output range** and Type **C1** for **Output range** to denote the upper left cell of the output range, and check the box **Chart output** to put the histogram on the same sheet of the workbook as the data. For now, you won't use the other options (**Pareto** and **Cumulative percentage**.) Click **OK**.

3. The output entries under **Bin** in the C column are the upper limits of the boundaries for each class, and the corresponding frequencies appear in cells in the D column. (Notice the numbers at the bottom of each bar in the histogram.)

4. Adjacent bars should touch in a histogram. To change the bar width, right-click one of the bars to bring up the **Format data series** dialog box. On **Series Options**, change the **Gap width** from **150%** to **0%**. Then click **Close**.

5. To create a slightly different histogram using "bin" intervals that you choose, type **My Bins** (or any other appropriate label) in cell E1. (You may need to move your histogram.) Because the data values range from 14 to 55, you can use intervals of width 5, beginning with the interval from 10 to 15. So enter the values **15, 20, 25, 30, 35, 40, 45, 50, 55** in cells E2:E10. (This creates the bins or classes of: $10 <$ points scored ≤ 15; $15 <$ points scored ≤ 20; $20 <$ points scored ≤ 25; and so on, to the last class, $50 <$ points scored ≤ 55.) Notice that the right endpoint of each interval or class is included in the interval but the left endpoint is not.

6. Use the procedure for creating a histogram described previously, but type **E1:E10** in the **Bin range** area and type **G1** for **Output range**.

Some Selected Enhancements to Your Histogram

1. Here is an additional way to get rid of the legend. To do this, click the legend (the word *frequency* on the chart). Select the **Layout** tab. Click **Legend** (in the **Labels** group) and select **None** to turn off the Legend Option.

2. You can resize both the plot area (the gray region containing the histogram) and the larger rectangle by clicking in the boundaries and dragging the appropriate handles horizontally, vertically, or diagonally. Click outside the chart to "deselect."

3. To change the chart title, click the title word *Histogram*. A rectangular gray border with handles will surround the word. Click the word *Histogram* again and type an appropriate title. When finished, click outside the graph.

4. Click the words *My Bins* at the bottom of the chart and type in an appropriate X-axis title.

5. The **More** interval with **0** counts shouldn't be there. In the workbook cell containing the word *More*, change the label to **60**. The histogram is dynamically linked to the data and the label on the X-axis changes to 60.

b. Explain what your histogram shows about points earned by Super Bowl winning teams.

Additional Questions

5. Retrieve the data set "EA1.3.3 Governors Salaries.xls" from the text website or from WileyPLUS.

    a. Choose appropriate classes and use Excel to create a histogram for the governors' salaries.

    b. Explain why you picked the classes you did.

    c. Explain what your histogram shows.

6. Consider again the data in the "EA1.3.2 Super Bowl.xls" data file.

    a. Create, by hand, a stemplot of these data.

b. How is this graph different from the histogram for the same variable you created previously?

c. Which plot gives a better sketch of the data and why?

d. What information from the table is not reflected in your graphs?

Summary

In this activity, you grouped data to create histograms by hand and using Excel. With Excel, you created histograms in two different ways: as a bar graph using the frequencies you computed and then adjusting the width of the bars so that there were no gaps in between them, and as a histogram using the histogram function in Excel, in which frequencies are computed automatically.

Bivariate Data

After completing this topic, you will be able to:

- Construct a scatterplot to describe the relationship between two quantitative variables.

- Identify and explain, in the context of the data, any trends depicted in a scatterplot.

- Describe the mathematical concept of a function and recognize when one variable is a function of another.

- Represent functions using words, tables, symbols, and graphs, and move from one mode of representation to another.

- Recognize and formulate directly proportional relationships.

There are many instances where we want to determine how two variables are related; for example, we might want to determine how gender and salaries in a particular industry are related. Because gender is a categorical variable and salary is a quantitative variable, we use comparative bar graphs or pie charts to help identify relationships. When we are working with two quantitative variables (that is, bivariate data sets), as in the exploration in Topic 1 involving the average verbal SAT score for each state and the percentage of high school seniors in the state taking the SAT test, we might want to use a **scatterplot** to see how the two variables are related.

In considering the relationship between two quantitative variables, we can sometimes identify one of the variables as the **explanatory variable**, or **independent variable**, and the other as the **response variable**, or **dependent variable**. The response or dependent variable generally *depends on* or is *explained by* the explanatory variable. For example, the time a student spends playing video games (the explanatory variable) might *explain* the student's grade point average (the response variable). A child's height (the response variable) *is explained by* his age (the explanatory variable). Here is a table showing how a particular boy's height changed between two years and five years of age:

Age in Years	2	3	5
Height in Inches	35	38	42.5

(Note that there are other quantities that might also explain the relationship between age and height. Here we are simplifying the analysis by looking at just one explanatory variable.)

Sometimes we will want to choose values of the explanatory variable and see how the response variable is affected. At other times it might not be obvious which is the explanatory variable and which is the response variable.

When creating a scatterplot, we will use a rectangular coordinate system and plot the explanatory variable on the horizontal axis and the response variable on the vertical axis. We denote points in a rectangular coordinate system as ordered pairs using parentheses, with the explanatory variable as the first coordinate and the response variable as the second coordinate, like this: (explanatory, response). If the choice is not obvious, we might plot either variable on the horizontal axis. The following graph shows a scatterplot of the three points from the boy's age-height table: (2, 35); (3, 38); (5, 42.5).

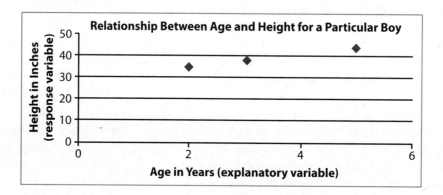

In Example 2.1, we use SAT data to identify the explanatory and response variables, create a scatterplot, and look for trends in the graph.

Example 2.1

Consider the critical reading SAT data for states and the percentage of high school seniors in each state who take the test. (You used these data in Topic 1, Exploration 12. A portion of the table is shown here.)

State	Average Critical Reading SAT	Percent Taking Test
AL	556	7
AK	518	48
AZ	519	25
AR	566	4

Create a scatterplot for these bivariate data and identify which variable should be the explanatory variable, which variable should be the response variable, and why you made this choice. Also note any patterns or trends that this plot reveals. Does a high average critical reading SAT score for a state mean the state has a sound education system?

Solution

Because the percentage of seniors taking the test might help "explain" the state's critical reading SAT score, we chose that as the explanatory variable and plotted it on the horizontal axis. For example, if a lower percentage of seniors took the test in a state, those students might be the higher-achieving students, which helps "explain" the state's score. Each point on the graph represents the data for one state (that is, one individual in this data set); the percent taking the test is the first coordinate, measured on the horizontal axis, and the average critical reading SAT score is the second coordinate, measured on the vertical axis.

On the following graph, the first point, with coordinates (7, 556), represents Alabama with 7% of students taking the test and an average critical reading SAT of 556. To plot the point, we count 7 units to the right for the first coordinate and 556 units up for the second coordinate. The next point, representing Arkansas with 48% of students taking the test and an average critical reading SAT of 518, is the second point given on the plot (48, 518).

The plot of the data for all states is shown here:

The scatterplot for all the states shows that, in general, higher critical reading SAT scores tend to occur in states in which a lower percentage of high school seniors take the test. Lower critical reading SAT scores tend to occur in states in which a higher percentage of seniors take the test. So we really cannot say that states with higher critical reading SAT scores necessarily do a better job of educating their high school students. A more plausible explanation is that in the states in which a smaller percentage of seniors take the SAT test, the stronger students are the ones who tend to take the test, thus resulting in a higher average critical reading SAT score for that state.

Note that the scales used on the two axes in Example 2.1 are not the same. There are some instances in which we will want to use the same scale on both axes and others in which, because of the nature of the variables, we won't. (See Exploration 7 at the end of this topic for an example where we use the same scale on both axes.) In Example 2.1, you should also note that the scale on the vertical axis does not start with 0, but with 450. In this data set, as in many data sets, the values of one or both of the variables are much larger than 0, so the plot shows the relationships much more clearly if the axes intersect at a point different from (0, 0). We need to be sure to mark the axes clearly.

Two variables have a **positive association** if larger values of one variable tend to occur with larger values of the other variable. The variables have a **negative association** if larger values of one variable tend to occur with smaller values of the other variable. From the data in the previous example, we see that states' average critical reading SAT score and the percentage of seniors taking the test are negatively associated. This negative association can be seen in the trend of the data in the scatterplot.

Example 2.2

Match the following four descriptions to the following four scatterplots, and explain why you chose the match. Note that in each plot, time is the explanatory variable and the years are given on the horizontal axis. The scale of the response variable is not marked on the vertical axis. Think about the data to match the descriptions to the pictures. Observe if the association between the variable described and year is a positive association, a negative association, or neither.

a.   U.S. life expectancy at birth, all races, both sexes (*Source:* The Centers for Disease Control, www.cdc.gov)

b.   U.S. gross federal debt in millions (*Source:* The President's Budget for Fiscal Year 2012, www.whitehouse.gov)

c. Active-duty military personnel, excluding reserves on active duty for training (*Source: Information Please, www.infoplease.com*)

d. U.S. cases of tuberculosis per 100,000 population (*Source:* The Centers for Disease Control, www.cdc.gov)

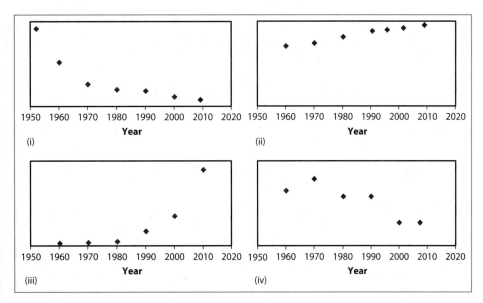

Solution

Graph (i) shows a decline in the value of the response variable as time increases, which is a negative association. Therefore, this graph represents the cases of tuberculosis per 100,000 population; this decline is a result of better healthcare, improved hygiene, and more advanced treatments for disease. Graph (ii) shows a fairly modest increase (a positive association) and so would be a plot of U.S. life expectancy. Graph (iii) also shows an increase in the response variable as time passes (a positive association), but the increase is more dramatic and thus would be a plot of U.S. gross federal debt over time. Graph (iv) shows an increase, followed by a decrease, and would be a plot of U.S. active-duty military personnel. This graph represents an association that is neither positive nor negative. (What accounts for the peak around 1970?)

In mathematics, a general pairing of quantitative variables is called a **relation** or **relationship**. In the next example, we examine a relationship based on the board game Scrabble, to see how the Scrabble point value of a word is related to its length.

Example 2.3

The following table gives the Scrabble point value for each letter in the alphabet:

A = 1	B = 3	C = 3	D = 2	E = 1	F = 4	G = 2	H = 4	I = 1
J = 8	K = 5	L = 1	M = 3	N = 1	O = 1	P = 3	Q = 10	R = 1
S = 1	T = 1	U = 1	V = 4	W = 4	X = 8	Y = 4	Z = 10	

For each of the following ten words, find its Scrabble point value: *case; categorical; chart; data; function; graph; increase; quantitative; scatterplot; variable*. Then make a scatterplot of the variables "number of letters in the word" and "Scrabble point value of the word," and explain what the scatterplot shows.

Solution

Here is a table listing the Scrabble point value for each of the words:

Word	Number of Letters	Scrabble Point Value
Case	4	6
Categorical	11	16
Chart	5	10
Data	4	5
Function	8	13
Graph	5	11
Increase	8	10
Quantitative	12	24
Scatterplot	11	15
Variable	8	13

We will plot "number of letters in the word" on the horizontal axis because it *explains* the Scrabble point value. (Another way to think about this is to realize that the "Scrabble point value of a word" *depends on* the number of letters in it.) For example, the point (12, 24) corresponds to the word *quantitative*. Note that the point (8, 13) corresponds to two words. The scatterplot shown here reveals that, in general, as the number of letters in a word increases, the Scrabble point value of the word tends to increase as well. However, there are words in our list that are exceptions to this tendency. For example, the word *graph* has 5 letters but its Scrabble point value is 11 points; however, the word *increase* has 8 letters, but it has a smaller Scrabble point value of 10 points.

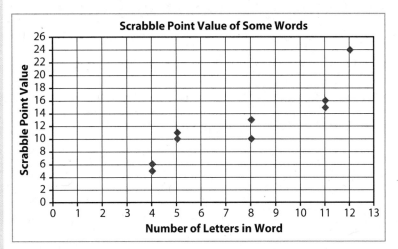

In ordinary language when we say that one entity is a function of another, we mean that the first thing is related to the second and is in some way dependent on it. As we saw in Example 2.3, the Scrabble point value of a word is related to the length of the word. But if we are using the term *function* mathematically, the Scrabble point value is not a function of the length of the word. In mathematics, we mean something precise when we use the term *function*.

A **function** is a relation in which each value of the explanatory variable is paired with exactly one value of the response variable. We will call the explanatory variable x and the response variable y. The variable $y =$ "Scrabble point value of a word" is not a function of the variable $x =$ "number of letters in the word" because there are values of the explanatory variable x paired with multiple values of the response variable. For the word *chart*, $x = 5$ and $y = 10$, which results in the point $(5, 10)$, but for the word *graph*, we have $x = 5$ and $y = 11$, resulting in the point $(5, 11)$. This collection of ordered pairs is not a function because two ordered pairs with the same first values (5 in this case) have different second values. Sometimes we will want to take data for which there is not a functional relationship between the explanatory and response variables and *fit* a function to it. We will look at this modeling process in Topic 6.

Functional relationships, or functions, can be represented in various ways. Some functions are represented by **tables** in which we give a list of the allowable values of the explanatory variable x, and for each value of x we give its associated value of y. (Note that because each value of x in the table has exactly one value of y associated with it, there will be one y value for each x value listed, and the x value will not be repeated.) The following table of data supports graph (i) given in Example 2.2. Examining the data, we can see that there is a unique value of y for each x value given in the table.

Year (x)	Cases of Tuberculosis per 100,000 Population (y)
1953	52.6
1960	30.7
1970	18.1
1980	12.2
1990	10.3
2000	5.8
2009	3.8

Functions can also be given using **symbols**. If we let h represent the hours worked in one week at a job that pays $6.15 per hour, then total wages w earned for the week could be represented symbolically as $w = 6.15h$. We could then show this function relationship in a table for selected values of h as follows (note that w is a function of h because for each h value, there is exactly one w value):

Hours h	5	10	12	15	20	30	40
Wages w	$30.75	$61.50	$73.80	$92.25	$123.00	$184.50	$246.00

If a function is given symbolically as a formula, we often use y to represent the response variable and x to represent the explanatory variable. For example, if the function is given by $y = 2x+1$, we can see that for each number we put in for x we will compute a unique y value. This means that y is a function of x. When representing functions using symbols, we often use letters that suggest the quantities they represent, instead of x and y. For example, we used h to represent hours worked and w to represent total wages. It really doesn't matter what letters we use, but h and w might help us remember the quantity each represents.

Example 2.2 shows four functions represented as **graphs**. Note that on each of these graphs, every value of the year (which is the explanatory variable) that appears on the graph is paired with only one value of the response variable. The graphs in Example 2.2 show functions with a finite number of points. In Example 2.4 we will show a function with infinitely many points. The graph of a function with infinitely many points might appear as a line, a curve, or a series of line segments and/or curve segments.

Words can also be used to describe a function. For example, the energy cost in calories to an individual engaging in an activity such as jogging is affected by a variety of factors that vary from person to person but weight is a critical factor. The calories used while doing a particular activity is the response variable, and the individual's weight is the explanatory variable. According to one model, a 110-pound person, for example, burns approximately 3.4 calories per minute playing table tennis, while a 150-pound person burns approximately 4.5 calories per minute and a 190-pound person burns approximately 5.9 calories per minute on the same activity. The next example takes a function given in words and asks us to represent it in table form, as a graph, and in symbols.

Example 2.4

According to one model, a 150-pound person uses approximately 5.4 calories per minute walking at a normal pace on an asphalt road. Represent the total number of calories used by a 150-pound person as the response variable c, and represent c as a function of the number of minutes spent walking m. Represent this function as a table, in symbols (that is, using the letters c and m and relating them in an equation), and as a graph. (*Source:* George Constable, ed., *Setting Your Weight*, Alexandria, VA: *Time-Life Books*, 1987, p. 32.)

Solution

When we describe a function in a table, we need to decide which values of the explanatory variable to include in the table. We choose the explanatory variable to be "minutes spent walking"; we'll use increments of 5 minutes, over an interval from 10 to 50 minutes. We'll then enter the corresponding values for the response variable "total calories used" into the table:

Minutes Spent Walking	10	15	20	25	30	35	40	45	50
Total Calories Used	54	81	108	135	162	189	216	243	270

The table gives total calories expended by a 150-pound person using increments of 5 minutes from 10 minutes to 50 minutes. In symbols, we represent the functional relationship between the minutes m and the calories c as $c = 5.4\,m$. A graph of this function is shown next:

In the function examined in Example 2.4, the ratio $\frac{c}{m} = 5.4$ for any pair (m, c) given in the table or any point represented on the graph. In a situation like this, when the ratio of the response variable over the explanatory variable is a fixed number or a constant, we say that the response variable is **directly proportional** to the explanatory variable. The constant is called the **constant of proportionality**. For the function in Example 2.4, $\frac{c}{m} = 5.4$ and the constant of proportionality is 5.4. If we multiply both sides of the equation $\frac{c}{m} = 5.4$ by m, we write $c = 5.4m$. As we did in Example 2.4, this is the way we typically write an equation relating directly proportional variables. The next example investigates other directly proportional functions.

Example 2.5

For each of the relationships described below, use the indicated letters to represent the variables and write an equation that gives the functional relationship between the variables.

a. Total number of miles m traveled in h hours if traveling at a constant speed of 55 miles per hour

b. Total cost c in dollars of p pounds of bananas if bananas cost $.39 per pound

c. Length in centimeters c of a ribbon that is i inches long (Recall that an inch is equal to 2.54 centimeters.)

d. Length in inches i of a belt that is c centimeters long

Solution

These are all functions in which the response variable is directly proportional to the explanatory variable.

a. The function is $m = 55h$. Note that the units on the left are miles and the units on the right are miles per hour times hours, which also gives miles.

b. Here, the function is $c = 0.39p$. The units on the left are dollars and on the right are dollars per pound times pounds, which also gives dollars.

c. Because 1 inch = 2.54 centimeters, i inches = $i \times 2.54$ centimeters. So the length in centimeters of a ribbon that is i inches long is $c = 2.54i$.

d. We start with the equation 1 inch = 2.54 centimeters and divide both sides by 2.54 to get 1 centimeter = $\frac{1}{2.54}$ inches. Then c centimeters = $c \times \frac{1}{2.54}$ inches. The length in inches of a belt c centimeters long is $i = \frac{c}{2.54}$.

In the next example we use proportional relationships to analyze the size of a television screen.

Example 2.6

Traditional television screens have a width-to-height ratio of 4:3. Manufacturers give the size of a television by the length of its diagonal. For example, a 32-inch television is one with a diagonal measure of 32 inches.

a. Write an equation that gives the width of a television screen in terms of its diagonal.

b. Give the dimensions (width and height) of a 32-inch television screen.

Solution

a. If h represents the height and w the width, we know that the width-to-height ratio is $\frac{w}{h} = \frac{4}{3}$. We can use this equation to write h in terms of w: Multiply both sides of the equation by $3h$ to get $3w = 4h$. Then, dividing by 4, we have $h = \frac{3}{4}w$.

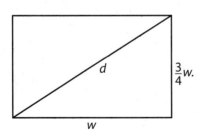

The diagonal d is the hypotenuse of the triangle with sides w and $\frac{3}{4}w$. By the Pythagorean theorem, the square of the hypotenuse is the sum of the squares of the sides. So $d^2 = w^2 + \left(\frac{3}{4}w\right)^2 = w^2 + \frac{9}{16}w^2 = \left(1 + \frac{9}{16}\right)w^2 = \left(\frac{16}{16} + \frac{9}{16}\right)w^2 = \frac{25}{16}w^2$. Hence, $d^2 = \frac{25}{16}w^2$. To solve for w, we multiply both sides of the equation by $\frac{16}{25}$ and get $\frac{16}{25}d^2 = w^2$. Because d and w represent lengths, they are positive numbers. We take square roots on both sides of the equation and obtain $\sqrt{\frac{16}{25}}d = w$; thus, $w = \frac{4}{5}d$.

b. If the diagonal of a television screen measures 32 inches, then its width is $w = \frac{4}{5}d = \frac{4}{5}(32) = \frac{128}{5} = 25.6$ and its height is $\frac{3}{4}w = \frac{3}{4}(25.6) = 19.2$. We conclude that a 32-inch television screen is 25.6 inches wide and 19.2 inches high.

Summary

In this topic, we looked at how a scatterplot could be used to show the relationship between two quantitative variables, and we investigated the trends that could be seen using this type of graph. Some pairs of quantitative variables are positively associated, others are negatively associated, and some show no particular relationship. We considered how to express functional relationships between two variables using words, symbols, tables, and graphs. We also explored directly proportional functions; in each directly proportional function involving two variables, the ratio of the variables is a constant.

Explorations

1. Colonial population estimates (in round numbers) are given in the following table for the decades before the establishment of the U.S. Census in 1790. Create a scatterplot for this data table and describe the trends shown by your graph.

Year	Population	Year	Population
1610	350	1700	250,900
1620	2,300	1710	331,700
1630	4,600	1720	466,200
1640	26,600	1730	629,400
1650	50,400	1740	905,600
1660	75,100	1750	1,170,800
1670	111,900	1760	1,593,600
1680	151,500	1770	2,148,100
1690	210,400	1780	2,780,400

Source: Time Almanac 2004, p. 175.

2. For each of the following pairs of variables, sketch a rough plot (as a series of points) that could reasonably represent the relationship between the explanatory and response variables. Indicate which is the explanatory variable and which is the response variable, and explain why you made the sketch you did. Also determine if your plot shows a positive association between the variables, a negative association, or neither.

a. Monthly grocery bill of a household; number of people in the household

b. Hours per day spent watching television; a college student's grade point average

c. Time it takes to run 100 yards; the runner's age

d. Number of minutes elapsed since being taken out of the oven; temperature of a pizza

3. For each of the relationships described below, determine if the variables are directly proportional. Write a sentence or two for each to justify your answers.

   a. Total number of words a typist can type and the number of minutes spent typing if he or she types 72 words per minute

   b. The temperature of an asphalt road at several times on a hot summer day and air temperature at those same times

   c. Total number of miles you travel if you average 26 miles per gallon and the gallons of gas used

   d. The total days of vacation earned after y years of work if you earn two vacation days for every three months you work

4. For each of the relations described below, use the indicated letters to represent the variables and write an equation that gives the functional relationship between the variables. (In some cases it may help you to draw a picture of the situation.)

   a. The length in yards y of a fence that is m meters long (One meter is equivalent to 1.0936 yards.)

   b. The cost c per course when a student takes four courses and pays t dollars in tuition

   c. The area of a rectangular rug with width w and length twice the width

   d. The number of 1-foot square tiles t needed to construct a 2-foot-wide path around a square garden of side x feet long

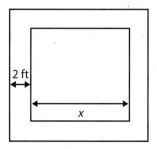

5. Wide-screen television screens have a width-to-height ratio of 16:9.

   a. Write an equation that gives the width of a television screen in terms of its diagonal.

   b. Give the dimensions (width and height) of a 32-inch television screen (a screen with a diagonal of 32 inches).

   c. Give the dimensions of a 45-inch television screen (a screen with a diagonal of 45 inches).

6. For each student in your class, collect the following data: student's gender; his or her height in inches; his or her hand-span measurement, also in inches. (Before collecting these measurements, decide how you will define the hand-span measurement.)

   a. Sketch a scatterplot of the data collected from the class and discuss any trends. Which variable did you choose for the horizontal axis and which variable did you choose for the vertical axis? Does it matter?

   b. Discuss any problems associated with collecting these measurements.

   c. How can you indicate a student's gender on the scatterplot?

7. The following table gives information from a sample of college students: gender; number of children in family of origin; and number of children in their ideal family, in which they may someday be a parent.

Gender	No. of Children in Family of Origin	No. of Children in Ideal Family
F	2	2
M	3	2
M	4	3
F	2	2
F	4	3
M	5	5
F	3	3
F	2	3
F	2	2
F	4	4
M	3	3
F	4	0
F	3	3
F	3	4
F	1	2
M	2	2
M	3	3
M	2	2
M	1	1
M	2	0

a. Sketch a scatterplot of the data collected from the students and discuss any trends. (Use the same scale on both axes.) Which variable did you choose for the horizontal axis and which variable did you choose for the vertical axis? Does it matter?

b. Describe the role of the diagonal line $y = x$ and what it helps you see about the data.

c. Do you think the data for males and the data for females should be considered separately? Why or why not?

8. The following table of data gives hockey star Eric Lindros' career numbers for the eight seasons he played with the Philadelphia Flyers:

Year	Games	Goals	Assists	Points
1993–94	65	44	53	97
1994–95	46	29	41	70
1995–96	73	47	68	115
1996–97	52	32	47	79
1997–98	63	30	41	71
1998–99	71	40	53	93
1999–2000	55	27	32	59
Totals	**486**	**290**	**369**	**659**

Source: The Internet Hockey Database, www.hockeyDB.com.

a. Sketch a scatterplot of the two variables "number of games played in a season" and "number of points scored by Lindros," and discuss any trends.

b. Which variable did you choose for the horizontal axis and which variable did you choose for the vertical axis? Does it matter?

c. Determine if there appears to be a relationship between goals and assists. What might explain this?

9. Here is a table relating age and height for a particular girl between two years and ten years of age. Create a scatterplot for these data and describe any trends. Is height a function of age? Give a reason for your answer.

Age in Years	2	3	4	5	6	8	10
Height in Inches	33	36	37	41	44	49	53

2-1

Estimating Dates: Scatterplots

In this activity, you will estimate the dates of several major events and then use a scatterplot to compare your estimates with the actual dates these events occurred. You will analyze relationships between two variables that can be read from a scatterplot.

You will start with an estimation experiment. Work with one or two classmates and fill in the following table, giving your group's best estimate of the year in which each event occurred.

1. For each of the following twentieth-century events, estimate the year in which the event occurred. Record only the last *two* digits of the year; because you know all dates are in the 1900s, you don't need to record the "19." Do not look up the actual dates.

Event	Estimated Year Event Occurred
Oprah Winfrey born	
Martin Luther King assassinated	
Original Woodstock music festival held	
Nineteenth Amendment established women's suffrage	
Long-distance telegraphic radio signal sent across the Atlantic	
Bart Simpson's character made his debut	
Aerosol can invented	
Franklin D. Roosevelt elected to a third term of office	

"Scotch" tape invented	
Hank Aaron inducted into the Baseball Hall of Fame	
Pearl Harbor bombed	
President Nixon resigned	
Schindler's List won the Academy Award for best picture	
San Francisco destroyed by earthquake and fire	
The World Wide Web was developed	

2. Obtain the actual dates for those events from your instructor.

a. Create a scatterplot of the data, with the horizontal axis *x* representing "year event occurred" and the vertical axis *y* representing "estimated year event occurred." (You might want to record these actual dates immediately to the left of your estimates in the previous table.)

b. Why is it appropriate to use the variable "year event occurred" as the explanatory variable and "estimated year event occurred" as the response variable?

c. How close were your predictions of events that occurred in the 1980s to the actual dates?

d. What would the scatterplot look like if you had guessed the correct year for each event?

e. Sketch the line $y = x$ on your graph. Does it appear that you overestimated more than you underestimated, or that you underestimated more than you overestimated or neither? How did you determine this from your graph?

f. Use Excel to redraw the scatterplot you created previously. To do so, you need to enter the data in two columns. One column should contain the values of the explanatory variable and the second column the values of the response variable.

Instructions to Create a Scatterplot

1. Get the file "EA2.1 Events and Dates.xls" from the text website or WileyPLUS. In column C of the Excel worksheet, enter your estimates of the year each of the events occurred. (Enter only the last two digits of the year.) Add an appropriate column title in cell C1.

2. To create a scatterplot of the x-y data with x representing "year event occurred" and y representing "estimated year event occurred," select the two columns of data, including the labels. Go to the **Insert** tab and choose **Scatter** from the **Charts** group and click on the first type of graph (the "dots").

3. Select the graph by clicking inside the border of the graph. Go to **Chart Layouts** and select Layout 1 (the first one in the first row).

4. Enter a relevant chart title and titles for the two axes. Also, click on the legend and delete it by pressing the **Delete** key. Note that you can find the coordinates of any point on your completed graph by moving the cursor to the point. If you want to change the scale on either axis, point to the numbers on the axis and right-click; then select **Format Axis**. Then, using the **Axis Options**, select **Fixed** for **Minimum** and **Maximum** and enter the appropriate minimum and maximum values for your axis. Finally, click **Close**.

g. Change your estimated values so the values of "estimated year event occurred" and the values of "year event occurred" are equal, and look at the corresponding scatterplot. Describe the graph. Does your description agree with your answer in part (d)?

Additional Questions

3. The following data give the number of strikes or lockouts involving more than 1,000 workers, and the percentage of the total labor force belonging to a union between the years 1950 and 2010:

Year	Strikes and Lockouts	Union Membership Percentage
1950	424	31.5
1960	222	31.4

1970	381	27.4
1980	187	23
1990	44	16.1
1995	31	14.9
2000	39	13.4
2005	22	12.5
2010	11	11.9

Source: The Bureau of Labor Statistics, www.bls.gov.

a. What kinds of information, related to these data, would you want to get from a graph?

b. Create a scatterplot that relates the number of strikes and lockouts with the percentage of the total labor force with union membership.

c. Which variable did you use as the explanatory variable? Why?

    d. Explain what your graph shows.

4. Create two other scatterplots using data from the previous table. Explain what each graph shows.

Summary

In this activity, you learned how to use Excel to create a scatterplot. In creating scatterplots, you decided which variable should go on the horizontal axis (that is, which variable is the explanatory variable) and which should go on the vertical axis (that is, which is the response variable). You also discovered how the line $y = x$ can help you see relationships in paired data. You will use these skills to visualize relationships between variables in bivariate data sets.

2-2

State Governors' Salaries and Per Capita Income: More on Scatterplots

In this activity, you will use scatterplots to investigate the relationship between the governor's salary and the average per capita income in the state. You will also analyze differences between scatterplots and other types of graphs.

1. The following table gives the average per capita income and the governor's salary in 2010 for each state in a group of nine northeastern states:

State	Per Capita Personal Income 2010 in $	Governor's Salary 2010 in $
Connecticut	56,001	150,000
Maine	37,300	70,000
Massachusetts	51,552	140,535
New Hampshire	44,084	113,834
New Jersey	50,781	175,000
New York	48,821	179,000
Pennsylvania	41,152	174,914
Rhode Island	42,579	117,817
Vermont	40,283	142,542

a. Explain how you could use a bar graph to represent and interpret one or more aspects of the data. Which aspects of the data would the bar graph help interpret?

b. Would a histogram help you interpret the data, or an aspect of the data? Explain.

c. Which type of graph would help you answer this question: Is there any relationship between per capita income in a state and the governor's salary?

d. Create a scatterplot of the data using the two quantitative variables, "per capita income" and "governor's salary."

e. Which variable did you use on the *x*-axis? Why?

f. Which variable did you use on the *y*-axis? Why?

g. Explain what your graph shows about the relationship of the two variables.

2. To analyze this relationship further, you will create a scatterplot using data on all 50 states. To do so, you will first create a two-column table that records the governor's salary and the average per capita income for each state.

Instructions to Create a Table from Other Tables

1. Retrieve the file "EA2.2.1 Per Capita Income.xls" from the text website or WileyPLUS. This file contains a list of all states and the per capita personal income for each for 2010. (*Source:* Bureau of Business and Economic Research, http://bber.unm.edu.)

2. Open the file "EA2.2.2 Governors Salaries.xls" from the text website or WileyPLUS, and then follow instructions in (3) below to copy and paste the list of governors' salaries into the worksheet "EA2.2.1 Per Capita Income.xls."

3. First, ensure that both files contain all the states listed in the same order. Then put the cursor at the top of column B, "Governors' Salaries," and click to highlight the entire column. Click on the **Copy** icon (two pieces of paper partly overlapping) from the **Clipboard** group, the first group on the left, under the **Office Button**.

4. Return to the per capita income worksheet. Place the cursor at the top of column C and then select **Paste** from the **Clipboard** group to insert the governors' salaries into column C of the per capita worksheet. You should now have three columns in this worksheet.

a. Create a scatterplot of this data set. You should change the scale of the horizontal (or x) and vertical (or y) axes so the data are easier to read. (To change the scale on an axis, point to the numbers on that axis and right-click; then select **Format axis**. Using the **Axis Options**, select **Fixed** for **Minimum** and **Maximum**, and enter the appropriate minimum and maximum values for the axis.) What is a reasonable minimum value of x to use? What about y?

b. Are there any trends or patterns to this data? Explain.

c. Are there any data points that appear to be "away from" the rest of the data? If so, which one(s) and what makes them stand out?

Additional Questions

3. The file "EA2.2.3 Children in Poverty.xls," which you will find on the text website or WileyPLUS, contains the table shown here. This table gives the percentage of children younger than 18 who were living below the poverty level in the United States from 1976 to 2006. Use Excel to create a scatterplot for this data table and write a paragraph describing the trends shown by your graph. Does this scatterplot represent a function? Why or why not?

Year	Percentage of Children	Year	Percentage of Children
1976	16	1992	22.3
1977	16.2	1993	22.7
1978	15.9	1994	21.8
1979	16.4	1995	20.8
1980	18.3	1996	20.5
1981	20	1997	19.9
1982	21.9	1998	18.9
1983	22.3	1999	16.9
1984	21.5	2000	16.2
1985	20.7	2001	16.3
1986	20.5	2002	16.7
1987	20.3	2003	17.6
1988	19.5	2004	17.8
1989	19.6	2005	17.6
1990	20.6	2006	17.4
1991	21.8		

Source: U.S. Census Bureau, www.census.gov

4. The table here (also available in the "EA2.2.4 US Farms.xls" file on the text website or WileyPLUS) gives the total number of acres (in thousands) of land devoted to farming in the United States and the farm population (in thousands) from 1900 to 2008.

Year	Farm Acreage (in thousands)	Farm Population (in thousands)
1900	841,202	29,835
1910	881,431	32,077
1920	958,677	31,974
1930	990,112	30,529
1940	1,065,114	30,547
1950	1,161,420	23,048
1960	1,176,946	15,635
1970	1,102,769	9,712

1980	1,039,000	6,051
1990	986,850	4,801
1995	962,515	N.A.
2000	945,080	N.A.
2008	920,000	3,282

Source: The New York Times Almanac 2011, p. 316.

Note that some data are not available. You need to decide what to do in that situation when answering the following questions.

a. Create an appropriate scatterplot that you can use to analyze any trends about the land used for farming from 1900 to 2008. Describe these trends.

b. Create a second scatterplot and use it to describe the relationship between the number of acres used for farming and the total farm population in the United States.

Summary

In this activity, you considered the differences among a bar graph, a histogram, and a scatterplot and the information conveyed. You learned to copy and paste in Excel. You looked at trends in a sample compared to trends in a full data set and identified unusual observations. You also investigated trends over time and relationships between variables.

Graphs of Functions

After completing this topic, you will be able to:

- Determine when a graph represents a function.

- Use the graph to identify if the values of a function are increasing or decreasing.

- Use the graph to find absolute and relative maximum and minimum values of a function.

- Recognize, from the graph of a function, when the rate of change is increasing or decreasing.

- Apply the information from a graph to help analyze a particular situtation.

W e use functions to model relationships between two variables with available data to better understand the relationship and, in many cases, to predict values. The graph of a function is an important tool for analyzing the function's behavior.

The graph of a function consists of all points with coordinates (a, b), where b is the value of the response or dependent variable that corresponds to the value a of the explanatory or independent variable. In many situations, the values of the independent variable can take on infinitely many values in a given interval (for example, when the independent variable is time), so the graph consists of a continuous curve, rather than isolated points. The following graph shows the temperature in a northeastern U.S. city during 24 hours starting at noon on a winter day.

The graph of a function describes the function "at a glance." It provides a quick way of identifying important properties of the function such as whether it is increasing or decreasing and where it peaks. Because each value of the explanatory variable is paired with exactly one value of the response variable, a vertical line drawn anywhere on the graph will intersect the graph in no more than one point. This gives us a test, called the **vertical line test**, to determine whether a graph is that of a function or not.

A function is **increasing** if the values of the response (dependent) variable increase when the corresponding values of the explanatory (independent) variable increase. A function is **decreasing** if the values of the response variable decrease when the corresponding values of the explanatory variable increase.

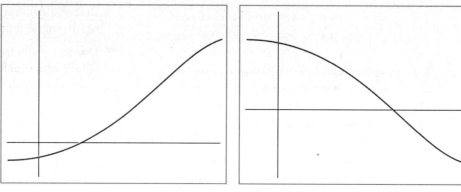

Increasing Function   **Decreasing Function**

Because the independent variable is represented on the horizontal axis with values increasing to the right, and the dependent variable is represented on the vertical axis with values increasing upward, the graph of an increasing function rises when traced from left to right. Similarly, the graph of a decreasing function falls when traced from left to right. Most functions are increasing over some intervals of the independent variable and decreasing over others.

The graph of the temperature in a northeastern city over 24 hours starting at 12 noon on a winter day shows that the temperature decreased during the first 17 hours and then increased for approximately 5 hours. The temperature stayed approximately constant during the last 2 hours.

Points of special interest on the graph of a function are the highest and lowest points, which give the (absolute) **maximum** and (absolute) **minimum** values of the function, respectively; that is, they give the largest and smallest values of the response variable.

Example 3.1

The following graph represents the national student loan default rate, as a percentage of borrowers, from 1987 to 2008:

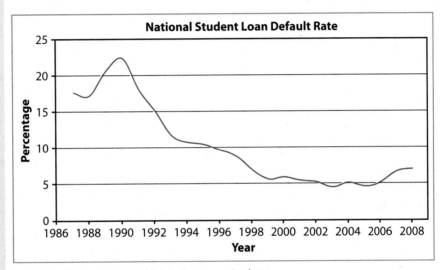

Source: U.S. Department of Education, www2.ed.gov.

Use the graph to answer the following questions.

a. Is this the graph of a function? Why?

b. During which years was the default rate increasing?

c. During which years was the default rate decreasing?

d. What was the default rate in 2000?

e. What was the maximum default rate between 1987 and 2008? When did it occur?

Solution

a. Because any vertical line intersects this graph at no more than one point, the graph is that of a function.

b. We estimate the interval of time when the default rate is increasing by reading, on the horizontal axis, the time interval over which the graph of the function rises from left to right: We estimate that the default rate was increasing from 1988 to 1990, from 1999 to 2000, from 2003 to 2004, and from 2005 to 2008.

c. The default rate was decreasing from 1987 to 1988, from approximately 1990 to 1999, from 2000 to 2003, and from 2004 to 2005.

d. The default rate in 2000 is given by the value of the dependent variable (represented in the vertical axis) that corresponds to the value 2000 of the independent variable (represented on the horizontal axis). Using this graph we estimate that the default rate in 2000 was 6%.

e. Observing that the highest point of the graph is the point corresponding to the value 22.5 of the dependent variable and 1990 of the independent variable, we see that the maximum default rate was 22.5%, which occurred in (approximately) the year 1990.

In addition to the maximum and minimum, some points of special interest on the graph of a function are all those where the function changes from increasing to decreasing, and those where the function changes from decreasing to increasing. These are "turning points." The value of the response variable at a point where the function changes from increasing to decreasing is a **relative maximum** (or **local maximum**) value of the function. The value of the response variable at a point where the function changes from decreasing to increasing is a **relative minimum** (or **local minimum**) value of the function. The graph here shows a function with two relative maximum values (M_1 and M_2) and one relative minimum (m).

Relative Minimum Value: *m*. Relative Maximum Values: M_1 and M_2.

Example 3.2

Using the function in Example 3.1, give the maximum value, the minimum value, and any relative maximum or minimum values that exist. Also indicate the year in which they occur. Explain why these values are of interest.

Solution

The maximum value of the function is approximately 22.5%, as given in the solution of part (e) of Example 3.1. It occurred in 1990, which means the highest default rate occurred in 1990 when approximately 22.5% of borrowers defaulted. This value of 22.5% is also a relative maximum, and a point where the function changes from increasing to decreasing. There are two other relative maximum values that are less noticeable. They are 6%, which occurred in the year 2000, and 5%, which occurred in 2004. For the government analyst, these points are of interest because they show a trend change and might indicate broader economic improvements.

The minimum value is approximately 4.5% (in 2003). This value is also a relative minimum because at the point (2003, 4.5) the function changes from decreasing to increasing. Other relative minimum values are approximately 17% (in 1988), 5.5% (in 1999), and 4.6% (in 2005). The points (1988, 17), (1999, 5.5), (2003, 4.5), and (2005, 4.6) would be of interest because the default rates stopped decreasing and started to increase. Because this is an unwanted change, the analyst would study what may have occurred at that time to cause such a change.

The intervals of the horizontal axis over which a function is increasing or decreasing may be quite small. Sometimes we need to overlook small changes to see the general trend. In Topic 6, we will see how to model more formally such a general trend.

Example 3.3

The following graph shows the fluctuations in annual mean temperature in New York City's Central Park for the years 1869 to 2010. Disregarding small oscillations, explain the general behavior of annual mean temperature in Central Park, giving the maximum and minimum values.

Source: National Weather Service Forecast Office, www.erh.noaa.gov/okx/climate.html.

Solution

During the time period from 1869 to 2010, the annual mean temperature in Central Park ranged from a minimum of approximately 49.5° around 1888 to a maximum of 57° in 1998. The relative minimum values are getting larger as time increases, which can be seen if we look at the "dips" in the graph around 1888, 1893, 1904, 1917, 1926, 1940, 1958, 1967, 1978, and several between 1990 and 2010. Similarly, the peaks in the graph also follow a somewhat upward trend.

Another useful piece of information about a function that we can observe from its graph directly is the rate at which the values of the response (dependent) variable are changing per unit change in the explanatory (independent) variable. This tells us whether the function values are increasing or decreasing rapidly or slowly.

If x_1 and x_2 are two values of the explanatory variable and y_1 and y_2 are the corresponding values of the response variable, the **average rate of change** of y per unit change in x over the interval from x_1 to x_2 is the ratio $\frac{y_2 - y_1}{x_2 - x_1}$. We will refer to this quantity as the rate of change of y from x_1 to x_2.

Example 3.4

Estimates of cigarette consumption in the United States and of the numbers of cigarettes exported each year are given in the following table. Compare the rates of change of the two given functions over the intervals of time from

a. 1970 to 1980

b. 1990 to 1995

c. 1995 to 1996

d. 2003 to 2006

Year	U.S. Consumption of Cigarettes (in billions)	Exports of Cigarettes (in billions)
1960	484.4	20.2
1970	536.4	29.2
1980	631.5	82.0
1990	525.0	164.3
1995	487.0	231.1
1996	487.0	243.9
1997	480.0	217.0
1998	465.0	201.3
1999	435.0	151.4
2000	430.0	148.3
2001	425.0	133.9
2002	415.0	127.4
2003	400.0	121.5
2004	388.0	118.7
2005	376.0	113.3
2006	371.0	111.3

Source: U.S. Department of Agriculture, Economics, Statistics and Market Information System, http://usda.mannlib.cornell.edu.

Solution

a. Because 536.4 billion cigarettes were consumed in the United States in 1970 and 631.5 billion were consumed in 1980, the rate of change in cigarette consumption from 1970 to 1980 was $\frac{631.5 - 536.4}{1980 - 1970} = \frac{95.1}{10} = 9.51$ billion cigarettes per year. The rate of change in number of cigarettes exported during the same interval of time was $\frac{82 - 29.2}{10} = 5.28$ billion cigarettes per year. Both the consumption and the export figures increased during this period, with the rate of increase in national consumption being almost twice the rate of increase in exports.

b. From 1990 to 1995, the rate of change in number of cigarettes consumed in the United States was $\dfrac{487 - 525}{1995 - 1990} = \dfrac{-38}{5} = -7.6$ billion cigarettes per year. The negative sign here reflects the fact that the number of cigarettes consumed decreased. The rate of change in the number of cigarettes exported during the same period of time was $\dfrac{231.1 - 164.3}{5} = 13.36$ billion cigarettes per year. During this five-year period, consumption decreased at a rate of 7.6 billion cigarettes per year, while exports increased at a rate of 13.36 billion per year.

c. From 1995 to 1996, national consumption remained the same (the rate of change was $\dfrac{0}{1} = 0$), while exports increased at a rate of 12.8 billion per year.

d. During the three-year period from 2003 to 2006, national cigarette consumption decreased at a rate of 9.67 billion cigarettes per year, while exports decreased at a rate of 3.4 billion cigarettes per year.

Note that when the function is decreasing, the rate of change is negative because if $x_1 < x_2$ then $x_2 - x_1$ is positive and $y_2 - y_1$ is negative; that is, the values of the response variable decrease when the values of the explanatory variable increase. If the function is increasing, then $y_2 - y_1$ and $x_2 - x_1$ are both positive when $x_1 < x_2$ and so the rate of change is positive.

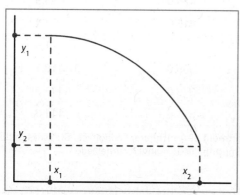

$y_2 - y_1$ is negative.

$y_2 - y_1$ is positive.

We can compare rates of change over different intervals by observing how steep the graph of the function is over each interval. If the rate of change is positive, a steeper graph means the rate of change is greater. We know this because for the same interval length $x_2 - x_1$, greater values of $y_2 - y_1$ give greater values of the quotient $\dfrac{y_2 - y_1}{x_2 - x_1}$.

Example 3.5

The following graphs are of the two functions given in the table in Example 3.4. Use the graph to answer the following questions.

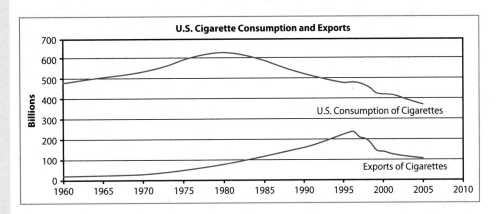

U.S. Cigarette Consumption and Exports

a. Give a time interval when the rate of change of U.S. cigarette consumption was negative, while the rate of change of exports of cigarettes was positive.

b. Give an interval of time when the number of cigarettes consumed in the United States increased faster than the number of cigarettes exported. Which of the two functions has the larger rate of change over this interval?

c. Let R_1, R_2, and R_3 be the rates of change of U.S. cigarette consumption during the ten-year periods 1960–1970, 1970–1980, and 1980–1990, respectively. Without calculating these rates, write them in descending order.

Solution

a. The graph shows that cigarette consumption decreased while exports increased from 1980 to approximately 1995. An interval when the rate of change of U.S. cigarette consumption was negative and the rate of change of exports was positive is, for example, the interval 1980–1995 (or any smaller interval contained within that interval).

b. The number of cigarettes consumed increased faster than the number of cigarettes exported during the interval 1970–1980. During that interval, the graph of the cigarette consumption function is steeper than the graph of the exports function; therefore, the rate of change of cigarette consumption is greater than the rate of change of exports.

c. R_3 is the smallest; it is negative because the graph is decreasing over the interval 1980–1990. The other two rates are positive, and R_2 is greater than R_1 because the graph is steeper over the interval 1970–1980 than it is over the interval 1960–1970. Placing the rates in descending order, we have R_2, R_1, R_3.

We conclude this topic with another observation we can make from the graph of a function. The way the graph is curved, upward or downward, indicates whether the *rate of change* of the function is increasing or decreasing.

We say that a function is **concave upward** when its graph is a curve bent upward. These are two such graphs:

Increasing and Concave Upward

Decreasing and Concave Upward

The following two graphs are curved downward. The functions they represent are **concave downward**.

Increasing and Concave Downward

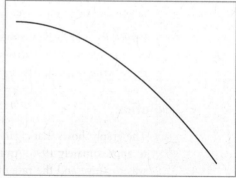

Decreasing and Concave Downward

If the graph of a function is curved upward, then the rate of change of the function is increasing. In the graph in Example 3.5, the graph of the function showing the number of cigarettes exported is increasing and curved upward over the interval 1980–1990. If we look at the rates of change over intervals of equal length, say, two-year intervals, within the 1980–1990 period, we can see that these rates are positive and growing larger as we move to the right because the graph is getting steeper.

If the graph of a function is decreasing and curved upward, then the rate of change (which is negative) is getting larger (or "less negative") because the graph is becoming less steep.

This is the case, for example, for the function that shows U.S. cigarette consumption over the time interval 1986–1993 (see the graph in Example 3.5), where cigarette consumption is decreasing so the rate of change is negative, but consumption is leveling off, so the rate at which it is decreasing is getting larger (approaching 0).

When the graph of a function is curved downward, the rate of change is decreasing. If the function is increasing and the graph curves downward, then the rate of change is positive and getting smaller; that is, the graph is rising but less steeply as we move from left to right. When the function is decreasing and curved downward, then the rate of change is negative and getting smaller (more negative); that is, the curve is falling more steeply as we move from left to right.

In many situations, it is important to look at whether the rate of change is increasing or decreasing; for example, if during an epidemic the number of new cases of sick people is increasing at an increasing rate, health officials would see it as a good sign when the rate of increase of number of new cases starts to decrease. This would mean that their control methods are working and that the end of the epidemic is nearer.

Example 3.6

The following graph shows the approximate number of daily visitors (in millions) to the two most popular social networks, Facebook and MySpace, from April 2007 to April 2008. Describe the changes in the number of visitors to each network and make a prediction about the number of visitors in the months of May and June of 2008. Which of the two networks would you expect to have more visitors in June 2008?

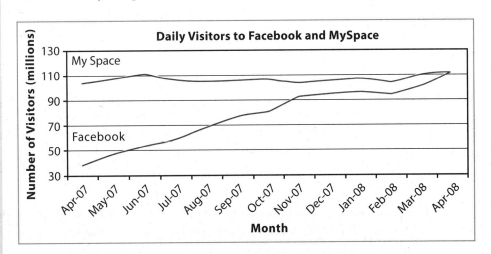

Solution

The number of visitors to MySpace was about 105 million in April 2007 and remained fairly constant throughout the year. The highest number of visitors was approximately 112 million and occurred in June 2007 and again in April 2008.

In April 2007, the number of visitors to Facebook was much smaller than the number of visitors to MySpace, but grew quickly and in April 2008 both networks had the same number of visitors. The number of visitors to Facebook increased from April 2007 to January 2008 and from February 2008 to April 2008. It decreased slightly from January to February of 2008.

Although both curves show that the number of visitors is increasing from February to April 2008, the rate of increase decreased for MySpace and increased in the case of Facebook. Assuming that the trend continues, we predict that the number of visitors to Facebook would be larger than the number of visitors to MySpace in the months following April 2008.

In the following example, we construct a graph from a verbal description of trends.

Example 3.7

The number of students enrolled each year in elementary and secondary schools in the United States is given by a function that satisfies the following: The number of students increased from 40 million in 1960 to 51.2 million in 1970. From 1970 to 1983, enrollment decreased slowly until 1977, when enrollment was 49 million, and then decreased more rapidly from 1977 to 1985, when it reached a minimum of 44.9 million. Enrollment then increased rapidly (the rate at which it was increasing was also increasing) to 51.5 million in 1996. From 1996 until 2006, it continued increasing, but with a decreasing rate of change. Enrollment was 55.3 million in 2006 and has remained at similar levels for the rest of the decade. Using this information, draw a possible graph of the function that describes student enrollment from 1960 to 2010.

Solution

To satisfy the given description, we need to draw a graph in which enrollment increases from 1960 to 1970, decreases from 1970 to 1985, and increases again from 1985 to 2006.

There is a local maximum of 51.2 million in 1970 and a local minimum of 44.9 million in 1985. Based on the data given, other points that we need to include are (1960, 40), (1977, 49), and (1996, 51.5). To satisfy the other conditions, we draw a function that is concave downward from 1970 to 1985 with a decrease that is not so rapid at first, but gets more rapid. Indicating that the rate of increase from 1985 to 1996 is increasing, we draw the curve concave upward; that is, curved upward. Because the rate of increase decreases from 1996 to 2006, the graph is concave downward but still increasing from 1996 to 2006.

Here is a possible graph:

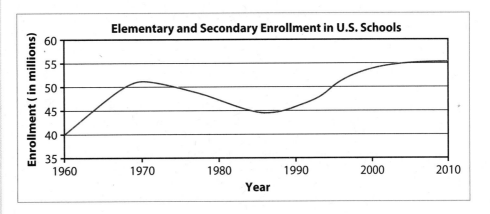

Summary

In this topic, we analyzed graphs of functions to identify relative (local) maximum values, relative (local) minimum values, intervals where the function is increasing and where it is decreasing, where the graph is concave upward and where it is concave downward. We interpreted these characteristics, easily seen on the graph of a function, in terms of the values of the function and the function's rates of change. We also looked at practical implications.

Explorations

1. The following graph, created using data from a 1999 article in *The New York Times*, represents the television audience of nightly news programs from three major broadcast networks from 1980 to 1998.

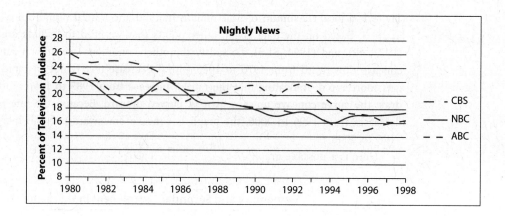

a. Give an interval of time when all three networks were losing viewers.

b. Give an interval over which the audiences of two of the three networks were increasing while the audiences were decreasing for the third network.

c. Give an interval of time when the audiences of two of the three networks were decreasing while the audiences of the third network were increasing.

d. For the function that gives the percentage of viewers of *NBC Nightly News*, give each relative minimum and the year it occurred.

e. For the same function as in part (d), give each relative maximum and the year it occurred.

2. The following table gives the U.S. budget surpluses or deficits for the years 1960 through 2010. Surpluses and deficits are given in millions of dollars.

Year	Surplus or Deficit (−)	Year	Surplus or Deficit (−)	Year	Surplus or Deficit (−)
1960	301	1977	−53,659	1994	−203,186
1961	−3,335	1978	−59,185	1995	−163,952
1962	−7,146	1979	−40,726	1996	−107,431
1963	−4,756	1980	−73,830	1997	−21,884
1964	−5,915	1981	−78,968	1998	69,270
1965	−1,411	1982	−127,977	1999	125,610
1966	−3.698	1983	−207,802	2000	236,241

Year	Surplus or Deficit (−)	Year	Surplus or Deficit (−)	Year	Surplus or Deficit (−)
1967	−8,643	1984	−185,367	2001	128,236
1968	−25,161	1985	−212,308	2002	−157,758
1969	3,242	1986	−221,227	2003	−377,585
1970	−2,842	1987	−149,730	2004	−412,727
1971	−23,033	1988	−155,178	2005	−318,346
1972	−23,373	1989	−152,639	2006	−248,181
1973	−14,908	1990	−221,036	2007	−160,701
1974	−6,135	1991	−269,238	2008	−458,555
1975	−53,242	1992	−290,321	2009	−1,412,686
1976	−73,732	1993	−255,051	2010	−1,555,582

Source: The World Almanac and Book of Facts 2011, p. 67.

a. Give the rate of change of the surplus/deficit function from 1960 to 1965, from 1965 to 1970, from 1975 to 1980, and from 1985 to 1990.

b. Use the information obtained in part (a) of this Exploration to decide whether the graph of the function will generally rise or fall over the interval of times mentioned. Does the information tell you that the graph will always rise or fall over each of those intervals? Explain.

c. What is the rate of change from 1998 to 2002? How does it compare with the rate of change over the period 1994 to 1998?

d. Give the rate of change of the surplus/deficit from 2002 to 2006 and compare it with the rate of change from 2006 to 2010.

3. The following graph gives the surplus/deficit function from 1975 to 2010:

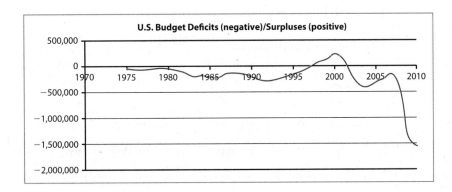

a. Give the intervals where the surplus/deficit function is increasing and the intervals where it is decreasing.

b. Give the minimum and the maximum value of the function and the time when it occurred.

c. Give a relative maximum and a relative minimum [other than the values you gave in part (b) of this Exploration] and the time each occurred. Explain what these values show about the surplus/deficit.

d. Give an interval where the function is increasing at an increasing rate and explain how you can tell this from the graph.

e. Give an interval where the function is decreasing at an increasing rate (so the rate of change of the function is becoming less negative). Explain what this shows about the surplus/deficit.

f. Based on the graph, what do you predict will happen with the values of the surplus/deficit function after 2010? Give a reason for your answer.

4. Explain what this graph from Example 3.5 shows overall about U.S. consumption and exports of cigarettes:

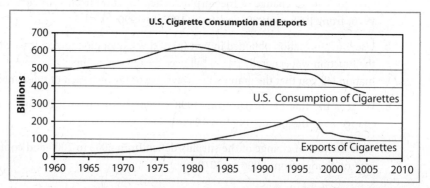

5. The following graph represents the federal debt from the years 1960 to 2016, with estimated values for 2011 through 2016:

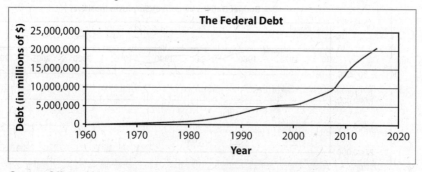

Source: Office of Management and Budget, www.whitehouse.gov.

a. Give the time intervals (if there are any) over which the rate of change of the debt function is positive.

b. Give the time intervals (if there are any) over which the rate of change of the debt function is negative.

c. Estimate the intervals over which the function is concave upward and those intervals where it is concave downward.

d. Estimate the intervals over which the rate of increase of the federal debt is growing, and the intervals where the rate of increase of the federal debt is shrinking.

e. Describe the changes in the values of the federal debt.

6. This graph shows the annual number of U.S. elementary and secondary school students who were enrolled in private schools:

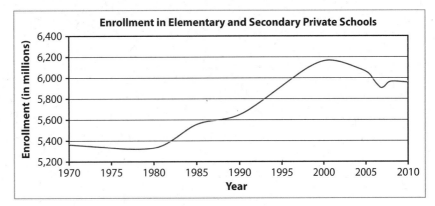

Source: U.S. Department of Education, National Center for Education Statistics, 2011, http://nces.ed.gov.

a. Give the time intervals (if there are any) over which the rate of change of the function is positive.

b. Give the time intervals (if there are any) over which the rate of change of the function is negative.

c. Estimate the intervals over which the function is concave upward and those intervals over which it is concave downward.

d. Estimate the intervals over which the rate of increase of the number of private school students is growing, and the intervals over which the rate of increase of the number of private school students is shrinking.

7. The National Football League average salary was approximately $20,000 in 1960 and rose slowly but at an increasing rate. In 1975, the average salary was approximately $50,000 and, in 1980, approximately $100,000. The rate at which salaries continued to rise increased quite sharply until 1991 when the average salary was approximately $780,000. Then, the

average salary increased at a small constant rate until 1997, when it started to increase at a rate of approximately $100,000 per year for several years. Draw a possible graph of the function that represents the National Football League average salary from 1960 to 2000.

8. In the 1983–1984 academic year, the number of students per computer in U.S. public schools was 125. The number of students per computer decreased to 75 in the 1984–1985 academic year and continued to decrease to 4.9 students per computer in the 2001–2002 academic year. Suppose we also know that the rate at which the number of students per computer decreased each year was increasing during this time period. Draw a possible graph of the function that represents the number of students per computer in U.S. public schools during this time period.

9. The following table gives the yearly Major League Baseball (MLB) television revenue from 1976 to 1996:

Year	MLB Television Revenue (millions of $)	Year	MLB Television Revenue (millions of $)
1976	50.01	1986	321.60
1977	52.21	1987	349.80
1978	52.31	1988	364.10
1979	54.50	1989	246.50
1980	80.00	1990	659.30
1981	89.10	1991	664.30
1982	117.60	1992	363.00
1983	153.70	1993	616.25
1984	268.40	1994	716.05
1985	280.50	1995	516.40
		1996	706.30

Source: Michael Haupert, "The Economic History of Major League Baseball," in *EH.Net Encyclopedia*, edited by Robert Whaples, August 2003, http://eh.net/encyclopedia/article .haupert.mlb.

a. Find the rate of change of TV revenue for each of the following four-year periods: 1976–1980, 1980–1984, 1984–1988, 1988–1992, and 1992–1996.

b. Give the change in TV revenue for each one-year period between 1984 and 1988. Are any of these numbers equal to the rate of change in TV revenue over the four-year period 1984–1988? What is the relationship between the change per year and the rate of change over the four-year interval?

c. Suppose that the rate of change of TV revenue over the four-year period 1996–2000 equals the annual change for each year during that period. What would the graph of the function over that period of time look like? Explain.

10. The next table gives the average ticket price of Major League baseball games from 1995 to 2002. Use this information to answer the following questions.

Year	Average Ticket Price ($)
1995	10.76
1996	11.32
1997	12.06
1998	13.58
1999	14.45
2000	16.22
2001	17.20
2002	17.85

Source: Michael Haupert, "The Economic History of Major League Baseball," in *EH.Net Encyclopedia,* edited by Robert Whaples, August 2003, http://eh.net/encyclopedia/article.haupert.mlb.

a. Give the rate of change of average ticket price over each one-year period between 1995 and 2002.

b. When was the average ticket price decreasing? When was it increasing?

c. Use the data obtained in part (a) of this Exploration to decide where the graph of the average ticket price function will be concave upward and where it will be concave downward.

d. Graph the average ticket price function. Does your graph agree with your answers for parts (b) and (c)? Explain.

11. The following table gives the average ticket price of Major League baseball games for some years between 2002 and 2011.

Year	Average Ticket Price ($)
2002	17.85
2006	22.21
2007	22.69
2008	25.40
2011	26.91

a. Find the rate of change of ticket price for the period 2002–2006.

b. Assuming that the rate of change for each one-year period between 2002 and 2006 is equal to the rate of change you found in part (a), find the ticket price in 2003, in 2004, and in 2005.

c. Find the rate of change of ticket price for the period 2006–2011 and use it to estimate the ticket price in 2009 and in 2010. Explain.

d. Use only the data for the years 2007 and 2008 to estimate the ticket price in 2009. How does it compare with your estimate in part (c)?

12. The following graph shows the top Major League baseball salary for each of the years 1988 to 2011. Write a paragraph to explain what this graph shows. In particular, identify when the graph is concave upward and when it is concave downward and explain what that tells you about the top Major League baseball salaries during that time period. Add any other information you can get from the graph.

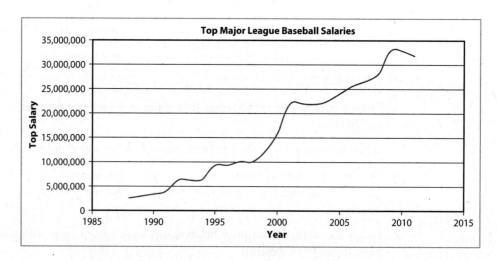

13. Describe what the graph of a function on an interval from $t = 0$ to $t = 10$ might look like if the function is neither concave up nor concave down on that interval. Are there other possible graphs?

14. The graph below shows the percent of people in the United States living in poverty in three different age groups over the years 1966 to 2009. Write a clear explanation of the story this graph tells.

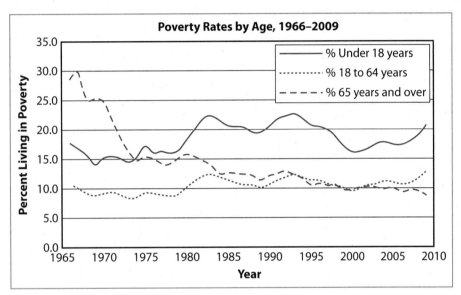

Source: U.S. Census Bureau, www.census.gov.

3-1

Temperature Patterns: Functions and Line Graphs

In this activity, you will work with examples in which curves obtained by joining known points of the graph of a function can help you understand the data. These graphs are called **line graphs**. The temperature in a region varies over the year and is a function of the time of year. You will look at average temperature data as a function of month in various cities and investigate patterns in the data. You will also create a graph involving minimum wage data.

1. On the following axes, sketch a curve that you think shows how the average temperature in Fairbanks, Alaska, changes over the course of a year. (Think about when the temperature would be a maximum, when it would be a minimum, when it would be increasing, and so on.)

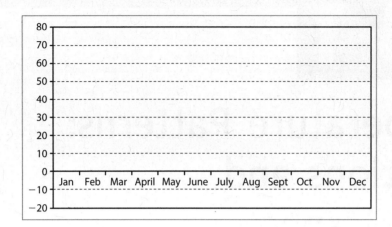

Next you'll see how your graph compares to the actual data, and how the average temperatures over the year for four cities in different parts of the world compare.

2. Retrieve the Excel file "EA3.1.1 Avg Temp Four Cities.xls" from the text website or WileyPLUS. (*Source:* World Climate, www.worldclimate.com.)

   a. Use the instructions that follow to create line graphs.

Instructions to Create Line Graphs

1. Highlight the block of data for the months Jan through Dec, including the column and row headings but not the Year column (column N), and go to the **Insert** tab. Select **Line** from the **Charts** group and select the first entry for **Chart sub-type**. With the graph still "selected," from the **Chart Layouts** group, select Layout 7 and insert appropriate titles for your graph.

2. Note that if you point (between vertical grid lines) on the "unselected" graph to one of the functions, the data values are given.

3. Create a second graph using the same block of data, but this time use **Column** as the **Chart type**.

   b. How does the line graph for Fairbanks, Alaska, compare with the one you created in Question 1?

   c. Explain what the graphs you created show.

   d. Which of the two types of graphs (line or column) do you think is preferable for these data and why?

   e. Use the line graphs you created previously to answer the following:

      i. Identify for each of the four cities when during the year the maximum average temperature occurs.

      ii. Identify for each of the four cities when during the year the minimum average temperature occurs.

      iii. Identify for each of the four cities when during the year the average temperature is increasing.

      iv. Identify for each of the four cities when during the year the average temperature is decreasing.

      v. Over the interval Jan−May, which city's average temperature increased the fastest and how does the graph show that?

    vi. Identify where the graph of the average temperature of Fairbanks, Alaska, is concave upward and where it is concave downward.

3. Retrieve the data set "EA3.1.2 Min Wage.xls" from the text website or WileyPLUS, or enter the following data on minimum wage in years in which it increased into an Excel worksheet.

Year	1974	1975	1976	1978	1979	1980	1981	1990	1991	1996	1997	2007	2008	2009
Wage ($)	2.00	2.10	2.30	2.65	2.90	3.10	3.35	3.80	4.25	4.75	5.15	5.85	6.55	7.25

    a. Use Excel to create and label two graphs of these data: a scatterplot (make sure you select **Scatterplot** instead of **Line** in the **Charts** group) with just the points shown, and a scatterplot using a line to connect the points (you may choose to show the points or not, and you may choose a smooth line or data points connected by line segments).

    b. Which variable should be on the horizontal axis of these graphs and why?

    c. Explain why your line graph (the one with points connected) is not really appropriate for these data.

4. Retrieve the data set "EA3.1.3 Normal Avg Temp.xls" from the text website or WileyPLUS.

    a. Select three cities in different regions of the United States that appear in this file and create an appropriate graph of the data. (*Source:* National Oceanic & Atmospheric Association, www.noaa.gov/climate.html.)

Instructions to Copy and Paste Data

1. For one of your cities, highlight the row of data for the city, go to the **Clipboard** group, and click on the **Copy** icon. (It looks like two overlapping sheets of paper.) Or, you can process **Ctrl** and the letter **C** to copy.

2. Then go to a new worksheet and select a new row from that sheet (leave the top row for labels, as shown in the Excel file "EA3.1.3 Normal Avg Temp.xls"). Click on **Paste** in the **Clipboard** group (or press **Ctrl** and the letter **V**). Repeat with your other chosen cities.

3. If you already have data in the first row, you can insert a blank row as follows: Click on any cell in the first row. Then select **Insert** from the **Cells** group and then **Insert Sheet Rows**. When you have finished copying the data from your three cities, you should have the data and labels, similar to those in the first file used, in your new worksheet.

b. Interpret your graph and explain what the graph shows about the average temperature over the year in your three cities. (Your explanation should include a lot more detail than the title of the graph conveys and should include information about when, over the course of the year, the temperature is increasing, when it is a maximum, and so on.)

Summary

In this activity, you compared bar graphs and line graphs, and used graphs to find maximum and minimum values of functions, intervals where a function is increasing, where it is decreasing, and where it is concave upward or concave downward. You learned how to use Excel to create line graphs, and how to copy and paste data and insert rows in an Excel worksheet.

Rates of Change and Concavity

In this activity, you will investigate the link between the graph of a function and rate of change and concavity.

1. Here are a graph and the table of data giving cases of tuberculosis over time. (This graph also appears in Example 2.2 of Topic 2.)

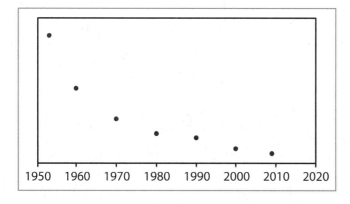

Year	Cases of Tuberculosis
1953	52.6
1960	30.7
1970	18.1
1980	12.2
1990	10.3
2000	5.8
2009	3.8

    a. Use the table to find the average rate of change in cases of tuberculosis per year.

        i. From 1953 to 1960:

ii. From 1960 to 1970:

iii. From 1970 to 1980:

iv. From 1980 to 1990:

v. From 1990 to 2000:

vi. From 2000 to 2009:

b. Explain how these rates of change confirm that the graph represents a decreasing function.

c. Explain how these rates of change confirm that the graph is concave upward from 1953 to 1990.

2. Here are a graph and the table of data giving life expectancy at birth over time. (This graph also appears in Example 2.2 of Topic 2.)

Year	Life Expectancy at Birth
1960	69.7
1970	70.8
1980	73.7
1990	75.4
1995	75.8
2000	76.8
2007	77.9

a. Use the table to find the average rate of change in life expectancy at birth per year.

i. From 1960 to 1970:

ii. From 1970 to 1980:

iii. From 1980 to 1990:

    iv. From 1990 to 1995:

    v. From 1995 to 2000:

    vi. From 2000 to 2007:

b. Explain how these rates of change confirm where the function represented by the graph is increasing and where it is decreasing.

c. Explain how these rates of change confirm where the graph is concave upward and where it is concave downward.

3. Here are a graph and the table of data giving U.S. military personnel over time. (This graph also appears in Example 2.2 of Topic 2.)

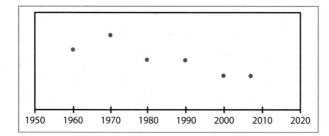

Year	Military Personnel
1950	1,459,462
1960	2,475,438
1970	3,064,760
1980	2,050,627
1990	2,043,765
2000	1,384,338
2007	1,380,082

a. Use the table to find the average rate of change of number of military personnel per year.

   i. From 1950 to 1960:

   ii. From 1960 to 1970:

   iii. From 1970 to 1980:

   iv. From 1980 to 1990:

v. From 1990 to 2000:

vi. From 2000 to 2007:

b. Explain how these rates of change confirm where the function represented by the graph is increasing and where it is decreasing.

c. Explain how these rates of change confirm where the graph is concave upward and where it is concave downward.

4. A large city has a group of homeless shelters that offer food and a bed to people in need. The director of the shelters has prepared some graphs showing the number of guests staying in the city's shelters each year over the past 15 years.

a. She observes that the graph is increasing over the 15-year period of time, and that the graph is concave downward during the first five years of the 15-year period of time and concave upward over the remainder of the time period. Explain in detail what these observations tell us about the population of the shelters over that period of time.

b. Sketch a possible graph of the population of the city's shelters over the past 15 years.

c. For funding the shelters, explain the consequences of what your graph and the data reveal.

Summary

In this activity, you looked at different sets of data and the corresponding graphs. You computed rates of change and verified that the rate of change indicates whether a function is increasing or decreasing. You also explored the relationship between concavity and whether the rate of change is increasing or decreasing.

Multiple Variable Functions

5	0	-5	-10
-5	-11	-16	-22
-10	-16	-22	-28
	-19	-26	-32
	-22	-29	-35
	-24	-31	-37

In previous topics, we looked at situations where we assumed, for simplicity, that one variable (the response or dependent variable) was determined completely by just one other variable. What we often do in examples like these is assume that other variables do not affect the value of the response variable (at least not enough to matter). We will now look at situations in which one response variable depends on several explanatory variables, and we will examine the nature of this dependence in a variety of situations.

We first explore some examples in which a response variable depends on several explanatory variables.

After completing this topic, you will be able to:

- Identify multiple explanatory variables that influence a response variable.

- Work with multiple variable functions using words, symbols, numbers in tables, and graphs.

- Recognize how the response variable changes when all but one explanatory variable is held constant.

- Analyze various quantities relating to real-world applications and understand how they are modeled by multiple variable functions.

Example 4.1

For each of the following response variables, identify at least three explanatory variables that influence the given variable, and for each explanatory variable identified, determine whether it is a categorical variable or a quantitative variable.

a. The cost of a 1,000-mile car trip

b. The length of time a traffic light is set to remain yellow

c. The time it takes to travel from Boston to Miami

d. A college student's grade point average

Solution

There are many variables that may influence each response variable; we give some possibilities but encourage other answers as well.

a. Three variables that influence the cost of a 1,000-mile car trip are the price of gasoline, the kind of car driven, and how much the driver pays in road tolls. The kind of car is a categorical variable, whereas the price of gasoline and amount paid in tolls are quantitative variables.

b. The length of time a traffic light is set to remain yellow depends on the speed limit of the road on which it is located, how many other traffic lights there are nearby on the same road, and how heavy the traffic load is on the road. The first two variables are quantitative. How heavy the traffic load is on the road would be a quantitative variable if we measured "how heavy the traffic load is" in terms of how many cars typically travel on the road. If we measured "how heavy the traffic load is" by using categories such as light, medium, and heavy, then it would be a categorical variable.

c. The time it takes to travel from Boston to Miami is a function of the mode of transportation, when you travel (morning, afternoon, evening, or night), and how many stops you make. "How many stops" is a quantitative variable; the other two are categorical.

d. Explanatory variables that influence a college student's grade point average are how much time the student spends studying, what types of courses he or she is enrolled in, and how much time he or she spends watching television or playing video games. "Types of courses" is a categorical variable; the other two are quantitative variables.

Functional relationships in which there are multiple explanatory variables can be given in the same four ways that functions with one explanatory variable are given: using words, tables, symbols, and charts or graphs. The following examples illustrate these modes.

Example 4.2

An adult's basal metabolic rate (BMR) is roughly equivalent to the number of calories burned in a day. The formula (which may be found at www.health.gov) to help people between the ages of 20 and 90 calculate their BMR can be divided into the following six steps:

1. Multiply your weight in pounds by 4.4 (women) or 6.2 (men).

2. Multiply your height in inches by 4.7 (women) or 12.7 (men).

3. Add the answers from Steps 1 and 2.

4. Multiply your age in years by 4.7 (women) or 6.8 (men).

5. Subtract the answer for Step 4 from the answer for Step 3.

6. Add 655 (women) or 666 (men).

The final answer is your BMR.

a. Compare the BMR for a man and a woman, each of whom is 5 feet 7 inches tall, weighs 145 pounds, and is 34 years old.

b. Write two symbolic formulas to calculate BMR, one for men and one for women.

Solution

a. We first need to convert 5 feet 7 inches to inches; 5 feet is 60 inches, so the man and woman are 67 inches tall. For the woman, we perform the following computations:

1. $4.4 \times 145 = 638$

2. $4.7 \times 67 = 314.9$

3. $638 + 314.9 = 952.9$

4. $4.7 \times 34 = 159.8$

5. $952.9 - 159.8 = 793.1$

6. $793.1 + 655 = 1,448.1$

For the man, the steps are similar:

1. $6.2 \times 145 = 899$

2. $12.7 \times 67 = 850.9$

3. $899 + 850.9 = 1,749.9$

4. $6.8 \times 34 = 231.2$

5. $1,749.9 - 231.2 = 1,518.7$

6. $1,518.7 + 666 = 2,184.7$

The man's BMR is higher because most men have larger muscles. These numbers are averages, of course, and are based on studies using large numbers of men and women. In addition to the amount of lean tissue in one's body affecting BMR, an individual's activity level will also affect his or her BMR.

b. We'll use the six steps to put together a general symbolic formula for women and another one for men:

$$BMR \text{ for women} = 4.4 \times \text{weight} + 4.7 \times \text{height} - 4.7 \times \text{age} + 655$$

$$BMR \text{ for men} = 6.2 \times \text{weight} + 12.7 \times \text{height} - 6.8 \times \text{age} + 666$$

In writing formulas, it is common to use just one letter to represent variables and to omit the sign for multiplication. Letting w represent weight in pounds, h represent height in inches, a represent age in years, and B represent BMR, we can write the formulas as

$$B_{\text{women}} = 4.4w + 4.7h - 4.7a + 655$$

$$B_{\text{men}} = 6.2w + 12.7h - 6.8a + 666$$

(Note that we use the subscript on B for gender.)

For quantities that are functions of more than one variable, we sometimes want to investigate the behavior of one variable at a time as we hold the other variables constant.

Example 4.3

a. Graph the BMR for a woman who is 5 feet 6 inches tall and weighs 130 pounds, as a function of age. Describe what happens to her BMR for every 10 years she ages. Does this description change for a woman of different height and weight?

b. Graph the BMR for a man who is 5 feet 10 inches tall and weighs 150 pounds, as a function of age. Describe what happens to his BMR for every 10 years he ages. Does this change for a man of different height and weight?

Solution

a. If we let $w = 130$ and $h = 66$, the formula for B_{women} becomes

$$B_{\text{women}} = 4.4 \times 130 + 4.7 \times 66 - 4.7 \times a + 655 \quad \text{or} \quad B_{\text{women}} = 1{,}537.2 - 4.7a$$

Here is the graph of this function:

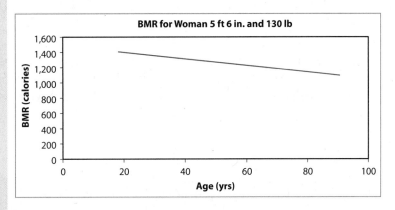

As the woman ages from 30 to 40, her BMR changes from 1,396.2 to 1,349.2, a decrease of 47. Additional computations will show the same decrease as she ages from 43 to 53, and for any 10-year change in age. If the woman were a different height and weight, the constant term in the formula (1,537.2) would be different, but the other term ($-4.7 \times a$) would remain the same. Therefore, the change in BMR for any 10-year period would still be a decrease of 47. This is a linear function (the graph is a straight line) and -4.7 is the slope of the line. It tells us that for every one-year increase in age, the BMR decreases by 4.7. And, as we saw, for a 10-year increase in age, BMR decreases by $4.7 \times 10 = 47$. (Linear functions are discussed in more detail in Topic 5.)

b. For a man who is 5 feet 10 inches tall and who weighs 150 pounds, the formula for BMR is

$$B_{men} = 6.2 \times w + 12.7 \times h - 6.8 \times a + 666$$
$$= 6.2 \times 150 + 12.7 \times 70 - 6.8 \times a + 666, \qquad \text{so } B_{men} = 2,485 - 6.8a$$

Here's the graph:

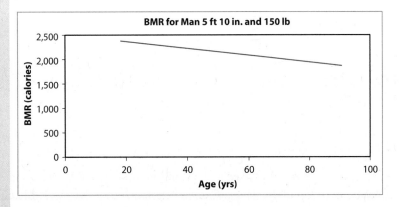

For each 10-year increase in age, a man's BMR decreases by 68. This will be true for a man who is 5 feet 10 inches tall and weighs 150 pounds, as well as for any other man, using this model.

Often in winter months, weather reporters will give temperature and wind chill equivalent temperature because how cold we feel when we go out into winter weather depends not only on the outside temperature, but also on wind speed. The term *wind chill* is used to describe the equivalent temperature felt by exposed skin for a particular temperature of the surrounding air and wind speed. The following table gives wind chill equivalent temperatures for various combinations of air temperature, in degrees Fahrenheit, and wind speeds, in miles per hour. Note that the effect of the wind is negligible if its speed is less than 4 miles per hour, and wind speeds greater than 45 miles per hour do not significantly affect body heat further.

Wind Chill Equivalent Temperature

Wind Speed (mph)	Thermometer Temperature (°F)																	
	40	35	30	25	20	15	10	5	0	−5	−10	−15	−20	−25	−30	−35	−40	−45
5	36	31	25	19	13	7	1	−5	−11	−16	−22	−28	−34	−40	−46	−52	−57	−63
10	34	27	21	15	9	3	−4	−10	−16	−22	−28	−35	−41	−47	−53	−59	−66	−72
15	32	25	19	13	6	0	−7	−13	−19	−26	−32	−39	−45	−51	−58	−64	−71	−77
20	30	24	17	11	4	−2	−9	−15	−22	−29	−35	−42	−48	−55	−61	−68	−74	−81
25	29	23	16	9	3	−4	−11	−17	−24	−31	−37	−44	−51	−58	−64	−71	−78	−84
30	28	22	15	8	1	−5	−12	−19	−26	−33	−39	−46	−53	−60	−67	−73	−80	−87
35	28	21	14	7	0	−7	−14	−21	−27	−34	−41	−48	−55	−62	−69	−76	−82	−89
40	27	20	13	6	−1	−8	−15	−22	−29	−36	−43	−50	−57	−64	−71	−78	−84	−91
45	26	19	12	5	−2	−9	−16	−23	−30	−37	−44	−51	−58	−65	−72	−79	−86	−93
50	26	19	12	4	−3	−10	−17	−24	−31	−38	−45	−52	−60	−67	−75	−81	−88	−95
55	25	18	11	4	−3	−11	−18	−25	−32	−39	−46	−54	−61	−68	−75	−82	−89	−97
60	25	17	10	3	−4	−11	−19	−26	−33	−40	−48	−55	−62	−69	−76	−84	−91	−98

Source: Mount Washington Observatory, www.mountwashington.org

Example 4.4   ■ ■ ■ ■ ■ ■ ■ ■

a. Suppose wind speed is 25 miles per hour (mph). Graph wind chill equivalent temperature as a function of thermometer temperature and describe this graph.

b. Suppose the temperature is fixed at 30°F and the wind speed is increasing. If wind speed increases by 10 mph from 15 to 25 mph, how much does the wind chill equivalent temperature change? If wind speed increases from 35 to 45 mph, how much does the wind chill equivalent temperature change?

c. Suppose the temperature is 20°F. Graph the wind chill equivalent temperature as a function of wind velocity and describe this graph.

d. Suppose the weather forecaster says that the wind chill equivalent temperature is −43°F and the temperature is −10°F. How fast is the wind blowing?

Solution

a.

The graph shows a fairly steady and constant increase in wind chill equivalent temperature as the thermometer temperature increases over the interval from −45 to 40°F.

b. If wind speed increases from 15 to 25 mph, the wind chill equivalent temperature drops from 19 to 16° so it decreases by 3°. If wind speed increases from 35 to 45 mph, the wind chill equivalent temperature decreases by 2°.

c. The graph shows that as wind velocity increases, the wind chill equivalent temperature decreases if the temperature is fixed at 20°F. The decreases are greater for the lower wind velocity values and level off as wind velocity increases.

d. If the temperature is −10°, we look at the column corresponding to that temperature and locate the wind chill equivalent temperature of −43. Then we trace that row to the left to find the corresponding wind velocity, which is 40 mph.

Summary

In this topic, we investigated response variables that depend on more than one explanatory variable. We worked with multiple variable functions given in words, in symbols, as graphs, and in table format. We explored basal metabolic rate (BMR) and wind chill equivalent temperature to see how the response variable is affected when all but one explanatory variable is held constant. We identified a linear relationship between BMR and age, which sets the stage for Topic 5.

Explorations

1. In each of the following situations, identify at least three explanatory variables that influence the given response variable. For each explanatory variable, identify if it is quantitative or categorical.

   a. The amount of time a student spends writing a research paper for a particular course in which he or she is enrolled

   b. A one-month electric bill for a family's residence

   c. The amount of profit made by a business during a one-year period

2. The estimated number of calories burned in one day also depends on a person's activity level. If a person is sedentary, his or her activity factor is approximately 1.3; if moderately active, the activity factor is 1.4; if very active, the activity factor is 1.7. Then,

   Total calories needed in one day = BMR × activity factor

Find the daily calorie needs for the woman and man described in Example 4.2 if they are sedentary. By how much will their calorie needs change if they are moderately active? By how much will their calorie needs change if they are very active?

3. One way to find out whether your weight is "reasonable" for your height is to calculate your body mass index (BMI). A BMI of 20 to 25 is normal. BMI is a function of height and weight and is determined as follows: To calculate a person's BMI, divide his or her weight in kilograms by the square of his or her height in meters. (To convert into kilograms, multiply pounds by 0.4536; to convert into meters, multiply inches by 0.0254.) Thus, you can write an equation to calculate BMI:

$$\text{BMI} = \frac{0.4536 \times \text{wt. in pounds}}{\left(0.0254 \times \text{ht. in inches}\right)^2}$$

a. Describe how this equation can be used to calculate BMI for a friend who does not understand the description of BMI.

b. Abraham Lincoln was 6 feet 4 inches in height and weighed about 180 pounds, according to one historical account. From pictures you have seen of Abraham Lincoln, do you think his BMI would fall in the normal range? Calculate Lincoln's BMI.

c. Find the BMI for a man who weighs the same amount as Lincoln did but who is 5 feet 8 inches tall. Comment on how this compares to Lincoln's BMI.

d. A person who is 5 feet 8 inches tall reduces his or her weight from 200 pounds to 140 pounds. Create a table that contains weights at 10-pound intervals and find this person's BMI at each of those weights. By how much is BMI reduced when weight is reduced from 200 to 170 pounds? By how much is BMI reduced when weight is reduced from 170 to 140 pounds? What can you say about the formula to calculate BMI for fixed height?

4. For improving cardiovascular fitness, exercise physiologists recommend exercising so your heart rate is in the "target zone." The upper and lower limits of this target zone can be represented as a function of two variables: a person's age and resting heart rate (in beats per minute). These limits are found as follows:

Lower limit = (220 − age in years − resting heart rate) × 0.6 + resting heart rate
Upper limit = (220 − age in years − resting heart rate) × 0.8 + resting heart rate

a. Find the upper and lower limits of the target zone for a 20-year-old individual whose resting heart rate is 68 beats per minute.

b. How do these limits change for a 40-year-old individual with the same resting heart rate of 68 beats per minute?

c. Compare the upper and lower limits of the target zone for two 20-year-old individuals with resting heart rates of 65 and 75 beats per minute, respectively.

5. A person's blood alcohol level is a function of his or her gender, body weight, number of drinks consumed, and amount of time spent drinking. The following two tables, obtained from the Pennsylvania State Police, PA Liquor Control Board, give the approximate blood alcohol level for various values of these variables. Note that one drink is 1¼ ounces of 80 proof liquor, 12 ounces of beer, or 5 ounces of table wine.

Male	Body Weight in Pounds							
Drinks	100	120	140	160	180	200	220	240
0	0.00	0.00	0.00	0.00	0.00	0.00	0.00	0.00
1	0.04	0.03	0.03	0.02	0.02	0.02	0.02	0.02
2	0.08	0.06	0.05	0.05	0.04	0.04	0.03	0.03
3	0.11	0.09	0.08	0.07	0.06	0.06	0.05	0.05
4	0.15	0.12	0.11	0.09	0.08	0.08	0.07	0.06
5	0.19	0.16	0.13	0.12	0.11	0.09	0.09	0.08
6	0.23	0.19	0.16	0.14	0.13	0.11	0.10	0.09
7	0.26	0.22	0.19	0.16	0.15	0.13	0.12	0.11
8	0.30	0.25	0.21	0.19	0.17	0.15	0.14	0.13
9	0.34	0.28	0.24	0.21	0.19	0.17	0.15	0.14
10	0.38	0.31	0.27	0.23	0.21	0.29	0.17	0.16

Female		Body Weight in Pounds							
Drinks	90	100	120	140	160	180	200	220	240
0	0.00	0.00	0.00	0.00	0.00	0.00	0.00	0.00	0.00
1	0.05	0.05	0.04	0.03	0.03	0.03	0.02	0.02	0.02
2	0.10	0.09	0.08	0.07	0.06	0.05	0.05	0.04	0.04
3	0.15	0.14	0.11	0.10	0.09	0.08	0.07	0.06	0.06
4	0.20	0.18	0.15	0.13	0.11	0.10	0.09	0.08	0.08
5	0.25	0.23	0.19	0.16	0.14	0.13	0.11	0.10	0.09
6	0.30	0.27	0.23	0.19	0.17	0.15	0.14	0.12	0.11
7	0.36	0.32	0.27	0.23	0.20	0.18	0.16	0.14	0.13
8	0.40	0.36	0.30	0.26	0.23	0.20	0.18	0.17	0.15
9	0.45	0.41	0.34	0.29	0.26	0.23	0.20	0.19	0.17
10	0.51	0.45	0.38	0.32	0.28	0.25	0.23	0.21	0.19

a. Describe how blood alcohol levels compare for a male and a female, each weighing 160 pounds, for increasing numbers of drinks; that is, describe the effect of the variable "gender" on blood alcohol levels for a weight level of 160 pounds. Use appropriate graphs.

b. On the same set of axes, sketch a graph of blood alcohol level as a function of number of drinks consumed for an "average" woman weighing 120 pounds and an "average" man weighing 160 pounds. Describe the graphs.

c. For a male consuming four drinks, sketch a graph of blood alcohol level as a function of weight. Repeat for a female who consumes four drinks and explain what the graphs show.

6. A student's final grade in a course is often a function of various student performance measures such as homework, class participation, hour exams, and final exam, as well as the weights the professor assigns to each of these measures. Suppose one professor uses the following system:

$$\text{Final grade} = 0.4 \times \text{hour exam average} + 0.25 \times \text{final exam score} + 0.25$$
$$\times \text{homework average} + 0.1 \times \text{class participation grade}$$

and a second professor uses this system:

$$\text{Final grade} = 0.6 \times \text{hour exam average} + 0.15 \times \text{final exam score} + 0.15$$
$$\times \text{homework average} + 0.1 \times \text{class participation grade}$$

a. Which professor's grading system would you prefer if your grades were as follows: hour exam average = 84; final exam grade = 68; homework average = 90; class participation grade = 95?

b. If you scored 10 points higher on the final exam, what effect would that have on your final grade under each professor's grading system?

c. If you scored 6 points lower on your hour exam average, what effect would that have on your final grade under each professor's grading system?

d. Suppose you earned the following grades in a class: hour exam average = 78; final exam grade = 88; homework average = 70; class participation grade = 80. What weights would you assign to each of these components to set up a grading system? Assume each component must count for at least 5% of the grade (that is, have a weight of at least 0.05), and explain why you chose the weights you did.

7. Just as low temperatures and high winds together can reduce effective temperature, the combination of high temperatures and high humidity can produce adverse conditions. The following equation gives one model for the apparent temperature (AT), sometimes called *heat index,* as a function of air temperature in degrees Fahrenheit and relative humidity as a decimal between 0 and 1.

$$AT = 2.70 + 0.885 \times \text{air temperature} - 78.7 \times \text{relative humidity} + 1.20$$
$$\times \text{air temperature} \times \text{relative humidity}$$

a. Suppose the air temperature is 90°F. Sketch a graph of the apparent temperature as a function of relative humidity. Describe your graph and what it tells you about humidity levels when the temperature is 90°.

b. Suppose the air temperature is 60°F. Sketch a graph of the apparent temperature as a function of relative humidity. Describe your graph and what it tells you about humidity levels when the temperature is 60° and how this compares to your graph and description in part (a) of this Exploration.

c. Fill in the following table. Describe how apparent temperature changes when the air temperature increases 15°F from 60 to 75° and from 75 to 90° at 20% relative humidity versus 50% relative humidity versus 90% relative humidity. What is the rate of change of apparent temperature over each of these intervals?

Temperature	Relative Humidity		
	0.20	0.50	0.90
60°			
75°			
90°			

8. In addition to the well-known Body Mass Index (BMI), a new measure is currently used by some researchers to estimate possible health risks. The Body Adiposity Index (BAI) approximates a person's percent body fat and is calculated using only the person's height and hip circumference:

$$BAI = \frac{1000 \times (\text{hip size in centimeters})}{(\text{height in centimeters})\sqrt{\text{height in centimeters}}} - 18$$

a. Find the BAI for a person who is 160 centimeters tall and whose hip size is 105 centimeters.

b. Find the BAI for a person who is 160 centimeters tall and whose hip size is 95 centimeters.

c. What is the percent change in body fat when the hip size of a 160-centimeter-tall person drops from 105 to 95 centimeters?

9. Write a formula that gives the BAI (see Exploration 8) when the height and hip size are in inches. (Use 1 inch = 2.54 centimeters.) Then use your formula to calculate the percent body fat of a person who is 6 feet tall and has a 38-inch hip size.

10. Credit scores are one way that lenders measure the risk associated with giving credit to an individual. FICO scores are the credit scores most lenders use to determine credit risk. There are three credit bureaus (Equifax, Experian, and TransUnion) so an individual has three FICO scores. (These scores are called "FICO scores" because most credit scores are obtained from software developed by a company named Fair Isaac and Company.)

    a. What variables do you think should be part of an individual's credit score?

    b. Research FICO scores to determine what variables do contribute and what variables do not contribute to an individual's credit score.

4-1

Blood Alcohol Levels and Credit Cards: Working with More Than Two Variables

In this activity, you will look at examples in which there is a need to graph or analyze expressions and functions that involve the interactions of more than two variables. In the first part of this activity, you will work with "blood alcohol level" as the response variable, and in the second part you will explore how some variables affect the balance on a credit card.

Blood alcohol level tables are given in Exploration 5 of Topic 4 and are contained in the file "EA4.1.1 Blood Alcohol Level.xls." You will compare the blood alcohol level of 140-pound males and 140-pound females based on the number of drinks consumed. To do so, you will create a graph that shows two scatterplots (one relates the number of drinks consumed with the blood alcohol level of a 140-pound male, the other relates the same variables for a 140-pound female) on the same axes.

1. Retrieve the Excel file "EA4.1.1 Blood Alcohol Level.xls" from the text website or WileyPLUS.

   a. Using these tables and the copy-and-paste functions of Excel, create a three-column table in the same worksheet. The first column should be "Number of Drinks," and the second and third columns should be "Blood Alcohol Level of a 140-lb Male" and "Blood Alcohol Level of 140-lb Female," respectively.

b. Create and label appropriate scatterplots (in a single graph) using the table you created. Explain what your scatterplots show about blood alcohol levels of 140-pound males and females.

c. What variables are involved in this graph and what other variables might you want to include in order to understand the issue more fully?

2. How fast the balance on your credit card decreases depends on the original amount charged, any additional purchases charged to the card, the monthly interest rate, and how much you pay off each month. Suppose you have charged $1,000 on a particular credit card and do not make any additional charges on the card. Suppose also that the annual interest charge on this card is 16.8%. Then the monthly charge, as a percent, is $16.8/12 = 1.4$. The credit card company requires that you pay off a minimum balance of $20 each month. Then you can calculate the next month's balance using the formula:

Balance next month = balance this month + 0.014 × balance this month − 20

a. Explain what each part of the "balance next month" formula represents.

b. Set up a spreadsheet to use the formula to calculate how much you will owe after one month, after six months, and after one year, if you pay off only the minimum amount of $20 each month and don't charge anything else on your card. Here are instructions to set up the spreadsheet. Using the spreadsheet, you will be able to answer the questions that follow.

Instructions to Create a Spreadsheet and Enter a Formula

1. Go to a new sheet in your Excel workbook. In cell A1, enter the label **Month**, and in cells 2 and 3 of column A, enter the numbers **1** and **2**, respectively. Select cell **A2**, then press the **Shift** key and while holding it down, press the downarrow key to select cell **A3** (cell A2 won't look highlighted but cell A3 will, and you will see a bold line around the two boxes). Point the cursor to the lower-right corner of the highlighted cells until it changes to a bold plus sign, and then press and hold the left mouse button and drag down to cell 13 of column A. You should have the numbers 1 through 12 in cells 2 through 13 of column A.

2. Label column B with **Balance This Month** and column C **Balance Next Month**. Enter **1000** in cell B2 (without a $ sign).

3. You will enter a formula in cell C2 that will compute the balance next month for the first month. To enter a formula, use the "equals" sign. In cell C2, entry **=B2+0.014*B2-20**. Note that this is the formula given previously to calculate "balance next month" because cell B2 contains the "balance this month."

4. For month 2, you want to transfer the amount in cell C2 (that is, the "balance next month" for month 1) to cell B3, which is the "balance this month" for month 2. To do this in an efficient way, type **=C2** in cell B3. Now, to calculate the "balance next month" for month 2, you can either retype the formula for "balance next month" in cell C3, or put the cursor on cell C2 and then move it to the lower-right corner of the cell and drag the fill handle (the bold "plus" sign) to cell C3 to auto-fill the cell.

5. To extend these same formulas to the rest of the months, highlight cells B3 and C3, move the cursor to the lower-right corner of the highlighted cells, and drag the bold plus sign down to cell C13 (that is, auto-fill down to cells B13 and C13).

6. There is a nice feature of Excel that allows you to see what formulas you used in creating the spreadsheet. Press the **Ctrl** button together with the accent key ` (in many computers, at the upper left corner of the keyboard). This puts the spreadsheet in **Audit mode**. In Audit mode you can see all the formulas that were entered. Press **Ctrl** and the accent key ` together again to get out of Audit mode.

7. If you want to round the numbers to two decimal places (which is appropriate in this situation because you are dealing with money), highlight the cells you wish to round, and click on the small arrow at the bottom right of the **Number** group on the **Home** tab. In the **Format Cells** dialog box that appears, for **Category**, choose **Number** and for **Decimal places**, enter **2** and click **OK**.

c. How much progress did you make in paying off your debt in month 1?

d. What is the "balance next month" for month 2?

e. How much will you owe after six months? After one year?

f. If you continue to pay $20 each month and don't charge anything else on your card, how long will it take to pay off the whole amount? (To answer this question, highlight cells A13, B13, and C13 and auto-fill.)

3. Suppose you charged $1,300 and you pay $20 per month and don't charge anything else on your card. Use Excel to set up a table to determine how much will remain to be paid on your card after one year. Explain what your table shows.

4. Suppose you want to pay off only $14 each month. If the initial balance is $1,000, you don't charge anything else on your card, and the interest is 16.8% per year, find how much you will owe after the first month, after six months, and after one year.

5. What happens if you owe $1,000 and you can pay only $13 each month?

6. What conclusions can you draw from the previous computations?

Additional Questions

7. We collected information on family size from a group of 43 students. We asked students how many children there were in the family they were raised in and how many children they would like to have when they establish a family of their own. We organized the data in a file titled "EA4.1.2 Class Data Num Child.xls"; retrieve this file from the text website or WileyPLUS.

   a. You'll notice that each case has three variables associated with it. What are the cases and what are the variables?

   b. Sort the data by gender, highlight the portion of the data in columns B and C for female cases, and create and label a scatterplot for the female cases. Now highlight the portion of the data in columns B and C for male cases, and create and label another scatterplot for male cases. Explain what your two scatterplots show. Are the two scatterplots basically the same, or are there differences? Do your graphs show all data points? Explain.

c.   What other kinds of responses might you want to gather from your respondents to understand any possible relationship between these two variables more fully?

Summary

In this activity, you learned how to enter a formula in Excel and drag down to do a series of calculations. You also learned how to use Audit mode in Excel to check formulas. You looked at explanatory variables that affect blood alcohol level. You also looked at various scenarios and explored how the original amount charged on a credit card and how much is paid off each month affect the balance remaining on the card after a period of time.

Proportional, Linear, and Piecewise Linear Functions

After completing this topic, you will be able to:

- Identify proportional (directly and inversely), linear, and piecewise linear functions.

- Graph directly proportional, linear, and piecewise linear functions.

- Write equations to model situations using proportional, linear, or piecewise linear functions.

- Find rates of change for functions that are proportional, linear, or piecewise linear.

I n this topic, we will study several types of functions that establish a fairly straightforward relationship between the response variable and the explanatory variable—linear and piecewise linear functions. Many situations are modeled by these types of functions.

Directly proportional functions are functions in which the response variable is equal to a constant c times the explanatory variable; that is, the response variable is directly proportional to the explanatory variable. In symbols, this is represented as $y = c \times x$, or $y = cx$, where y and x represent the response and explanatory variables, respectively, and c is a constant. You saw examples of such functions in Topic 2. In one such example (Example 2.4), we represented the relationship between minutes spent walking, m, and the approximate total number of calories used by a 150-pound person, c, as $c = 5.4m$. The graph of a directly proportional function is a (straight) line that passes through the point $(0, 0)$. Any function whose graph is a line is a **linear function**. Directly proportional functions are special cases of linear functions. In linear functions, the relationship between the response and explanatory variable is given symbolically by $y = c \times x + d$, or $y = cx + d$, where c and d are constants.

Example 5.1

A sports nutritionist recommends that exercisers drink 2 cups of water or other liquid before a race or workout and 1 cup for every 15 minutes they exercise during a workout to avoid dehydration.

a. Suppose a runner takes 3 hours and 15 minutes to complete a marathon. How many cups of fluid should the runner consume for this event?

b. Let F represent cups of fluid and N represent number of 15-minute intervals spent exercising, and write an equation that can be used to find the recommended number of cups of fluid to drink while exercising.

c. Explain the significance of each of the numbers that appears in the equation and sketch a graph of the equation.

Solution

a. Three hours and 15 minutes corresponds to 13 intervals of 15 minutes. So the runner needs 13 cups (one for each 15-minute interval) plus the 2 initial cups, which is 15 cups of fluid.

b. In part (a) of this example, we computed: Cups of fluid $= 1 \times$ (number of 15-minute intervals spent exercising) $+ 2$. We can shorten this equation to $F = N + 2$.

c. The number 2 in the equation represents the number of cups of fluid recommended at the start of the race; that is, when $N = 0$. The other number that appears in the equation is 1, which is the number of cups of fluid recommended for each 15-minute interval spent exercising. We graph $N =$ "number of 15-minute time intervals spent exercising" on the horizontal axis because it is the variable that "explains" $F =$ "number of cups of fluid recommended."

Recommended Fluid Consumption

Data that fall in an exact linear pattern are characterized by the fact that no matter which two points you choose, the change in the dependent variable divided by the change in the independent variable will be a constant, called the **slope of the line**. This is the rate of change of values of the function (see Topic 3). This rate of change is given by the number c in the expression $y = c \times x + d$.

(To see that this is true, we compute $\dfrac{y_2 - y_1}{x_2 - x_1}$, where x_1 and x_2 are two different values of the explanatory variable and y_1 and y_2 are the corresponding values of the response variable. Because $y_2 = c \times x_2 + d$ and $y_1 = c \times x_1 + d$, we have

$$\frac{y_2 - y_1}{x_2 - x_1} = \frac{(c \times x_2 + d) - (c \times x_1 + d)}{x_2 - x_1} = \frac{c \times x_2 + d - c \times x_1 - d}{x_2 - x_1} = \frac{c \times x_2 - c \times x_1}{x_2 - x_1}$$

$$= \frac{c \times (x_2 - x_1)}{x_2 - x_1} = c$$

If the constant c is positive, the function is increasing, so the line rises as we move from left to right. If c is negative, the function is decreasing and therefore the line falls from left to right. The magnitude of c determines how steep the line is; the greater the magnitude, the steeper the line.

In the following example, we compare two linear functions and their rates of change.

Example 5.2

In Example 4.3 in Topic 4, we saw that the basal metabolic rate (BMR) for a woman who weighs 130 pounds and is 66 inches tall is related to the woman's age according to $B_{women} = 1{,}537.2 - 4.7 \times a$, where a is the age in years and B_{women} is the BMR for a woman a years old. If we calculate (following the pattern described in Topic 4) the BMR for a man who also weighs 130 pounds and is 66 inches tall, we obtain $B_{men} = 6.2 \times 130 + 12.7 \times 66 - 6.8 \times a + 666 = 2{,}310.2 - 6.8 \times a$.

a. Describe the graphs of the two functions and give the rate of change per year for B_{women} and for B_{men}.

b. Graph both functions on the same graph and explain their differences.

Solution

a. The two given functions are linear functions because they are both of the form $B = c \times a + d$, where B, the response variable, is B_{women} in one function and B_{men} in the other. The constants are $c = -4.7$ and $d = 1{,}537.2$ for women, and $c = -6.8$ and $d = 2{,}310.2$ for men. The graphs are lines of slope -4.7 and -6.8 for women and men, respectively. The rate of change of B_{women} is -4.7 points per additional year of age and the rate of change of B_{men} is -6.8 points per additional year of age.

b.   Here is the resulting graph:

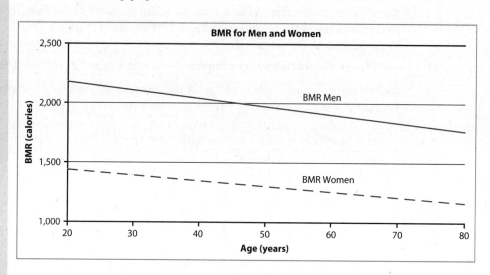

Both lines decline as we move from left to right, which is indicated by the negative slopes. One difference between the two lines is that the graph of the BMR for women is not as steep as that for men. This is because both slopes are negative and the slope of BMR for men is smaller (negative but larger in magnitude) than the slope of the BMR for women.

Linear functions are the only functions with a constant rate of change. To recognize linear functions, we compute the rate of change over several intervals. If we can see that the rate of change is always the same, then we can conclude the function is linear.

If we have confirmed that the relationship between two variables x and y is linear, we can find the equation of the line $y = cx + d$ that gives the response variable (y) as a linear function of the explanatory variable (x). To do this, we need to find the specific constants c and d for this linear relationship.

The value c is the slope of the line, or the constant rate of change we found when we confirmed that the data are linear. We can find the value of d in several ways. If we know the value of y when $x = 0$, this is also the value of d, because d is the y-intercept of the line. (We can confirm this by substituting $x = 0$ into the equation of the line; when $x = 0$, $y = c \times 0 + d = 0 + d = d$.)

Alternatively, after we have found the slope of the line, we can find d by taking one value of x from our data and the corresponding value of y and substituting these values for x and y into the equation $y = cx + d$. We know c and we have values of x and y, so the only unknown in our equation is d. We then solve for that unknown value d.

For example, suppose we know that a certain type of bacteria grows at a constant rate of 50 per hour; that is, the constant rate of change $c = 50$. Two hours after starting an experiment, there are 7,000 individuals in a particular dish. We can first write the equation for this linear relationship, $y = cx + d$, where y is the number of individuals in the dish and x is the number of hours, as $y = 50x + d$. Substituting $y = 7,000$ individuals when $x = 2$ hours gives $7,000 = 50 \times 2 + d$. Solving for d, we get $d = 7,000 - 100 = 6,900$. The equation of the line is $y = 50x + 6,900$.

Example 5.3

The following is a portion of a table that indicates the postal rate for an international letter from the United States to Canada as of July 2011:

Weight Up to (in oz)	Rate (in $)	Weight Up to (in oz.)	Rate (in $)
1.0	0.80	5.0	
2.0	1.08	6.0	
3.0	1.36	7.0	
4.0		8.0	

Source: U.S. Postal Service, www.usps.gov.

a. If we consider only integer numbers of ounces, verify that the values given in the table define the postal rate as a linear function of the letter weight.

b. Give the equation that relates postage (p) and letter weight (w).

c. Assuming the same pattern continues, fill the missing entries in the given table.

Solution

a. We observe from the table that for a 1.0-ounce change in weight, the cost increases by $0.28. The change in the dependent variable divided by the change in the independent variable is always $\frac{0.28}{1.0} = 0.28$ dollars per ounce, which tells us that the function is linear and its graph is a line with slope 0.28.

b. We now know that the postage and the letter weight are related through a linear equation: $p = c \times w + d$, where $c = 0.28$. To find the value of the constant d, we note that the postage corresponding to a weight of 1.0-ounce is $0.80, so $0.80 = 0.28 \times 1.0 + d$; hence, $d = 0.8 - 1.0 \times 0.28 = 0.52$. Thus, the linear relationship between postage and weight, for integer values of letter weight, is given by $p = 0.28 \times w + 0.52$.

c. We complete the table using this equation:

Weight Up to (in oz)	Rate (in $)	Weight Up to (in oz.)	Rate (in $)
1.0	0.80	5.0	1.92
2.0	1.08	6.0	2.20
3.0	1.36	7.0	2.48
4.0	1.64	8.0	2.76

A function whose graph consists of pieces of different lines is called a **piecewise linear** function. The function that associates income tax owed with salary earned is such a function, as the following example shows.

Example 5.4

The following table gives the year 2010 Federal Tax Rate Schedule for a single individual based on his or her taxable income:

Taxable Income	What You Pay
0 to $8,375	10% of the taxable income
Over $8,375 but not over $34,000	$837+15% of the excess over $8,375
Over $34,000 but not over $82,400	$4,681.25+25% of the excess over $34,000
Over $82,400 but not over $171,850	$16,781.25+28% of the excess over $82,400
Over $171,850 but not over $373,650	$41,827.25+33% of the excess over $171,850
Over $373,650	$108,421.25+35% of the excess over $373,650

Source: The New York Times Almanac 2011, p. 182.

a. Give the tax an individual should pay if his or her taxable income is:

i. $5,000

ii. $21,000

iii. $54,000

iv. $250,000

v. $385,000

b. Describe symbolically the function that gives the taxes to be paid for a given taxable income of i dollars.

c. Graph the function and describe the graph.

d. Give the rate of change in income tax if taxable income changes from $9,000 to $21,000.

e. Give the rate of change in income tax if taxable income changes from $21,000 to $54,000.

f. Give the rate of change in income tax if taxable income changes from $320,000 to $350,000.

Solution

a. All tax and income amounts are given in dollars. In the following calculations, we omit the dollar sign ($) for the sake of simplicity.

   i. Since 5,000 is between 0 and 8,375, the corresponding tax is 10% of 5,000; that is, $0.10 \times 5,000 = 500$.

   ii. To find the tax that corresponds to a taxable income of 21,000, we need to find 15% of the difference between $21,000 - 8,375$ and add it to 837. The tax is then $837 + 0.15 \times (21,000 - 8,375) = 2,730.75$.

   iii. The tax corresponding to a taxable income of 54,000 is $4,681.25 + 0.25 \times (54,000 - 34,000) = 9,681.25$.

   iv. For an income of 250,000, the tax is $41,827.25 + 0.33 \times (250,000 - 171,850) = 67,616.75$.

   v. For an income of 385,000, the tax is $108,421.25 + 0.35 \times (385,000 - 373,650) = 112,393.75$.

b. To describe this function, we need to use six different expressions, one for each given income range. Note that if i represents the taxable income and t the corresponding tax amount, the function is as follows:

If i is between 0 and 8,375, then $t = 0.1 \times i$.
If i is over 8,375 and not over 34,000, then $t = 837 + 0.15 \times (i - 8,375)$.
If i is over 34,000 and not over 82,400, then $t = 4,681.25 + 0.25 \times (i - 34,000)$.
If i is over 82,400 and not over 171,850, then $t = 16,781.25 + 0.28 \times (i - 82,400)$.
If i is over 171,850 and not over 373,650, then $t = 41,827.25 + 0.33 \times (i - 171,850)$.
If i is over 373,650, then $t = 108,421.25 + 0.35 \times (i - 373,650)$.

c. The graph of this function consists of pieces of six different lines. A portion of a line of slope 0.10 lies over the interval on the horizontal axis from 0 to 8,375; a portion of a line of slope 0.15 lies over the interval from 8,375 to 34,000; a portion of a line of slope 0.25 lies over the interval from 34,000 to 82,400; a portion of a line of slope 0.28 lies over the interval from 82,400 to 171,850; a portion of a line of slope 0.33 lies over the interval from 171,850 to 373,650; and a portion of a line of slope 0.35 lies over the values of the horizontal axis beyond 373,650. Here is the graph:

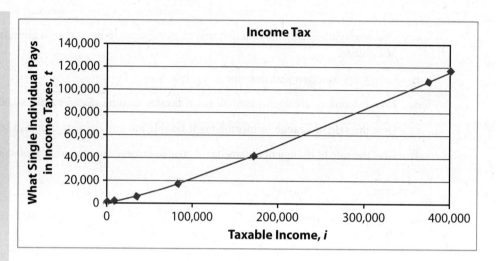

d. The rate of change from 9,000 to 21,000 is the rate of change of the function over that interval. That is, the slope of the line $t = 837 + 0.15 \times (i - 8,375)$ or 0.15. (We could also compute this rate by dividing the change in taxes by the change in income, but we have already seen that it equals the slope of the line.)

e. Over the interval from 21,000 to 54,000, the function consists of portions of two different lines, so we need to compute the rate of change by calculating the ratio of the change in the dependent variable divided by the change in the independent variable. The values of t corresponding to the income values of 21,000 and 54,000 are 2,730.25 and 9,681.25, respectively; the rate of change is $\frac{(9,681.25 - 2,730.75)}{(54,000 - 21,000)} \approx 0.211$, or 21.1 cents per dollar change in taxable income.

f. Over the interval from 320,000 to 350,000, the function satisfies the linear equation $t = 108,421.25 + 0.35 \times (i - 373,650)$. The rate of change is the slope of this line, which is 0.35.

The following is another example of a piecewise linear function. Each piece consists of a horizontal segment with a "jump" between them.

Example 5.5

The instructions on the use of the allergy medicine Triaminic® Syrup contain the following dosage table:

Age	Weight	Dosage
Under 6 yrs	Under 48 lb	Consult a doctor
6 to under 12 yrs	48–95 lb	2 tsp
12 yrs to adult	96+ lb	4 tsp

a. How much medication should we give a 10-year old boy who weighs 88½ lb? One who weighs 95 lb 7 oz? One who weighs 99 lb 2 oz?

b. Describe in words and in symbols, and graph the function that relates the dosage as the response variable with the weight as the explanatory variable.

Solution

a. The dosage for a 10-year old who weighs 88½ lb is 2 tsp because his weight falls in the 48–95 lb range. A child who weighs 95 lb 7 oz is between two categories, so the decision is not that clear. We prefer to err on the safe side and will give this child 2 tsp. (Some people may decide to round up to the closest pound and, in this case, would administer the higher dose. Alternatively, we could decide to give the child who weighs 95 lb 7 oz a dosage of 3 tsp, a dosage between 2 and 4 tsp.) The child who weighs 99 lb 2 oz gets 4 tsp of medicine because his weight is more than 96 lb.

b. There is no assigned value when the weight is less than 48 lb, so we say that the function is not defined for weights of less than 48 lb. As explained previously, we will administer the higher dose of 4 tsp only when the boy's weight reaches 96 lb, so from 48 lb to 96 lb (not included), the assigned value of the response variable is always 2 tsp. The corresponding value when the boy's weight is 96 lb or more is 4 tsp. We may describe this function in symbols as follows: $d = \begin{cases} 2 \text{ if } 48 \le w < 96 \\ 4 \text{ if } \quad\ w \ge 96 \end{cases}$, where d represents dose in teaspoons and w represents weight in pounds. The graph of this function consists of all points $(w, 2)$ for $w < 96$ and all points $(w, 4)$ for $w \ge 96$. Here is the graph:

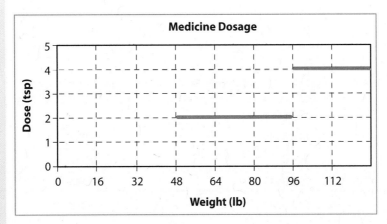

A function that has a symbolic expression $y = \dfrac{c}{x}$, where c is a constant, y represents the dependent variable, and x represents the independent variable, is an **inversely proportional** function. These are not linear functions as the following example shows.

Example 5.6

The average velocity v of a moving object is given by the distance d traveled divided by the time t elapsed. In symbols, $v = \dfrac{d}{t}$. The following table gives the average speed of the winners of the Daytona 500 auto race for several years:

Year	1960	1965	1970	1975	1980	1985	1990	1995	2000	2005	2010
Avg. mph	124.74	141.539	149.601	153.649	177.602	172.265	165.761	141.71	155.669	135.173	137.284

Source: The World Almanac and Book of Facts 2011, p. 960.

a. For each of the years given, calculate the time it took the winner to complete the 500-mile race. Create a table that displays the velocity and the corresponding time.

b. Graph the function that gives the time as the response variable and the average velocity as the explanatory variable for the 500-mile race. Include the points obtained in part (a) of this example.

c. Give the rate of change in time for an increase in average velocity from 130 to 140 mph, from 140 to 150 mph, and from 150 to 160 mph. How does this function differ from a linear function?

Solution

a. Because the distance traveled is 500 miles, the relationship between the distance and the time is given by $v = \dfrac{500}{t}$ or, solving for t, $t = \dfrac{500}{v}$, where t represents time in hours and v represents average velocity in miles per hour. So for $v = 124.74$, $t = \dfrac{500}{124.74} \approx 4.008$. Using this formula, we obtain the following times for the given velocities:

Average Speed (mph)	Time (hrs)
124.74	4.008
141.539	3.533
149.601	3.342
153.649	3.254
177.602	2.815

Average Speed (mph)	Time (hrs)
172.265	2.903
165.761	3.016
141.710	3.528
155.669	3.212
135.173	3.699
137.284	3.642

b. This is the graph of the function $t = \dfrac{500}{v}$. We have included all the points given in the table along with additional ones.

c. The time corresponding to 140 mph is $\dfrac{500}{140} \approx 3.571$ hrs, and the time corresponding to 130 mph is $\dfrac{500}{130} \approx 3.846$ hrs. The rate of change in time when the velocity increases from 130 to 140 mph is then $\dfrac{(3.571 - 3.846)}{(140 - 130)} = \dfrac{-0.275}{10} = -0.0275$ hrs per each mile per hour change in velocity. This means that the time is reduced, on average, 0.0275 hrs = $0.0275 \times 60 = 1.65$ minutes per 1 mph increase over the interval from 130 to 140 mph. Similarly, we calculate the rate of change for an increase in velocity from 140 to 150 mph and from 150 to 160 mph, obtaining approximately -0.0238 and -0.0208 hrs per mph, respectively. We see that the rate of change of this function is not constant, but it decreases as the values of the independent variable increase. This distinguishes it from a linear function, which has a constant rate of change.

Summary

In this topic, we analyzed directly and inversely proportional functions. A directly proportional function has an equation of the form $y = cx$ and is a particular case of a linear function. Inversely proportional functions are given by equations of the form $y = \dfrac{c}{x}$ and are not linear. We also discussed linear functions, which have equations of the form $y = cx + d$, and piecewise linear functions, which are described by different linear equations over different intervals of x.

Explorations

1. A wireless phone service charges $14.99 as a monthly charge and $0.30 per minute for calls.

   a. How much would you pay in a month when you made 10 calls of 3 minutes each?

   b. Let C represent your total monthly cost; assuming you only made calls, did not receive any calls, and the total number of minutes you were on the phone is m, write the equation to find your monthly bill.

   c. How can you tell that the relationship between C and m follows a linear pattern?

   d. Graph the function that gives the total monthly cost and give the slope of the line.

2. A household electricity bill for one month in which 2,494 kWh (kilowatt per hour) were used includes, among other charges, a distribution charge. The bill reads:

Current Charges	
Charges for – UTILITIES	
Residential Rate: RS for Jun 9 – Jul 12	
Distribution Charge:	
Customer Charge	6.47
200 KWH at 1.79600000¢ per KWH	3.59
600 KWH at 1.59400000¢ per KWH	9.56
1,694 KWH at 1.47200000¢ per KWH	24.94
Transmission Charge:	
2,494 KWH at 0.37700000¢ per KWH	9.40
Transition Charge:	
200 KWH at 1.79800000¢ per KWH	3.60
600 KWH at 1.59400000¢ per KWH	9.56
1,694 KWH at 1.47300000¢ per KWH	24.95
Generation Charge:	
Capacity and Energy	
200 KWH at 4.82600000¢ per KWH	9.65
600 KWH at 4.23800000¢ per KWH	25.43
1,694 KWH at 3.88600000¢ per KWH	65.83
PA Tax Adjustment Surcharge at 0.11818100%	0.23
Total UTILITIES Charges	**$193.21**
Pay This Amount No Later Than Aug 2, 2002	$193.21
Account Balance	**$193.21**

a. Give the total distribution charge when electricity consumption is:

   i. 150 kWh

   ii. 755 kWh

   iii. 920 kWh

b. Describe the function that the utility company uses to find the total distribution charge.

c. Sketch and describe the graph of the function.

d. Give the rate of change of the distribution charge if the electricity consumed changes from 150 to 160 kWh.

e. Give the rate of change of the distribution charge if the electricity consumed changes from 195 to 205 kWh.

f. Give the rate of change of the distribution charge if the electricity consumed changes from 300 to 310 kWh.

3. An aerobics instructor says she drinks 1 cup (8 ounces) of water before she starts her classes and then she drinks 10 ounces every 10 minutes.

   a. How much water does she drink when she teaches 2 one-hour classes in a row?

   b. Let W represent ounces of water and N the number of 10-minute intervals the aerobics instructor spends exercising, and write an equation to find the number of fluid ounces the instructor consumes during exercise.

   c. Explain the significance of each of the numbers that appears in the equation and sketch the graph of the function.

   d. Is the aerobics instructor following the advice of the sports nutritionist of Example 5.1?

4. The following is the recommended dose of a junior-strength painkiller, according to the child's weight:

Weight (lb)	Number of Tablets
Less than 48	Consult Physician
48–59	2
60–71	2½
72–96	3
Over 96	4

a. How much medication should you give a child who weighs:

   i. 62 lb 3 oz?

  ii. 71 lb 9 oz?

 iii. 98 lb?

Explain how you obtained your answers.

b. Describe and sketch the graph of the function that relates the dosage as the response variable with the weight as the explanatory variable.

c. Find the rate of change of dosage if the child's weight increases from:

  i. 49 to 62 lb

 ii. 72 lb 2 oz to 92 lb

d. Is the function linear? Is it piecewise linear?

5. The number of violent crimes reported in the United States in 1995 was 1,798,790 and decreased to 1,682,280 in 1996. Assume that the function that relates the number of violent crimes as the dependent variable, with the year that the number of crimes occurred as the independent variable, is linear.

  a. Write an equation for the line.

  b. If this equation were valid forever, how many years would it take for violent crime to disappear?

  c. Is the assumption that the function is linear, a reasonable assumption? Explain.

6. The following table contains the winning times in the women's 3,000-meter speed skating races in the Olympic games from 1960 to 2010. The time is given in minutes, seconds, and tenth of seconds. For example, the winner's time in 1960 was 5 minutes and 14.3 seconds or $5 \times 60 + 14.3 = 314.3$ seconds.

Year	Winner's Time
1960	05:14.3
1964	05:14.9
1968	04:56.2
1972	04:52.1
1976	04:45.2
1980	04:32.1
1984	04:24.8
1988	04:11.9
1992	04:19.9

Year	Winner's Time
1994	04:17.4
1998	04:07.3
2002	03:57.7
2006	04:02.4
2010	04:02.5

a. For each given year, find the winner's average velocity.

b. Give the function that relates the winner's time as the explanatory variable with the winner's average velocity as the response variable.

c. Graph the function described in part (b) from time 0 to 350 seconds.

d. Find the rate of change of average velocity when the time increases from 314 to 324 seconds and from 334 to 344 seconds.

e. Is the function described in part (b) linear? Explain.

f. Is the function described in part (b) increasing, decreasing, or neither increasing nor decreasing over the interval of time from 0 to 350 seconds.

g. Is the function described in part (b) concave upward or downward over the given interval? What does this say about the velocity function?

7. The table shown in Example 5.3 gives the postage rate for letters whose weights are integers (in ounces). When the weight is not an integer number of ounces, the post office rule is to round up to the nearest ounce to calculate the postage.

a. What would the postage be for a letter weighing 1.6 ounces?

b. Sketch a graph of the function for all weights from 0 to 4 ounces. Explain why the horizontal axis is used for the weight of the letter.

8. It was reported that Google websites (an aggregation of sites including YouTube and Blogger) were the most-visited world websites in June 2010, with the number of visitors (in thousands) at 943,791. (*Source: The World Almanac and Book of Facts 2011*, p. 372.)

a. Assume that the number of June visitors grew by 7% from June 2010 to June 2011. How many visitors, in thousands to the nearest thousand, visited Google websites in June 2011?

b. Assuming that the growth in June visitors to Google websites is linear, how many visitors can Google websites expect in June 2012? In June 2014?

c. Assuming that the growth in June visitors to Google websites is linear, find an equation that relates the number of visitors to Google websites and June year. Be specific about what any variables that you use represent.

d. Use your equation from part (c) to find the number of visitors Google websites can expect to experience in June 2020. Is your answer reasonable? How did you determine that it is or is not reasonable?

9. The U.S. Postal Service offers insurance that may be purchased to insure the value of standard mail, package services, and first-class or priority mail items. The costs to insure items (in addition to postage) are shown in the following table:

Declared Value	Insured Mail Fee
$0.01 to $50.00	$1.75
$50.01 to $100.00	$2.25
$100.01 to $200.00	$2.75
$200.01 to $300.00	$4.70
$300.01 to $400.00	$5.70
$400.01 to $500.00	$6.70
$500.01 to $600.00	$7.70
$600.01 to $5,000.00	$7.70 plus $1.00 per each $100 or fraction thereof over $600 in desired coverage

Source: The World Almanac and Book of Facts 2011, p. 379.

a. How much would it cost to insure a package valued at $950.00? A package valued at $3,500?

b. Describe and sketch the graph of the function that relates the insured mail fee as the response variable with the declared value as the explanatory variable.

c. Suppose we only want to consider items with declared values in $100 amounts between $600 and $5,000, including $600 and $5,000. Let F represent the insured mail fee and let v represent the declared value of the item. Write the equation that relates F and v for $v = \$600, \$700, \$800, \ldots, \$5,000$.

10. An airport has three rental car companies that rent a particular type of car. Company I offers a $23.50 daily rate with unlimited mileage (that is, there is no additional charge, regardless of how many miles are driven). Company II charges $19.99 per day and $0.15 for each mile driven; Company III charges $15.50 per day and $0.22 for each mile driven. Suppose you need to rent a car for 7 days.

a. Not counting the cost of gas, write an equation that relates the cost of renting a car from Company I (C_I) and the number of miles (m) driven during the week you rent the car.

b. Not counting the cost of gas, write an equation that relates the cost of renting a car from Company II (C_{II}) and the number of miles (m) driven during the week you rent the car.

c. Not counting the cost of gas, write an equation that relates the cost of renting a car from Company III (C_{III}) and the number of miles (m) driven during the week you rent the car.

d. Graph your three equations on the same set of axes. Describe what these equations show.

Rates of Change and Linear Functions

In this activity, you will investigate the link between the graph of a function and rate of change, particularly for data that describe a linear function. You will also investigate Major League baseball salary data and consider connections between rates of change and the graph of a function.

1. During the early years of the Olympics, the height of the winning pole vault increased approximately 8 inches every four years as shown in the following table:

Year	1900	1904	1908	1912
Height in inches	130	138	146	154

a. Explain how you can tell from the table that this is a linear function.

b. Let t be the number of years since 1900, and let H be the height of the winning pole vault. Verify that the equation $H = 130 + 2t$ gives the height of the winning pole vault as a function of t. What does the number 2 in this equation represent? What does the number 130 in this equation represent?

c. Could you use the equation given in part (b) to predict the height of the winning pole vault in the next summer Olympics? Explain.

2. The following table gives calories used per hour for individuals of different weights for three different activities:

Activity	Weight				
	100	120	150	180	200
Hiking	225	255	300	345	375
Cross-country skiing	525	595	700	805	875
Pleasure sailing	135	153	180	207	225

a. For each activity, find the rate of change of calories used per pound of weight from 100 to 120 pounds; from 120 pounds to 150 pounds; from 150 pounds to 180 pounds; from 180 pounds to 200 pounds.

b. What kind of a relationship is there between weight and calories used per hour for each activity, and what will the graphs of the calories used as a function of the individual's weight for each activity look like?

c. How will the graph of calories used for hiking compare with that of the other two activities?

d. Make a graph that shows all the data on the same graph. Is your answer from part (c) of this question correct?

e. Give an equation that relates C, calories used cross-country skiing, and w, an individual's weight.

3. Retrieve the Excel file "EA.5.1 Major League Salaries.xls" from the text website or WileyPLUS. The table in this file gives the average yearly salary of Major League baseball players from 1980 to 2010. (*Sources:* Michael Haupert, "The Economic History of Major League Baseball," in *EH.Net Encyclopedia*, edited by Robert Whaples, August 2003, http:eh.net/encyclopedia/article/haupert.mlb; Major League Baseball, http://mlb.com.)

   a. Find the average rate of change in average yearly salary per year from 1980 to 1981 and from 1981 to 1982, and record them here.

b. Now use the following Excel instructions to calculate the average rate of change in salary per year from one year to the next.

Instructions to Calculate Average Rate of Change

1. To enter the formula for rate of change of average salary, click cell C3, type =(**B3-B2**)/(**A3-A2**), and press **Enter**. [Does the number that appears in C3 agree with one of the two in part (a)?]
2. Click C3 and auto-fill to the end of your data.
3. Enter an appropriate title for column C in cell C1.
4. You can use **Audit mode** (**Ctrl** and ` together) to check that your formula for rate of change auto-filled correctly. (Press **Ctrl** and ` together again to exit **Audit mode**.)

c. What do the numbers in column C represent?

d. Notice that the rates of change for the years 1981 through 1986 are very similar to one another, compared to other values in column C. Explain why it would be reasonable to *approximate* the portion of the graph from 1980 to 1986 by a line.

e. Write an equation of a line that approximates the values of the given function from 1980 to 1986. To simplify the numbers, take the year 1980 as time $t = 0$, 1981 as time $t = 1$, and so on. Explain how you obtained your line.

f. Find the interval(s) of time when the average yearly salary was increasing. Record them here.

g. Find the interval(s) of time when the average yearly salary was decreasing. Record them here.

h. Within the years 1986–1997, find the interval(s) of time when the average rate of change in average salary per year is increasing. Give the intervals here.

i. Should the graph of the function that gives the average yearly salary of Major League players be concave downward or concave upward over the intervals of time indicated in part (h)?

j. Within the years 1986–1997, find the interval(s) of time when the average rate of change in average salary per year is decreasing. Give the intervals here.

k. Should the graph of the function that gives the average yearly salary of Major League players be concave downward or concave upward over the intervals of time indicated in part (j)?

l. Use the information obtained in parts (f–k) to draw a rough sketch of the function from 1986 to 1997. Include an appropriate scale on the vertical axis.

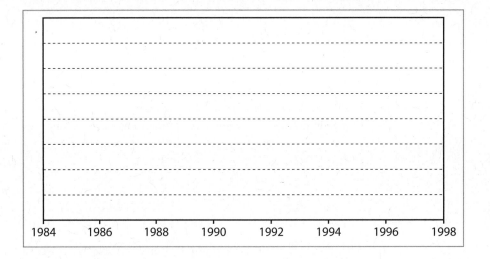

m. Use Excel to draw a graph of the average Major League salary function over time. (On your Excel graph, include all data given, starting in 1980.) How does this graph compare with your sketch?

Summary

In this activity, you analyzed a linear function that models the height of the winning pole vault in the Olympics and linear functions that model the calories spent in different exercise activities. You also used Excel to find average rates of change of Major League salaries for baseball players. You used these to draw a graph by hand and then used Excel to create a graph.

Modeling with Linear and Exponential Functions

Many times we want to summarize a set of data by constructing a function that in some way "fits" the data. The original data, however, might not represent a function in the mathematical sense, because at least one value of the explanatory variable appears with two (or even more) different values of the response variable. But the description of the situation or the data leads us to suspect that a straight line or an exponential function fits the situation. By analyzing the data and the circumstances, we evaluate if a linear (straight-line) model is appropriate or if some other type of curve would be a better model. In this topic, we look at linear and exponential models as examples. Although there are other ways to model data, we will not address them in this book.

After completing this topic, you will be able to:

- Recognize when data have an exact linear relationship.

- Visualize a regression line and use an equation of the line to predict values of the response variable for particular values of the explanatory variable.

- Determine when some quantity is growing or decreasing at an exponential rate.

- Understand the differences between linear and exponential growth.

Example 6.1

Each of the following tables represents a specific function. Determine if the relationship between the variables is a straight line.

a. This table represents the total popular votes, in thousands rounded up to the nearest thousand, cast for president of the United States in presidential election years 1968, 1972, and 1976.

Year	1968	1972	1976
Total Votes Cast (in thousands)	74,000	78,000	82,000

Source: The Federal Register, www.archives.gov/federal-register.

b. The following table gives the temperature in degrees Celsius and the corresponding temperature in degrees Fahrenheit.

Temperature (°C)	0	25	75	100
Temperature (°F)	32	77	167	212

c. The following table represents one model for recommended number of calories consumed per day as a function of a person's weight:

Weight (pounds)	120	140	180
Calories Per Day	1,800	2,000	2,300

Solution

a. We check to see if the slope, determined using successive pairs of points, is constant. We see that $\frac{(78,000 - 74,000)}{(1972 - 1968)} = 1,000$, and $\frac{(82,000 - 78,000)}{(1976 - 1972)} = 1,000$. Therefore, all three points lie on the same line and the relationship represented by these data is linear. Note that this is only a portion of the data representing all presidential elections, and the remaining data probably does not lie on the same line.

b. Again checking the slope between successive pairs of points, we see that $\frac{(77 - 32)}{(25 - 0)} = 1.8$, and the quotient of the difference in temperature in degrees Fahrenheit divided by the difference in temperature in degrees Celsius, using any two points in the table, will also be 1.8. For example, $\frac{(77 - 212)}{(25 - 100)} = \frac{(-135)}{(-75)} = 1.8$. Recall that it doesn't matter which point is used as the first point in calculating the differences for a rate of change. To calculate the slope of the line joining the first two points in the table, we could also compute $\frac{(32 - 77)}{(0 - 25)} = \frac{-45}{-25} = 1.8$. The relationship is linear.

c. For these data, $\frac{(2,000 - 1,800)}{(140 - 120)} = 10$, but $\frac{(2,300 - 2,000)}{(180 - 140)} = 0.753$, so these data are not linear.

The data in Example 6.1(a) show that the response variable, "total votes cast," experienced **linear growth** over the time interval from 1968 to 1976; that is, the function values increased at a constant rate over this time interval. In many cases, however, we have a data set such as the one given in Example 6.1(c), which is not exactly linear but which can be approximated or "fit" by a linear (straight-line) model. The line used most frequently to fit data that have an approximately linear relationship is called the **least-squares regression line**. A computer or calculator can be used to produce the equation of the least-squares regression line, sometimes called the **least-squares line** or the **regression line**. The regression line is the particular line, among all possible lines, that minimizes the sum of the squares of the vertical distances from the data points to the line. We can use the regression line to predict values of the response variable for appropriate values of the explanatory variable.

Example 6.2

The following table gives the median weekly earnings of full-time workers 25 years of age and older who have at least a Bachelor's degree, for the years between 1990 and 2010.

Year	Salary ($)	Year	Salary ($)
1990	638	2003	964
1992	696	2004	986
1994	733	2005	1,013
1996	758	2006	1,039
1998	821	2007	1,072
2000	891	2008	1,115
2001	921	2009	1,137
2002	941	2010	1,144

Source: U.S. Bureau of Labor Statistics, www.bls.gov.

a. Create a scatterplot of these data and verify that they do not represent an exact linear relationship.

b. Let s represent weekly salary, in dollars, and let t represent year. Sketch the line $s = 26.12t - 51,351$ on the scatterplot and describe what you see. (This is the regression line for these data. Although there are formulas to find the equation of this line, given a set of data points, we won't discuss those formulas here.)

c. Use the line to estimate the median weekly earnings for 1995 for workers with at least a Bachelor's degree.

d. Use the line to predict the median weekly earnings for 2012 for workers with at least a Bachelor's degree.

Solution

a. The data given in the table do not represent a linear relationship. Checking the rate of change quotients $\dfrac{\text{change in } s}{\text{change in } t}$ for successive pairs of points shows that the rate of change is not constant. We get different quotients, for example, $\dfrac{696 - 638}{2} = 29$, while $\dfrac{733 - 696}{2} = 18.5$ and $\dfrac{758 - 733}{2} = 12.5$. In the scatterplot shown here, the points look fairly close to a line, so it is appropriate to use a line as a model for these data.

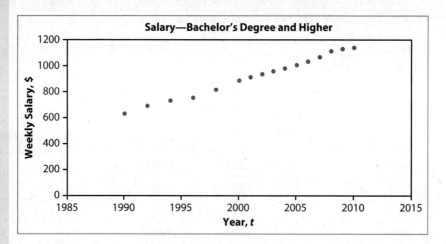

b. We sketched the given line on the following scatterplot. (To do so, we can plot any two points that satisfy the equation of the line and then draw the line through these two points.) We can see that some of the data points lie above the line and some lie below the line. In the equation of the line as given, the slope is 26.12. Note that this number is fairly close to the $\dfrac{\text{difference in salary}}{\text{difference in year}}$ calculated for several pairs of successive points in part (a) of the solution.

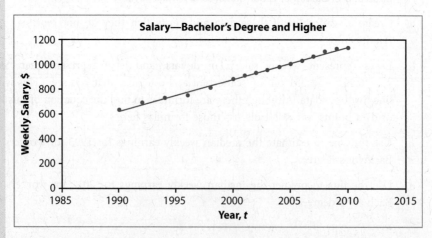

c. To use the line to estimate the median weekly earnings s for these workers in 1995, we let the explanatory variable t equal 1995 and compute $s = 26.12t - 51,351 = (26.12) \cdot 1995 - 51,351 = 758.4$. So we estimate that the median weekly earnings of a full-time worker with a Bachelor's degree or higher was $758.40 in 1995.

d. We can use the line to predict the median weekly earnings for these workers for the year 2012 because 2012 is not long after 2010, the last year for which we have this information. We let $t = 2012$ and use the equation of the line: $(26.12) \cdot 2012 - 51,351 = 1,202.44$. So we predict that the median weekly earnings for these workers is $1,202.44 in 2012. The point $(2012, 1,202.44)$ would lie on the line if we extended the line on the graph to the year 2012.

In some situations, given a data set in which the values represent growth (or decay) of some quantity, the values in the data set do not satisfy a linear relationship. We explore an exponential relationship in the next example.

Example 6.3

A family is in the process of furnishing their new home and is incurring some debt. At the beginning of the year, their debt was $500; the following table shows how their debt has grown during the first four months of the year. The explanatory variable is "time in months," using one-month increments, and the response variable is "amount of debt."

End-of-Month Number	Amount of Debt ($)
0	500.00
1	550.00
2	605.00
3	665.50
4	732.05

a. Add two columns to the table. In the third column of the table, compute the rate of change of debt with respect to time over each one-month period; that is, compute the difference of the current month's debt minus the previous month's debt. In the fourth column, compute the ratio of the current month's debt to the previous month's debt. (Note that for month 1, the previous month's debt would be the debt of $500 at the beginning of the year.)

b. Explain what these computations show.

Solution

a. The following table shows the two additional columns. To obtain each value in the third column, we compute a difference. The difference is a rate of change where the denominator of each rate is 1. To get each value in the fourth column, we compute a quotient.

End-of-Month Number	Amount of Debt ($)	Current – Previous Debt	Current/ Previous Debt
0	500.00	–	–
1	550.00	50.00	1.1
2	605.00	55.00	1.1
3	665.50	60.50	1.1
4	732.05	66.55	1.1

b. The values in the third column show that the difference of the current months's debt minus the previous month's debt is increasing. This means that the rate of change is not constant, so the growth of the debt is not linear. The fourth column shows that the ratio of the current month's debt to the previous month's debt is constant. We could write this relationship as $\dfrac{\text{current month's debt}}{\text{previous month's debt}} = 1.1$ or

current month's debt $= (1.1) \times$ previous month's debt.

The pattern shown in Example 6.3, in which the value of a quantity after the next time interval is a constant greater than 1 times the current value of the quantity, is characteristic of **exponential growth**. A graph can show how the family's debt will grow over a two-year period if it continues to increase by a factor of 1.1 each month. Notice that the graph is concave upward, which tells us that the rate of increase is increasing. This is exactly what the third column of the table in Example 6.3 tells us, too.

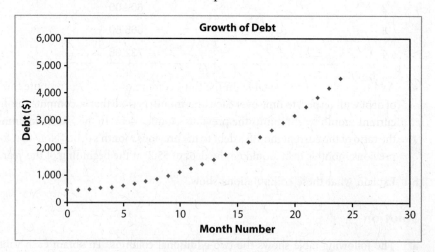

In the next example, we develop a formula that helps explain why this type of growth is called exponential growth.

Example 6.4

Find a pattern and develop a formula that gives the family's debt as described in Example 6.3 for any month m.

Solution

The debt in month 0 is \$500 and each succeeding month's debt is 1.1 times the previous month's debt. So let's use d_i to denote the family's debt in month i. Then $d_0 = \$500$, and $d_1 = (1.1) \cdot 500$, because the debt in month 1 is 1.1 times the debt in month 0. We know that the debt in month 2 is 1.1 times the debt in month 1, so $d_2 = (1.1) \cdot d_1 = (1.1) \cdot (1.1) \cdot 500 = (1.1)^2 \cdot 500$. We can continue this reasoning, and we see that the debt in month 3 is 1.1 times the debt in month 2: $d_3 = (1.1) \cdot d_2 = (1.1)(1.1)^2 \cdot 500 = (1.1)^3 \cdot 500$. Continuing the pattern, we note that for any value of m, the debt in month $m = d_m = (1.1)^m \cdot 500$. We can check to see that the formula is consistent with our graph; for example, in month 20, $d_{20} = (1.1)^{20} \cdot 500 \approx 3363.75$. It looks like the point on the graph is approximately (20, 3400). Notice that in the formula, the explanatory variable m appears as an exponent, so the term exponential growth makes sense.

Various phenomena grow exponentially. For instance, interest on investments involves exponential growth. Some analysts predict that the use of the Web to sell goods will grow exponentially. Population growth is another situation in which we often see exponential growth, as the next two examples show.

Example 6.5

The following table gives the estimated population (in millions) of Mexico for the years 1980 to 1986.

Year	Population in Millions
1980	67.38
1981	69.13
1982	70.93
1983	72.77
1984	74.66
1985	76.60
1986	78.59

Verify that the population of Mexico experienced exponential growth during the years 1980 through 1986, and explain why an exponential model for population growth is a reasonable one.

Solution

We check for exponential growth by dividing each year's population by the previous year's population and determining if this value is the same (that is, a constant) for the time period under consideration. For example, $\frac{\text{population in 1981}}{\text{population in 1980}} = \frac{69.13}{67.38} \approx 1.026$. Also, $\frac{\text{population in 1982}}{\text{population in 1981}} = \frac{70.93}{69.13} \approx 1.026$, and so on. We need to check each pair to make sure. Note that the *difference* in population in successive years is increasing from a growth of 1.75 million between the years 1980 and 1981 to an increase of 1.99 million between 1985 and 1986. Because individuals in the population reproduce, it is reasonable to predict that a population (unless there are other factors to be considered) will grow exponentially. Therefore, we might expect the population to increase at a rate proportional to the number of individuals now in the population. The population will grow faster and faster as time goes on. This is in contrast to a linear function, which always increases at the same rate.

Example 6.6

In the year 2000, a newspaper article pointed out that India's current population of approximately 1 billion may double to 2 billion in just another 100 years. The article predicted that India's population by 2050 will be 1.6 billion. Assume that the population growth of India is exponential and that the prediction of India's population in 2050 is correct; use this information to determine what the population of India will be at the end of this century, in 2100.

Solution

Using the information given, we assume that the population of India grows from 1.0 billion to 1.6 billion in the 50-year time period from 2000 to 2050. Thus the growth factor is $\frac{\text{population in 2050}}{\text{population in 2000}} = \frac{1.6}{1.0} = 1.6$. Therefore, assuming that this growth will continue, (population in 2100) $= 1.6 \times$ (population in 2050) $= 1.6 \times 1.6 = 2.56$ billion. If the growth continues at this rate, India will add more than 1 billion people in the next 100 years.

Example 6.7

The following table gives the number of mobile cellular subscriptions in the world for several years between 1990 and 2009.

Year	1990	1999	2000	2005	2006	2007	2008	2009
Subscriptions (in millions)	11	490	738	2,217	2,755	3,358	4,037	4,673

Source: The World Almanac and Book of Facts, 2011, page 375.

a. Was the growth in the number of subscriptions linear, exponential, or neither linear nor exponential from 1990 to 2009? Use one decimal place in your calculations.

b. Was the growth in the number of subscriptions linear, exponential, or neither linear nor exponential from 2005 to 2009? Use one decimal place in your calculations.

Solution

a. To see if the growth is linear, we calculate the rate of change from 1990 to 1999 and from 1999 to 2000. These are $\frac{490-11}{1999-1990} = 53.2$ (millions of subscriptions per year) and $\frac{738-490}{2000-1999} = 248$ (millions of subscriptions per year). These two rates are different, so the rate of change in the number of subscriptions is not constant. We conclude that the number of subscriptions did not grow linearly from 1990 to 2009.

To check for exponential growth, we need to consider data over intervals of time of equal length. Because there is a nine-year gap between the first two years in the table, we consider 1990–1999 and 1999–2008, the only two nine-year periods for which we have data, and calculate the ratios: $\frac{\text{subscriptions in 1999}}{\text{subscriptions in 1990}} = \frac{490}{11} = 44.5$; $\frac{\text{subscriptions in 2008}}{\text{subscriptions in 1999}} = \frac{4,037}{490} = 8.2$. These two ratios are different, so the growth was not exponential. We conclude that the growth was neither linear nor exponential from 1999 to 2009.

b. The rate of change from 2005 to 2006 is $\frac{2,755-2,217}{2006-2005} = 538$; and the rate of change from 2006 to 2007 is $\frac{3,358-2,755}{1} = 603$. These two rates are different, so the number of subscriptions did not grow linearly since 2005.

To check if the growth was exponential we calculate the ratio of the number of subscriptions each year over the number of subscriptions the previous year: $\frac{\text{subscriptions in 2006}}{\text{subscriptions in 2005}} = \frac{2,755}{2,217} = 1.2$; $\frac{\text{subscriptions in 2007}}{\text{subscriptions in 2006}} = \frac{3,358}{2,755} = 1.2$; $\frac{\text{subscriptions in 2008}}{\text{subscriptions in 2007}} = \frac{4,037}{3,358} = 1.2$; and $\frac{\text{subscriptions in 2009}}{\text{subscriptions in 2008}} = \frac{4,673}{4,037} = 1.2$. All these ratios are equal (after rounding to one decimal), so we conclude that the number of subscriptions grew (approximately) exponentially from 2005 to 2009.

If our data show that a quantity grows approximately exponentially over a period of time, we can find an exponential function to model the data. In the following example, we look for a function to model the data of the previous example over the years 2005 to 2009.

Example 6.8

Use the data given in Example 6.7 to write an equation that relates the number of mobile cellular subscriptions (s) and the time (t) over the years 2005 to 2009.

Solution:

We saw in Example 6.7 that the growth in subscriptions was approximately exponential from 2005 to 2009. We can use 1.2 as the ratio of the number of subscriptions one year divided by the number the previous year for that interval of time. Since the number of subscriptions in 2005 was 2,217 and $\frac{\text{subscriptions in 2006}}{\text{subscriptions in 2005}} = \frac{s \text{ in 2006}}{s \text{ in 2005}} = 1.2$, we can write the number of subscriptions in 2006:

(s in 2006) = (s in 2005) $\times$ 1.2 = 2,217 $\times$ 1.2. Repeating this process we have:

(s in 2007) = (s in 2006) $\times$ 1.2 = (2,217 $\times$ 1.2) $\times$ 1.2 = 2,217 $\times$ $(1.2)^2$;

(s in 2008) = (s in 2007) $\times$ 1.2 = 2,217 $\times$ $(1.2)^2$ $\times$ 1.2 = 2,217 $\times$ $(1.2)^3$;

(s in 2009) = (s in 2008) $\times$ 1.2 = 2,217 $\times$ $(1.2)^3$ $\times$ 1.2 = 2,217 $\times$ $(1.2)^4$.

To find the pattern, we organize this information in a table:

Year	2005	2006	2007	2008	2009
Subscriptions (s)	2,217	2,217 $\times$ 1.2	2,217 $\times$ $(1.2)^2$	2,217 $\times$ $(1.2)^3$	2,217 $\times$ $(1.2)^4$

We notice that each number in the second row is the product of 2,217 times a power of 1.2. Note that the value of s for the year 2005 is 2,217 = 2,217 $\times$ 1 = 2,217 $\times$ $(1.2)^0$ and the value of s in 2006 is 2,217 $\times$ 1.2 = 2,217 $\times$ $(1.2)^1$. If we represent by t the number of years since 2005, we see that $s = 2,217 \times (1.2)^t$.

If the value of a quantity after the next time interval is the current value of the quantity times a constant less than 1, we say that the quantity shows **exponential decrease**. The value of an appliance may decrease exponentially, and radioactive substances decrease exponentially (this type of decrease is called radioactive decay.)

Example 6.9

A refrigerator that cost $2,300 a year ago is now worth $1,978. If the price decreases exponentially, what will the refrigerator be worth two years from now?

Solution

Because the value decreases exponentially, we know that the ratio $\frac{\text{value one year}}{\text{value previous year}}$ is a constant. To find this constant we calculate $\frac{\text{value now}}{\text{value one year ago}} = \frac{\$1,978}{\$2,300} = 0.86$. So $\frac{\text{value one year from now}}{\text{value now}} = 0.86$ or value one year from now $= 0.86 \times \$1,978 = \$1,701.08$, and value two years from now $= 0.86 \times$ value one year from now $= 0.86 \times \$1,701.08 = \$1,462.93$.

Through several examples, we saw that a quantity that grows exponentially grows more quickly than one that grows linearly. The next example shows a situation in which growth is even faster than exponential growth.

Example 6.10

We want to test various brands of chocolate to see which is preferred by more subjects. To be fair, we want to vary the order in which the different brands are presented to the taste testers. If we have one brand to test, clearly there is only one way to arrange the order in which we present it to the testers.

a. How many orders of presentation are there if we have two brands to test? Three brands? Four brands?

b. Fill in the table with the values for the number of ways the brands can be presented for $n = 2$, 3, and 4 brands, and describe what you observe in this table.

Number of Brands Tested	Number of Orders of Presentation
1	1
2	
3	
4	
5	120
6	720
7	5,040
8	40,320
9	362,880
10	3,628,800

c. Sketch a graph of the number of orders of presentation as a function of brands tested. Explain why the growth in the number of orders of presentation is not exponential.

Solution

a. This is an example of a counting problem. If we think of lining up in a row the various brands of chocolate to be tested, we are actually counting the number of ways to rearrange objects in a row. If there are two brands, say A and B, either A can be first or B can be first. Thus, there are two orders of presentation. With three brands, say A, B, and C, any one of the three can be presented first. If A is offered first, then either B is next or C is next; if B is offered first, then A or C is next; and if C is offered first, then A or B is next. We can write out all the possibilities (that is, **enumerate** them) and see that there are six orders of presentation with three brands. With four brands, each of the four brands could be presented first. Then for each of those four choices, there are six orders of presentation for the remaining three brands (which we just figured out!). Thus, there are 4×6 or 24 orders of presentation with four brands.

b. The three missing numbers in the table are 2, 6, and 24. We observe that the number of orders of presentation is increasing rapidly, as the number of brands tested increases.

c.

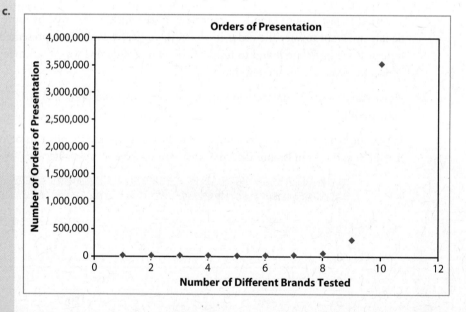

It's difficult to tell from the graph if the growth is exponential or not. We can see that if we connected the points on the graph, the graph would be concave upward. (What does this tell us?) When we look at the ratios of successive values of orders of presentation, we see that those values are increasing and are not constant. (You will be asked to verify this in Exploration 3 later in this topic.) For this to be an exponential function, those ratios must be the same.

Summary

In this topic, we investigated two types of models, linear and exponential models. We investigated how to recognize linear growth and exponential growth and explored some applications of each type of growth. We also saw how linear growth compares to exponential growth, looked at formulas for each type of growth, and examined why the terms *linear* and *exponential* are appropriate for each type of model.

Explorations

1. One linear model for computing an individual's "ideal" weight uses the following equations:

   Ideal weight for man of height h inches : $w_{man} = 160 + 6 \times (h - 65)$

   Ideal weight for woman of height h inches : $w_{woman} = 100 + 5 \times (h - 60)$

   a. Sketch the graph of each of these equations on the same set of axes with h on the horizontal axis and "ideal" weight on the vertical axis. Describe how the two equations differ.

   b. How do the "ideal" weights compare for a man and a woman who are both 5 feet 7 inches tall?

   c. What is a reasonable interval of values of h for which these equations make sense?

   d. How do you think these models were obtained?

2. The following table gives the median weekly earnings of full-time wage and salary workers 25 years and older who have had some college education or have an Associate's degree.

Year	Salary ($)	Year	Salary ($)
1990	476	2003	639
1992	484	2004	661
1994	499	2005	670
1996	518	2006	692
1998	558	2007	704
2000	596	2008	722
2001	617	2009	726
2002	629	2010	734

Source: U.S. Bureau of Labor Statistics, www.bls.gov.

   a. Sketch a scatterplot of these data.

   b. Let s represent salary and let t represent year. The least-squares regression line that best fits this data set is $s = 14.479t - 28,360$. Sketch this line on the graph with the scatterplot and describe what you see.

c. Compare this line with the line given in Example 6.2. What value for salary do you get from each of the lines if you substitute the value 2011 for the variable t, and what do those values tell you?

3. Verify that the ratios of successive values of orders of presentation in Example 6.10 are indeed increasing. Identify the pattern of these increasing values.

4. The following table gives the gross federal debt figures (in millions of dollars) for the United States every five years for the years 1960 to 2015:

Year	Gross Federal Debt (millions $)
1960	290,525
1965	322,318
1970	380,921
1975	541,925
1980	909,041
1985	1,817,423
1990	3,206,290
1995	4,920,586
2000	5,628,700
2005	7,905,300
2010	13,528,807
2015 (estimate)	19,775,536

Source: The President's Budget for Fiscal Year 2012, www.whitehouse.gov.

a. Sketch a scatterplot of these data. Does the graph appear to be similar to an exponential graph?

b. For each five-year value given in the table, compute the following ratio: gross federal debt in that year divided by the gross federal debt in the previous time period (five years earlier) and record these values. What do you observe about these ratios?

c. Is the growth of the gross federal debt exponential? How would you determine if the growth is exponential?

5. Cars depreciate in value as soon as you take them out of the showroom. A certain car originally cost $25,000. After one year, the car's value is $21,500. Assume that the value of the car is decreasing exponentially; that is, assume that the ratio of the car's value in one year to the car's value in the previous year is constant.

a. Find the ratio: $\dfrac{\text{value after one year}}{\text{original value}}$.

b. What is the car's value after two years? After ten years?

c. Approximately when is the car's value half of its original value?

d. Approximately when is the car's value one-quarter of its original value?

e. If you continue these assumptions, will the car ever be worth $0? Explain.

6. The U.S. Census Bureau International Database gave the population of the ten most-populated countries in the year 2000 and population predictions for 2050 for these countries: China had 1.2 billion people in 2000 and is predicted to have a population of 1.3 billion in 2050. The United States had a population of 284 million in 2000, with a prediction of 420 million by 2050.

   a. Assuming exponential population growth, predict the populations of China and the United States in the years 2100 and 2200.

   b. Assuming exponential population growth, approximately when will China's population double?

   c. What factors might make the assumption of exponential growth a faulty assumption?

7. World-wide food supply, or the ability to produce food, generally grows linearly while population tends to grow exponentially, if left unchecked. Explain why this is such a problem and illustrate with graphs.

8. Consider the formula $P = 67.38 \cdot (1.026)^t$. If we let P represent the population of Mexico in year t, where t is the number of years from 1980, confirm that this formula gives approximately the same population values as those given in the table in Example 6.5.

   a. Explain where the number 67.38 and the number 1.026 were obtained.

   b. What would the population in 1990 have been if growth had continued in this same pattern?

9. A newspaper article in *The Morning Call* on Friday, July 16, 2004, reported 2003 sales in billions of dollars and the number of stores for each of America's ten largest retailers. These data were used to sketch the following scatterplot:

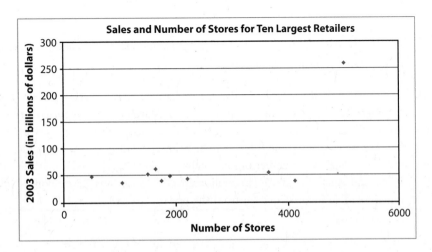

a. Is a line an appropriate fit for these data? Why or why not?

b. If you want to fit a line to these data, why should "number of stores" be the explanatory variable and "2003 sales" be the response variable?

c. Identify any unusual data value(s). What happens if you delete this (or these) value(s)?

10. The table contains data on U.S. online holiday sales from 2003 to 2009.

Year	Sales (in billions of dollars)
2003	18
2004	22
2005	27
2006	33
2007	37
2008	41
2009 (estimate)	45

Source: ReadWriteWeb, www.readwriteweb.com, with data from Forrester Research

a. Is the function that gives the U.S. online holiday sales from 2003 to 2009 linear, exponential or neither linear nor exponential. Explain.

b. If we consider only the data for years 2003−2006, could we use a linear or an exponential function to model the data? Explain. (Do your calculations to one decimal.)

c. If we consider only the data for years 2007−2009, could we use a linear or an exponential function to model the data? Explain.

d. Use the data in the table and an appropriate model to estimate the U.S. online holiday sales in 2002 and in 2001. Explain.

e. Use the data in the table and an appropriate model to estimate the U.S. online holiday sales in 2010 and 2012. Explain.

11. The article *"Number of U.S. Millionaires Soared in 2009: Spectrem Group"* posted by the *Huffington Post* (www.huffingtonpost.com) on March 9, 2010, reported a recent increase in the number of millionaires. The number of U.S. households whose net worth is $1million or more (not considering the value of their primary residency) was at an all-time high of 9.2 million in 2007. The number of millionaire families experienced a sharp decline and it was 27 percent smaller in 2008. The millionaire population was 7.8 million in 2009; this represented an increase of 16% from the previous year.

a. What was the U.S. millionaire population in 2008?

b. Assuming that the U.S. millionaire population grows linearly after 2008, estimate the millionaire population in 2011 and in 2012.

c. Assuming that the U.S. millionaire population grows exponentially after 2008, estimate the millionaire population in 2011 and in 2012.

d. Assume that the U.S. millionaire population follows a linear model from 2008 to 2025. Write an equation that gives the number of millionaire families m in terms of the time t, where t is the number of years since 2008 (that is, $t = 0$ corresponds to the year 2008, $t = 1$ corresponds to the year 2009, and so on).

e. Use the equation you found in part (a) to estimate the number of U.S. millionaire families in 2025.

f. Do you consider the estimate you found reliable? Explain why or why not.

12. The number of Facebook users worldwide was 676,733 in February 2011, and 692,998 in March 2011. (*Source:* Inside Facebook, www.insidefacebook.com)

a. Assuming that the increase is exponential, give the number of users in April and May of 2011.

b. Assuming that the number of Facebook users has been exponential since February 2011 and continues to be exponential until December 2015, write an equation that gives the number of Facebook users f in terms of t, the number of months since February 2011.

c. Use the equation you found in part (b) to give the number of Facebook users in March 2012 and in December 2015.

13. The following table gives the number of people in the United States that are living below the poverty level.

Year	Number of People (in thousands)
2006	36,460
2007	37,276
2008	39,829
2009	43,569

Source: U.S. Census Bureau, www.census.gov.

a. Decide whether a linear model or an exponential model is best for these data. Explain your answer.

b. Use an appropriate model and the given data to estimate the number of people living below the poverty level in the year 2012.

The Genie's Offer: Exponential Growth and Linear Growth

In this activity, you will explore differences between exponential and linear growth. You will also analyze an example where value is decreasing exponentially.

1. Suppose a magic genie offers you a choice: The genie will give you $1,000 on the first day of the year and will add $1,000 to what you have on each succeeding day of January until the end of the month. Or you may choose to receive 2 cents on January 1, and each day for the rest of the month, the genie will double the amount you had on the previous day.

   a. Which deal sounds better to you and why?

   b. Set up a spreadsheet to complete the following table. Use the appropriate formulas for the second and third columns, and fill in to the end of January.

Date in January	Total $ with the First Offer	Total $ with the Second Offer
1	1000	0.02
2	2000	0.04
3	3000	0.08
4		

(As a hint to help you fill in the spreadsheet, remember you can enter numbers **1** and **2** in the first two cells in a column and then drag down to fill in the rest of the integers. You can also enter a formula and drag it down a column and then use **Audit mode** to check the formulas entered. See Activity 4.1 if you need to refresh your memory.)

c. What does the table show?

d. At what point in the table is the amount in the third column greater than the amount in the second column and what does that mean?

e. Looking at your spreadsheet, identify a pattern and write an equation that shows how much money m you will have after d days if you take the genie's first offer of $1,000 on January 1 and an additional $1,000 each day after that.

f. Identify the type of function you wrote in part (e) of this question. How do you know it is that type of function?

g. Now you'll examine the genie's second offer to give you 2 cents on January 1 and then on each succeeding day to double the amount you had the day before. Identify the pattern developed below. Then analyze the pattern and develop an equation relating how much money m you will have on day d if you take the genie's second offer.

Money, in dollars, on day 1 = original amount of $ = 0.02

Money, in dollars, on day 2 = $ on Jan 2 = ($ on Jan 1) $*$ 2 = 0.02 $*$ 2

Money, in dollars, on day 3 = $ on Jan 3 = ($ on Jan 2) $*$ 2 = 0.02 $*$ 2 $*$ 2
= 0.02 $*$ 2^2

Money, in dollars, on day 4 = $ on Jan 4 = ($ on Jan 3) $*$ 2 = 0.02 $*$ 2^2 $*$ 2
= 0.02 $*$ 2^3

Money, in dollars, on day 5 = $ on Jan 5 = ($ on Jan 4) $*$ 2 = 0.02 $*$ 2^3 $*$ 2
= 0.02 $*$ 2^4

Money, in dollars, on day $d = m =$ _____. (What you fill in here should be an expression in terms of d.)

h. Fill in the blanks:

   i. On January 20, you have _____ times as much money as you had on January 1.

   ii. On January 31, you have _____ times as much money as you had on January 1.

i. Explain why it makes sense to call the function you wrote in part (g) of this question an **exponential function** and the kind of growth seen in the genie's second offer **exponential growth**.

j. On the same set of axes, graph the functions giving the amount of money you would have after d days with each of the genie's two offers. (You only need to graph up to day 25 to see the behavior of the two functions.) What does your graph show?

2. Appliances decrease in value as soon as you take them out of the store. A certain appliance originally cost $1,200. After one year the appliance is worth $1,080. Assume that the decline in value is exponential; that is, assume that the ratio of the appliance's value in one year to the appliance's value in the previous year is constant.

a. Thus, the ratio

$$\frac{(\text{value in year 1})}{(\text{value in year 0})} = \underline{\hspace{2cm}}; \text{or (value in year 1)} = \underline{\hspace{2cm}} * (\text{value in year 0})$$

will be the same for all succeeding pairs of years.

b. Set up a spreadsheet starting with the following table to compute the value of the appliance for years 2 through 20. You will need to enter the appropriate formula to compute the value of the appliance in year 2 and then drag it down the column for succeeding years.

Year	Value of Appliance in $
0	1200
1	1080
2	

c. What is the appliance's value after 10 years?

d. Approximately when is the appliance's value half of its original value?

e. Approximately when is the appliance's value one-quarter of its original value?

f. If you continue these assumptions, will the appliance ever be worth $0? Explain.

g. Use the letters v and y to indicate the appliance's value in dollars and the number of years from the purchasing date, respectively. Write a formula that gives v in terms of y. (Note that your formula must be set up so that when $y = 0$, $v = 1,200$, when $y = 1$, $v = 1,080$, and so on.)

h. Is v an exponential function of y? Why or why not?

i. Describe the similarities and differences between the formula for v in part (h) of this question and the formula for m in Question 1(g).

Summary

In this activity, you practiced creating formulas and scatterplots in Excel. You compared linear and exponential growth and explored the differences between these two fundamental types of growth. You also looked at a quantity that decreases exponentially and investigated patterns in several types of growth to find appropriate formulas.

6-2

Lines of Best Fit

It is often desirable to use a linear function to model a given set of data. In this activity, you will work with several data sets, and for each, you will look for a suitable line that approximates the given data. You will also work with the line that is generally used as the best-fitting line, the least-squares regression line. You will learn how to use Excel to find this line and its equation.

1. For the following scatterplots a, b, and c, draw an appropriate line to fit the data by "eye-balling" the graph and judging what line comes closest to all points. In the space following each graph, indicate whether the slope of the line you drew is positive, negative, or zero. For graph d, explain why a line would not be a good fit. (*Source: The Wall Street Journal Almanac 1999.*)

   a.

   Slope (negative, positive, or 0): _____

b.

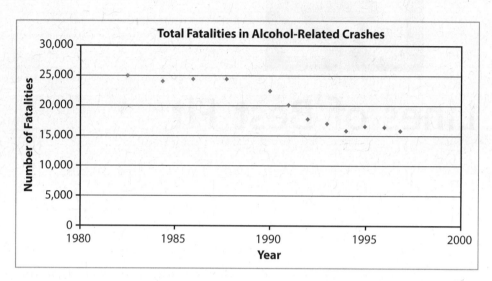

Slope (negative, positive, or 0): _____

c.

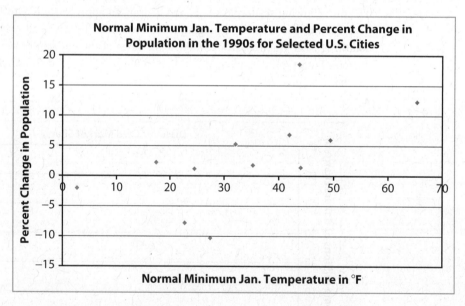

Slope (negative, positive, or 0): _____

d.

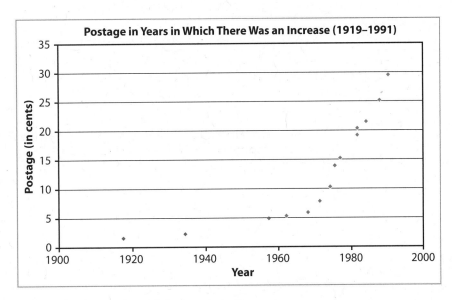

Why is a line *not* a good fit for graph d?

Because individual people might draw different lines, especially when the data is scattered as it is on graph c, we need a way to construct a line that doesn't depend on an individual's perception. The most commonly used method to construct such a line results in the **least-squares regression line** or just the **regression line**. This line, among all possible lines we could draw, is the one that makes the sum of the squares of the vertical distances of the data points from the line as small as possible.

The regression line is easy for a computer or calculator to find because it involves calculations using straightforward (but kind of messy) formulas.

2. The regression line is used to show how a response variable changes, on average, as an explanatory variable changes. You can use such a line to predict the value of the response variable for a particular value of the explanatory variable. The least-squares regression line, for the data given in Question 1(c), is shown on the following graph.

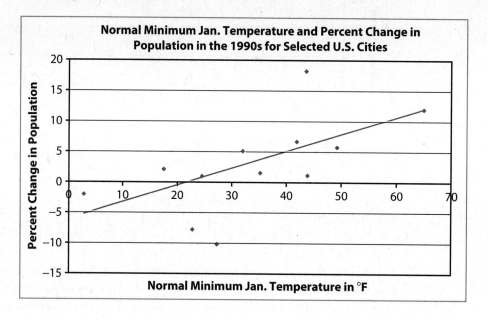

a. Draw in the vertical distance from each point in the data set to the regression line.

b. How many data points lie above the line? _____

c. How many data points lie below the line?_____

d. How can you tell from a scatterplot of the data, whether the slope of the regression line will be positive or negative?

3. Retrieve the data set "EA6.2.1 SAT Data.xls" from the text website or WileyPLUS.

   a. Create a scatterplot of "percent taking the test" and "critical reading SAT score." When creating the scatterplot, highlight only the two columns of data corresponding to the two quantitative variables; do not select the names of the states. Include appropriate titles, and change the scale on the vertical axis to go from 460 to 610. (See Activity 2.1 if you need a refresher on creating scatterplots and changing the scale.)

   b. Use the following instructions to find the regression line for these data.

Instructions to Find the Regression Line

1. There are several ways in Excel to find the least-squares regression line, but the easiest way is to point the cursor at one of the data points on the scatterplot and then right-click on the point to select it.

2. Select **Add trendline** from the menu box.

3. In the **Trendline Options** window, choose **Linear** for **Trend/Regression Type**, and **Automatic** for **Trendline Name**.

4. Click to place a check mark in the box **Display equation on chart**. Make sure there are no check marks in the other boxes. Then click **Close**. Your graph should display the regression line and its equation. You may need to click and drag the equation to a spot on the graph where you can read it clearly.

---

   c. Write the equation of your line and indicate what the variables x and y represent.

   d. What is the slope of the line and what does it represent? Interpret the slope in the context of the data.

   e. What is the y-intercept of the line and what does it represent? Interpret the y-intercept in the context of the data.

f. Use the line you found to predict the average critical reading SAT score for a state in which 60 percent of students take the exam. Where does this value appear on the graph? Mark it on a copy of the graph.

4. Retrieve the data set "EA6.2.2 Data Movies and DVDs.xls" from the text website or WileyPLUS. This file contains data collected from a sample of college students.

a. Create a scatterplot of the two quantitative variables and find the regression line for these data.

b. Write the equation of the line and indicate what the variables x and y represent in the equation.

c. Is there a clear choice of explanatory variable and response variable in this data set? Why or why not?

d. Describe what your scatterplot and line show.

e.   There is one clearly unusual data point in the data set—the male who estimated he saw 200 movies at a theater last year. Delete this case and look at the scatterplot for these adjusted data. Write the equation of this adjusted line and describe how the least-squares line changed when that point was deleted.

Summary

In this activity, you practiced creating scatterplots and learned how to find the regression line for a set of data using Excel. You interpreted the slope and y-intercept of a regression line in the context of the data set from which it was obtained, and used the regression line to predict values of the response variable. You also explored how an unusual data point can affect the regression line.

Logarithms and Scientific Notation

After completing this topic, you will be able to:

* Compare real-world quantities that are measured by logarithmic scales.

* Use properties of logarithms to solve problems.

* Use scientific notation and make estimates involving very large or very small quantities.

What do sound levels measured in decibels, intensity of earthquakes measured in points on the Richter scale, and acidity of liquids measured in pH levels, have in common? They are all measurements on a special scale called a logarithmic scale. In this topic, we look at logarithmic scales and discuss the definition and properties of the common logarithm.

A linear scale, like that on a yardstick, has the property that the distance between the 1- and 2-inch marks is the same as the distance between the 31- and 32-inch marks. In a similar way, the sequence 1, 4, 7, 10, ... is linear, or additive, because each term is three more than its predecessor. With a **logarithmic scale**, the unit steps increase in a multiplicative way. The sequence 1, 3, 9, 27, 81 ... is exponential, or multiplicative, because each term is three times its immediate predecessor. Logarithmic scales are used for some quantities that have a large range of variation, such as magnitude of earthquakes and levels of sound.

THE DECIBEL SCALE

The human ear can hear sounds that are as much as 100 trillion times louder than the faintest sounds. In the **decibel scale**, the least audible sound, which corresponds to a sound wave of intensity 10^{-12} watts/m^2, is assigned the value of 0; a sound that is 10 times louder than the least audible sound is assigned a value of 10; a sound $100 = 10^2$ times louder than the faintest sound is assigned a value of 20; a sound $1,000 = 10^3$ times louder than the faintest sound is assigned a value of 30; and so on. Thus, an increase of 10 decibels means a ten-fold increase in sound intensity or loudness.

Example 7.1

The sound of normal conversation measures approximately 60 decibels; the noise level inside a subway measures approximately 90 decibels.

a. How many times as loud as the faintest audible sound is normal conversation?

b. How many times as loud as the faintest audible sound is the noise inside a subway?

c. How many times as loud as normal conversation is the noise inside a subway?

Solution

a. Normal conversation at 60 decibels is $10 \times 10 \times 10 \times 10 \times 10 \times 10 = 10^6 = 1{,}000{,}000$ or 1 million times louder than the faintest audible sound.

b. The noise inside a subway is 90 decibels, so this noise is $10^9 = 1{,}000{,}000{,}000$ or 1 billion times as loud as the faintest sound.

c. The difference between the sound of normal conversation and the sound inside a subway is 30 decibels. Because each increase of 10 decibels corresponds to a ten-fold increase in loudness, an increase of 30 decibels corresponds to a sound $10 \times 10 \times 10 = 10^3 = 1{,}000$ times as loud. So the noise inside a subway is 1,000 times as loud as the sound of normal conversation. We can also obtain the answer by computing the ratio of the two sounds' loudness relative to the faintest sound, $\frac{10^9}{10^6} = 10^{9-6} = 10^3$.

The decibel scale is called logarithmic because the decibel measure and the loudness relative to the faintest audible sound are related through logarithms. Logarithms were first introduced by John Napier in 1614 to help simplify complicated calculations. The **common logarithm** (or logarithm with base 10) of a number n is the exponent r where $10^r = n$. That is, the logarithm of n is the power when n is written as 10 raised to a power. The common logarithm, also referred to as the **decimal logarithm** or just **logarithm**, of a number n is denoted by $\log n$.

The definition of the logarithm of a number tells us that, in general, $\log(10^p) = p$ for any number p. For example, $\log 1 = 0$ because $10^0 = 1$; $\log 10 = 1$ because $10^1 = 10$; and $\log 0.01 = -2$ because $10^{-2} = \frac{1}{10^2} = \frac{1}{100} = 0.01$.

For most numbers n, the computation of $\log n$ must be done using a scientific calculator or computer program. Before this technology was available, scientists and engineers used tables of logarithms or slide rules. For example, if we wanted to find the logarithm of $n = 50$, we would use a calculator and obtain $\log 50 \approx 1.699$, which means that $10^{1.699} \approx 50$. The function that relates the quantities n as the explanatory variable and r as the response variable through the equation $r = \log n$ is called the **logarithmic function with base 10**.

Example 7.2

Consider the logarithmic function with base 10, $r = \log n$.

a. Complete the following table of values of the function; use a calculator where necessary.

n	0.01	0.1	0.5	1	2	5	10	15	20	50	100
$r = \log n$	-2		-0.30								

b. Explain why when trying to evaluate log n for a value of $n \leq 0$, the calculator gives an error message.

c. Find the rate of change in r per unit change in n, when n increases from 0.1 to 0.5, from 1 to 5, from 10 to 15, and from 20 to 100.

d. Sketch and describe the graph of the function for values of n greater than 0.

Solution

a. We notice that $0.1 = \dfrac{1}{10} = 10^{-1}$, so $\log 0.1 = -1$. Also, $100 = 10^2$, so $\log 100 = 2$. We have already seen that $\log 1 = 0$ and $\log 10 = 1$. Using a calculator, we then approximate the remaining entries. Here is the completed table:

n	0.01	0.1	0.5	1	2	5	10	15	20	50	100
r	-2	-1	-0.30	0	0.30	0.7	1	1.18	1.30	1.7	2

b. For any value of r, the value of 10^r is positive. (Note that when r is negative, 10^r means the reciprocal of a positive power of 10. For example, $10^{-3} = \dfrac{1}{10^3} = \dfrac{1}{1{,}000}$ and $10^{-5} = \dfrac{1}{10^5} = \dfrac{1}{100{,}000}$. In general, when r is negative, $10^r = \dfrac{1}{10^s}$, where $s = -r$.) So if $r = \log n$ (as, for example, $-2 = \log n$) this means that $n = 10^r$ (for example, $n = 10^{-2} = 0.01$) and n must be positive.

c. The rate of change in r from 0.1 to 0.5 is $\dfrac{-0.30 - (-1)}{0.5 - 0.1} = \dfrac{-0.30 + 1}{0.4} = 1.75$ units per unit change in n. In the same manner, we compute the rate of change over the other given intervals. Here are the results:

The rate of change from 1 to 5 is $\dfrac{0.70 - 0}{5 - 1} = 0.175$ units per unit change in n.

The rate of change from 10 to 15 is $\dfrac{1.18 - 1}{15 - 10} = 0.036$ units per unit change in n.

The rate of change from 20 to 100 is $\dfrac{2 - 1.30}{100 - 20} = 0.00875$ units per unit change in n.

From these calculations, we can see that the rate of change decreases as n increases.

d. The following graph was obtained by using the points in the table completed in part (a) of this example. This function is always increasing (the rate of change is positive) and is concave downward (the rate of change is positive and decreasing, as the calculations in part (c) of this example show). Note that for positive values of n that are very close to 0, the values of r are large negative numbers, but r will never equal 0. For example, $\log(0.0001) = -4$, $\log(0.0000001) = -7$, and so on. There is no y-intercept on this graph.

In the next example, we investigate further the connection between logarithms and the decibel scale.

Example 7.3 ■ ■ ■ ■ ■ ■ ■ ■

Use logarithms to find a formula that relates the decibel measure of a sound with the loudness of the sound relative to the loudness of the faintest audible sound.

Solution

We assume that the loudness of the least audible sound, which measures 0 in the decibel scale, is 1. Then a sound that measures 10 decibels has loudness 10, which is 10 times 1; a sound that measures 20 decibels is 100 times louder than the least audible sound, so it has loudness 100, and so on. In the following table, we have included the decibel measure D for each of several sounds, the corresponding loudness L and the logarithm of L:

D	0	10	20	30	40	50	60
L	1	10	100	1,000	10,000	100,000	1,000,000
log L	0	1	2	3	4	5	6

Notice that the quantity $\log L$ differs from D by a factor of 10. The relationship between the two quantities, D and L, is given by $D = 10 \times (\log L)$. Thus D is directly proportional to $\log L$. (Note that D is not directly proportional to L, but to $\log L$.)

The logarithmic function $r = \log n$ is closely related to the exponential function with base 10. We study this relationship in the following example.

Example 7.4

Start with the function defined by the equation $r = \log n$ (also considered in Example 7.2), but use r as the explanatory variable and n as the response variable; that is, interchange the roles of r and n.

a. Express n in terms of r and determine what type of function is obtained.

b. Graph the function.

c. Using both horizontal and vertical axes from -2 to 10, sketch both this new function and the function $r = \log n$ on the same graph. Explain the relationship between the two graphs.

Solution

a. Because $r = \log n$ means $10^r = n$, n is obtained from r by raising 10 to the power r. The function $n = 10^r$ is an exponential function because the explanatory variable appears as an exponent in the formula. Because the base is 10, this function is the exponential function with base 10. This results in the table:

r	−2	−1	−0.30	0	0.30	0.7	1	1.18	1.30	1.7	2
n	0.01	0.1	0.5	1	2	5	10	15	20	50	100

b. To graph this function, we can use the table we created in part (a). Alternatively, we can use the table we used to graph the function in Example 7.2, but with the independent and dependent variables switched. We can see from the graph that this function is always increasing and is concave upward.

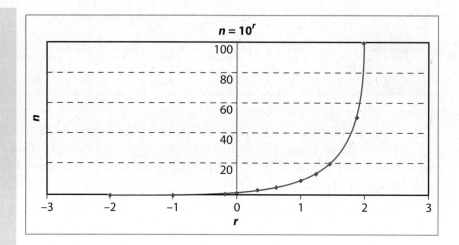

c. For this graph, we will use the names x and y for the independent and dependent variables, respectively. We can see that the graph of each of the functions can be obtained from the other function by reflection across the line $y = x$ (graphed as a dotted line). Thus, for example, the point $(0, 1)$ is on the graph of $y = 10^x$ and $(1, 0)$ is on the graph of $y = \log x$; the point $(0.7, 5)$ is on the graph of $y = 10^x$ and $(5, 0.7)$ is on the graph of $y = \log x$, and so on.

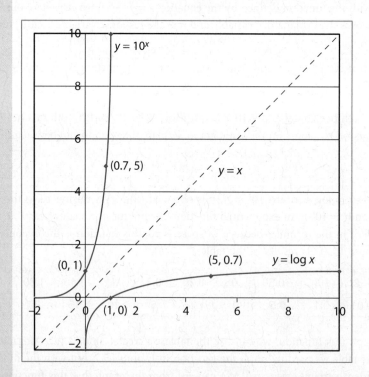

THE RICHTER SCALE

Another example of a logarithmic scale is the **Richter scale**. Named after the American geologist Charles F. Richter, it is used to measure the magnitude of earthquakes. Seismologists use records taken by seismographs to measure the total amount of motion produced by the vibrations during an earthquake. A number derived from these seismograph measurements is assigned to the earthquake and is called the magnitude of the earthquake. The magnitude of an earthquake, using the Richter scale, is determined from the logarithm of the amplitude of waves (that is, the measurement from the center of the wave to the highest point) recorded by seismographs. A unit change in the Richter scale, represents a tenfold increase in the amplitude of the earthquake. For example, the amplitude of waves recorded from an earthquake that measures 5 on the Richter scale is 10 times the amplitude of waves recorded for an earthquake that measures 4. If A represents the amplitude of waves produced by an earthquake and r indicates the magnitude of the earthquake on the Richter scale, then $A = 10^r$.

Example 7.5

In 1906, the American continents experienced two major earthquakes. The San Francisco earthquake, which occurred in April of that year, measured 8.3 on the Richter scale. In August of the same year, an earthquake that measured 8.6 on the Richter scale occurred in Valparaiso, Chile. How many times greater was the amplitude of the waves of the Valparaiso earthquake than that of the San Francisco one?

Solution

The amplitude of the waves of the San Francisco earthquake was $10^{8.3} \approx 199{,}526{,}231.5$, while the amplitude of the waves of the Valparaiso earthquake was $10^{8.6} \approx 398{,}107{,}170.6$. Because $\frac{398{,}107{,}170.6}{199{,}526{,}231.5} \approx 1.995$, the waves of the Valparaiso earthquake had an amplitude almost two times the amplitude of the waves of the San Francisco one.

Logarithms have been used extensively to simplify computations because they have the property that they transform products into sums, quotients into differences, and powers into products. That is, if a and b are two positive numbers, and r is any nonzero number, we have the following properties:

Property (1): $\log(ab) = \log a + \log b$

Property (2): $\log\left(\dfrac{a}{b}\right) = \log a - \log b$

Property (3): $\log(a^r) = \log a^r = r \log a$

Example 7.6

Use values from the table obtained in Example 7.2 to provide examples for each of the three properties just described. The table is repeated here.

n	0.01	0.1	0.5	1	2	5	10	15	20	50	100
r	-2	-1	-0.30	0	0.30	0.7	1	1.18	1.30	1.7	2

Solution

Because $20 = 2 \cdot 10$, property (1) gives $\log 20 = \log 2 + \log 10 = 0.30 + 1 = 1.30$. Note that $\frac{20}{2} = 10$ and from the table, we know that $\log 10 = 1$, $\log 20 = 1.30$, and $\log 2 = 0.30$. So by Property (2), we have $\log\left(\frac{20}{2}\right) = \log 20 - \log 2 = 1$. We illustrate Property (3) with 100. We know that $100 = 10^2$, so $\log 100 = \log 10^2 = 2 \log 10 = 2 \cdot 1 = 2$.

We can also use several of these properties in sequence. For example, from the table, $\log 50 = 1.7$. We see that $50 = 2 \cdot 5^2$, so $\log 50 = \log(2 \cdot 5^2) = \log 2 + \log(5^2) = \log 2 + 2 \cdot \log 5 = 0.3 + 2(0.7) = 1.7$.

Example 7.7

Use the properties of logarithms and the information that $\log a = 2.12$ and $\log b = 15.3$ to find the following values.

a. $\log(a \cdot b)$

b. $\log\left(\frac{b}{a}\right)$

c. $\log a^{-3}$

d. $(\log a)^{-3}$

e. $\log\dfrac{100,000}{b}$

Solution

a. By Property (1), $\log(a \cdot b) = \log a + \log b = 2.12 + 15.3 = 17.42$.

b. We use Property (2); $\log\left(\frac{b}{a}\right) = \log b - \log a = 15.3 - 2.12 = 13.18$.

c. Using Property (3), $\log a^{-3} = (-3) \cdot \log a = (-3)(2.12) = -6.36$.

d. We substitute for $\log a$: $(\log a)^{-3} = 2.12^{-3} = 0.105$.

e. By property (2), $\log \dfrac{100{,}000}{b} = \log 100{,}000 - \log b = \log 10^5 - \log b = 5 - 15.3 = -10.3$.

SCIENTIFIC NOTATION

Scientific notation is used mostly to express very large or very small numbers. A number is in scientific notation if it is written in the form

$$a \cdot 10^r, \text{ where } 1 \le a < 10 \text{ and } r \text{ is an integer.}$$

For example, 1,672,000 can be written in scientific notation as $1.672 \cdot 10^6$, and 0.000761 can be written in scientific notation as $7.61 \cdot 10^{-4}$. The number $16.54 \cdot 10^9$ is not in scientific notation because 16.54 is not between 1 and 10. To write $16.54 \cdot 10^9$ in scientific notation, we first write 16.54 as $1.654 \cdot 10$. We then write $16.54 \cdot 10^9$ as $1.654 \cdot 10 \cdot 10^9$ which is $1.654 \cdot 10^{10}$.

When a number is written in scientific notation, the logarithm of the number can be related easily to the logarithm of a number between 1 and 10. For example, $\log 1.672 \cdot 10^6 = \log 1.672 + \log 10^6 = \log 1.672 + 6$.

Some calculators use scientific notation in their output. For example, 4,260,000,000 would appear as 4.26E9 (or as 4.26e9), meaning $4.26 \cdot 10^9$ and 0.0000258 would appear as 2.56E-5, meaning $2.58 \cdot 10^{-5}$.

Example 7.8

The distance from a planet to the Sun varies as the planet moves in its orbit.

a. The closest distance (called the perihelion) from Mercury to the Sun is 2.86×10^7 miles and its farthest distance (aphelion) is 4.34×10^7 miles. What is the difference between the two measurements?

b. The farthest distance from Jupiter to the Sun is 5.07×10^8 miles. What is the difference between the farthest distance from Mercury to the Sun and the farthest distance from Jupiter to the Sun?

Solution

a. Because each term has a common factor of 10^7, we can factor it out and then subtract. The difference between the two measurements is $4.34 \times 10^7 - 2.86 \times 10^7 = (4.34 - 2.86) \times 10^7 = 1.48 \times 10^7 = 14{,}800{,}000$ miles.

b. For this calculation, we need the same term, 10^7, in each expression so we can factor it out. We first rewrite 5.07×10^8 as $5.07 \times 10 \times 10^7 = 50.7 \times 10^7$. Then, the difference between the farthest distance from Jupiter to the Sun and that of Mercury to the Sun is $5.07 \times 10^8 - 4.34 \times 10^7 = (5.07 \times 10 - 4.34) \times 10^7 = (50.7 - 4.34) \times 10^7 = 46.36 \times 10^7 = 463{,}600{,}000$ miles.

Scientific notation is useful to make estimates when large numbers are involved, because operating with powers of 10 is straightforward in the decimal numerical system that we use. In the next example, we see how to use scientific notation when estimating.

Example 7.9

According to its website (www.mcdonalds.com), as of February 2012, McDonald's has 33,000 restaurants in 119 countries around the world. Suppose that Australia, for example, has 792 McDonald's restaurants that serve approximately 1.5 million customers daily. Use this information to estimate how many McDonald's hamburgers are consumed each year around the world. State any assumptions you make.

Solution

We assume that, on average, every two customers who visit McDonald's eat one hamburger (some might eat chicken, fish, salad, or a breakfast meal but other customers consume more than one hamburger, so this assumption does not seem unreasonable). We first estimate the daily number of hamburgers. To do this we will assume that the average daily number of customers per restaurant all over the world is the same as the average daily number of customers in Australia. This is $\frac{1{,}500{,}000}{792} \approx 1{,}894$ customers per restaurant per day. Because there are 33,000 restaurants around the world, each serving 1,894 customers a day, or 947 hamburgers per day, a rough estimate of daily consumption of hamburgers is

$$3.3 \cdot 10^4 \cdot 9.5 \cdot 10^2 = 31.4 \cdot 10^6 = 31.4 \text{ million hamburgers per day}$$

In one year, the number of hamburgers consumed at McDonald's is approximately

$$31.4 \cdot 10^6 \cdot 360 = 3.1 \cdot 10^7 \cdot 3.6 \cdot 10^2 = (3.1 \cdot 3.6) \cdot 10^9 \approx 11 \text{ billion.}$$

We can use logarithms to solve for an unknown that appears in an equation as an exponent. When a variable appears as an exponent, as in the equation $10 = 5^x$, to solve for x we can take the logarithm of each side. (If two quantities are equal, their logarithms are also equal.) We set the logarithms equal and use the properties of logarithms to simplify. To solve $10 = 5^x$, we get $\log 10 = \log 5^x$, so $1 = x \cdot \log 5$. From Example 7.2, we find $\log 5 = 0.7$, so $x = \dfrac{1}{0.7} \approx 1.43$.

Example 7.10

One population model shows that a certain species of individuals (possibly bacteria or insects) lives in an isolated atmosphere. The equation that gives N, the number of individuals in the population as a function of t, time in days, is $N = 10 \cdot 1.2^t$. In this model, there are initially 10 individuals in the population, because when $t = 0$, $N = 10 \cdot 1.2^0 = 10 \cdot 1 = 10$. Use the facts that $\log 2 = 0.3$ and $\log 1.2 = 0.08$ to find the following:

a. How many individuals are in the population after 30 days?

b. When will the population have 200 individuals?

c. Suppose the population starts to decline (and this model would no longer work) when the population reaches 1,000 individuals. Use the model to predict when that will happen.

Solution

a. When $t = 30$, $N = 10 \cdot 1.2^{30} = 2{,}373.8$ individuals.

b. We set $N = 200$; $200 = 10 \cdot 1.2^t$. Next we apply the logarithm function to each side of this equation: $\log 200 = \log(10 \cdot 1.2^t)$. Simplifying each side of the equation, we get $\log 2 + \log 100 = \log 10 + t \cdot \log 1.2$. We substitute $\log 100 = 2$ and $\log 10 = 1$ and the given values for $\log 2$ and $\log 1.2$ to get: $0.3 + 2 = 1 + t \cdot 0.08$. Solving for t, we get $t = \dfrac{1.3}{0.08} = 16.25$. The population will have 200 individuals in 16.25 days.

c. From our answers to parts (a) and (b), we know the answer to this question will be between 16 and 30 days. We solve $1{,}000 = 10 \cdot 1.2^t$ for t. Applying the logarithm function to each side of this equation we get $\log 1{,}000 = \log 10 \cdot 1.2^t$, or $3 = \log 10 + t \cdot \log 1.2$. Substituting $\log 10 = 1$ and $\log 1.2 = 0.08$, we get $3 = 1 + t \cdot 0.08$, so $t = \dfrac{2}{0.08} = 25$. At time 25 days, the population starts to decline.

Summary

In this topic, we defined the common logarithm and applied rules of logarithms. We looked at two logarithmic scales: the decibel measure for sounds and the Richter scale for magnitude of earthquakes. We also did some calculations and made estimates using scientific notation.

Explorations

1. Explain why a logarithmic scale is useful when measuring sound.

2. The following graph appeared in *The Morning Call* on October 5, 1999. Use the information on the graph to answer the following questions.

HOW LOUD IS IT?

Typical range of common sounds, in decibels

Conversation

Chain saw
Diesel locomotive at 50 feet
Heavy truck at 50 feet
Power lawnmower
Automobile at 50 feet
Vacuum cleaner
Refrigerator

40 60 80 100 120

SOURCE: Environmental Protection Agency

a. How many times louder is the sound of the most silent chain saw than the sound of the loudest power lawnmower?

b. How does the sound of an approaching diesel locomotive compare to the sound of normal conversation when the locomotive is 50 feet away? (Consider the full range of possibilities.)

3. Consider the following statement: "An ambulance siren has a noise level of about 100 decibels when heard from 100 feet away. That is about 8 times as loud as normal conversation." Is this statement correct? Explain your answer.

4. Answer the following:

   a. Use the value $\log 2 = 0.30$ and the properties of logarithms discussed in this topic to find the following values: $\log 200$, $\log 0.002$, $\log 16 = \log 2^4$.

   b. Use a calculator to evaluate the logarithms in part (a) of this exploration. Did you get the same answer?

5. In February of 1997, northwestern Iran suffered an earthquake of magnitude 6.1 on the Richter scale. In May of the same year, an earthquake of magnitude 7.5 occurred in northern Iran. The first earthquake caused 1,000 deaths and the second earthquake caused 1,560 deaths.

   a. How many times greater was the amplitude of the waves recorded for the second earthquake than the first one?

   b. What was the ratio between the death tolls in the earthquakes?

   c. Name other factors that might increase the death toll of an earthquake other than its magnitude.

6. If an earthquake has magnitude 4.2 on the Richter scale, what is the magnitude on the Richter scale of an earthquake with an amplitude of waves that is 500 times greater? Explain your answer.

7. Pure water has a pH level of 7.0. The pH scale is a measurement scale that indicates how acidic or basic (that is, alkaline) a liquid is. Each one-unit decrease in pH indicates a ten-fold increase in acidity. Similarly, for pH values greater than 7.0, each one-unit increase indicates an acidity level one-tenth of the next lower integer value.

   a. Black coffee has a pH level of 5. How much more acidic is black coffee than pure water?

   b. Ammonia has a pH of 11. What is its acidity compared to pure water?

   c. Lemon juice has a pH of 2. How does its acidity compare with that of Milk of Magnesia with a pH of 10?

8. Suppose you know that $\log a = -1.3$ and $\log b = 7.6$. Find the following values.

   a. $\log(a \cdot b)$

   b. $\log\left(\frac{a}{b}\right)$

   c. $\log b^5$

   d. $\log(0.000001 \cdot a)$

   e. $\log \dfrac{b}{10,000}$

9. The annual waste in paper and paper products generated by a particular industry can be modeled by the equation $w = 15.35(1.03)^n$, where w is the waste in thousands of tons, and n is the year from 1960; that is, 1960 is year $n = 0$.

   a. Solve the equation $50 = 15.35(1.03)^n$ for n to find when this function estimates the annual waste at 50 thousand tons.

   b. Solve the equation $100 = 15.35(1.03)^n$ for n to find when this function estimates the annual waste at 100 thousand tons.

   c. How could the industry use this model?

10. The average distance from Earth to the Sun is $9.3 \cdot 10^7$ miles. The distance from Pluto to the Sun varies from 2.76×10^9 miles to 4.58×10^9 miles.

    a. What is the difference between the closest and farthest distances from Pluto to the Sun?

    b. Estimate how long it would take a rocket traveling at 2.9×10^3 miles per hour to reach the Sun.

    c. Find the exact answer to part (b).

11. In August 1999, a 36-year-old woman was awarded $23 million by a Texas jury who found that manufacturers of the diet drug Fen-phen were responsible for the woman's serious medical problems. There were 3,100 similar lawsuits against the company at the time. In July 1999, the company had reported second-quarter profits of $299 million.

    a. Estimate how much the company would have to pay out if the company loses all those lawsuits.

    b. Estimate how many years the company would take to pay off all that money. State any assumptions you make.

12. An angstrom is a unit of length equivalent to 0.0000001 millimeters.

    a. Write the number 0.0000001 as a power of 10.

    b. An electron is 10^{-12} millimeters in diameter. What is this diameter's length in angstroms?

13. A recent news report stated that, as a result of a recent downturn in the stock market, a man whose worth was $10 billion lost 99% of his wealth. How much is he now worth?

Richter Scale and Logarithms

In this activity, you will investigate earthquake data and explore the Richter scale as a measure of the intensity of an earthquake. You will consider how numbers on this scale compare with one another and study logarithms in the process.

1. How many times greater is the amplitude of the waves for an earthquake that measures 6.5 on the Richter scale than one that measures 6.0? How much farther away from your home is a restaurant that is 6.5 blocks away, as compared to one that is 6.0 blocks away? How are these measures similar and how are they different?

2. Retrieve the file "EA7.1.1 Deadly Earthquakes.xls" from the text website or WileyPLUS. (These data were obtained from the website http://earthquake.usgs.gov/.) This file contains the date, location, and magnitude on the Richter scale of earthquakes that occurred from 1975–2011 and involved the loss of 1,000 or more lives.

a. Sort the data in ascending order by magnitude and give the date and location of the strongest and weakest of the earthquakes on the list. (See Activity 1.2 if you do not remember how to sort data.)

Strongest:

Weakest:

b. How many times greater was the amplitude of the waves recorded for the strongest earthquake than for the weakest?

c. For each of the earthquakes listed, compute the amplitude of its waves using the following instructions. Recall that the amplitude A of an earthquake of magnitude m on the Richter scale is $A = 10^m$.

Instructions to Calculate Energy Released

1. In cell D2, enter **=10^C2** (the ∧ symbol indicates raised to a power) to calculate the amplitude of the waves of the first earthquake. Enter an appropriate title in cell D1.

2. To calculate the amplitude of the waves as a function of Richter scale magnitude for all earthquakes, use the "drag" feature of Excel to fill the column.

d. Create a scatterplot using columns C and D of your spreadsheet. (See Activity 2.1 if you do not remember how to create a scatterplot. Be sure to label your axes.)

   i. Explain what your graph shows.

   ii. What type of function (linear, exponential, or neither linear nor exponential) does your graph show? How do you know?

    iii. What variable is on the horizontal axis?

    iv. What variable is on the vertical axis?

  e. For each amplitude value A you found, compute its logarithm log A, using the following instructions.

Instructions to Calculate Logarithms

1. To have Excel compute these logarithms and enter them in column E, place the cursor in cell E2 and enter **=LOG(D2)**. Then drag down.
2. Enter an appropriate title for column E in cell E1.

  f. How are the values you just calculated in column E related to other values in your table?

  g. Create another scatterplot using columns D and E.
    i. Give the name and equation of the function just graphed.

    ii. What variable is on the horizontal axis?

    iii. What variable is on the vertical axis?

  h. Look at the two scatterplots you've created in this activity and describe how they are related. How could you obtain one from the other?

i. Use the following Excel instructions to write in scientific notation the amplitude of the waves of the first earthquake on the list.

Instructions to Write Scientific Notation

1. In cell F2, enter =**D2** to copy the number in cell D2 (which represents the amplitude of the waves of the earthquake) into cell F2.

2. Click on cell F2, go to the **Cells** group and select **Format**. Click on **Format Cells**.

3. In the **Format cells** window, click the **Number** tab and choose **Scientific** from the **Category** list. For **Decimal places**, enter **2**.

---

j. Write the number as it appears in cell F2 and also write it in standard scientific notation (using a power of 10).

k. What is the difference between the number in cell D2 and the number in cell F2?

l. Using the "drag" feature, enter the rest of the numbers in column D into column F in scientific notation. Estimate the ratio between the largest number and the smallest number in column F. What does this ratio say about the earthquakes given in the data set?

3. Retrieve the file "EA7.1.2 Earthquake Casualt.xls" from the text website or WileyPLUS. This file gives the same information as the file you used previously, except a new column has been added that shows estimated number of deaths.

    a. Create a scatterplot to show if there is any relationship between magnitude and number of deaths.

b. Explain in detail what your graph shows. (You might want to delete one or two "unusual data values" to see what the data shows. Be sure to explain what you did.)

Summary

In this activity, you compared strengths of major earthquakes from 1975 to 2011. You explored the relationship between the earthquake's magnitude on the Richter scale and the amplitude of the waves of the earthquake. You used Excel to draw graphs of the logarithmic and exponential functions involved and analyzed how they are related. You also used Excel to compute values of the common logarithmic function and to write numbers in scientific notation.

Estimations, Scientific Notation, and Properties of Logarithms

In this activity, you will use scientific notation to develop an estimate of a large quantity by breaking it down into small pieces. You will also use properties of logarithms to investigate and answer questions about an investment.

1. One of Ross Perot's policy recommendations in his 1992 Presidential campaign was a call for a $0.50 tax on every gallon of gasoline sold in the United States. Since that time, other lawmakers have supported additional gas taxes to help reduce consumption and raise money. This activity asks you to determine roughly how much revenue a $0.50 per gallon tax on gasoline would generate in a year. Before doing any calculations, take a guess as to how much revenue this proposal would generate in a year.

2. A useful strategy for estimating a quantity such as this one is to separate the problem into its component pieces and then to estimate each piece separately. These estimates can and should be fairly rough and are useful to provide a sense of the magnitude of a quantity.

   a. How many people are in the United States? (You might want to use the Internet or an almanac to get a sense of this number, if you don't already have an idea.) Write this estimate in scientific notation.

   b. Based on the number of people in the United States, estimate the number of automobiles in the United States. Write this estimate in scientific notation.

   c. Estimate the number of miles traveled by a typical car in a year. Write this estimate in scientific notation.

   d. Perform the appropriate operation on your answers to parts (b) and (c) of this question to obtain an estimate of the number of miles driven in the United States in one year. Use scientific notation.

   e. Estimate the number of miles that a typical car travels per gallon of gasoline.

   f. Perform the appropriate operation on your answers to parts (d) and (e) of this question to estimate the number of gallons of gasoline purchased in the United States in one year. Use scientific notation.

g. Use the answer to part (f) of this question to estimate the revenue that would be generated by Perot's $0.50 per gallon tax on gasoline.

h. What other quantities might you be able to estimate using this technique?

3. Now you will investigate an equation that relates to compound interest. Suppose you deposit $1,000 in a savings account that pays 4% annual interest, and the interest is compounded annually.

a. Assuming you don't withdraw any of the money or interest, before doing any computations, how much money do you think you will have after five years?

b. After one year your savings account will have $1,000 + 0.04 \times 1,000 = \$1,040$. Assuming you do not withdraw any money from the account, how much money will you have after two years? After three years?

c. Set up an Excel worksheet to help calculate money in the account after a period of time, using the following instructions.

Instructions to Calculate Interest

1. Open a new sheet in your Excel workbook and label column A as **years after initial deposit** and column B as **amount in the account**.

2. Enter numbers **0** through **20** in column A, starting in cell A2, and the corresponding amounts in column B. Use Excel to do the computations for you, by entering a formula in B3 that gives the amount in B3 in terms of the amount already in B2, and then drag to fill in the rest of the column.

d. What formula did you enter in cell B3?

e. What amount did you obtain for the account balance after 20 years?

f. The account balance after n years (still assuming an initial deposit of $1,000 and that no money is withdrawn during those years) can be given by a single formula in terms of n:

$$\text{Amount after } n \text{ years} = 1{,}000 \cdot (1.04)^n$$

In column C of your spreadsheet, enter the values obtained using this formula. (To do this, place your cursor in cell C2 and enter **=1000 * (1.04) ∧ A2**. Then drag down to year 20.)

g. How do the entries in columns B and C compare and why?

h. Using the previous formula, you can find the account balance after any number of years, without using a spreadsheet. You can also calculate the number of years it would take for the account balance to reach any given amount. For example, to find out how many years it would take for the account to reach $10,000, you need to find the value of n so that $10{,}000 = 1{,}000 \cdot (1.04)^n$.

Why is this equation the one we need to solve?

i. To solve for the number of years it would take to reach $10,000 in the account, we need to get n out of the exponent. If we apply the logarithm function to both sides of the equation, we get the following:

$$\log(10{,}000) = \log(10{,}000) + n \log(1.04)$$

What properties of logarithms did we use to get this equation?

j. Because $\log(10{,}000) = 4$ and $\log(1{,}000) = 3$, the previous equation is equivalent to $4 = 3 + n\log(1.04)$. Explain why $\log(10{,}000) = 4$ and $\log(1{,}000) = 3$.

k. Use algebra to solve (by hand) for n in terms of $\log(1.04)$.

l. Use a calculator or the computer to find an approximate value of $\log(1.04)$, and use it to approximate n to find the number of years it would take for the account to reach $10,000.

$n = $ _____

m. Use this method to find the number of years it would take the account to reach $100,000.

n. Drag down in your Excel sheet to verify your answers to parts (l) and (m). Do they agree with the answers Excel calculated?

Summary

In this activity, you used scientific notation to make a rough estimate of the revenue that would be generated by a $0.50 tax per gallon of gasoline. You used Excel to calculate compound interest and used logarithms to solve equations to determine how many years a savings account investment should be kept in the bank to obtain a certain return.

Indexes and Ratings

After completing this topic, you will be able to:

- Analyze trends in several commonly used indexes.

- Use and calculate indexes to understand and compare data.

- Investigate what might go into setting up a rating system.

W hat makes a city rank #1 for "livability"? How might such a ranking system be set up? Quantitative indexes and rating systems are used to give information about general trends and to allow us to make comparisons and judgments. We'll examine some frequently used indexes and rating systems and look at how to use them and what goes into setting them up.

The **Dow Jones Industrial Average (DJIA)** is a well-publicized index that reflects the value of stock prices. The DJIA includes 30 stocks that represent a variety of industries—financial, food, technology, retail, heavy equipment, oil, chemical, pharmaceutical, consumer goods, and entertainment. The DJIA is not a simple average but is adjusted to take into account the changes in price associated with stock splits in each of the included companies. The average is calculated by summing the prices of the 30 stocks and then dividing by a constant called the divisor that is adjusted periodically. (Originally, the divisor was 30, making the DJIA a simple average, but it has been reduced over the years and is currently less than 1.)

Example 8.1

The following graph of the DJIA shows the daily closing values from August 11, 2010, through August 11, 2011. Explain what trends this index shows for the time period covered in the graph. Why do you think the DJIA index uses 30 stocks rather than all stocks?

Source: yahoo! Finance, http://finance.yahoo.com.

Solution

The index shows that the stock market experienced an increasing trend through early May 2011 (with moderate dips at the end of August 2010, end of November 2010, and in late March 2011). Prices declined during May and most of June, and rose back to May levels during the last week of June. The DJIA remained fairly steady during the month of July but experienced a sharp decrease in early August. At the end of this one-year period, the DJIA was still higher than at the beginning.

Using 30 well-chosen, representative stocks (that is, a representative **sample** of the stock market) gives an indication of what is happening in the stock market as a whole, without having to include all stocks.

The **Consumer Price Index (CPI)** is an index number determined and published each year by the Bureau of Labor Statistics. It is used as one measure of inflation and measures the price of a market basket of approximately 300 goods and services purchased by consumers. (The market basket is weighted using the following percentages: 39.6% housing, 16.3% food and drinks, 17.6% transportation, 5.6% medical care, 4.9% apparel and upkeep, 6.1% entertainment, 5.5% education and communication, 4.3% other.) The CPI is not the only index used as an indicator of inflation, but it is a commonly used one. In the next example, we look at the CPI over time and then examine how we can use it to understand changes in the value of a dollar over time.

Example 8.2

The following table gives the CPI for the years 1978 through 2010:

Year	1978	1979	1980	1981	1982	1983	1984	1985	1986	1987	1988
CPI	65.2	72.6	82.4	90.9	96.5	99.6	103.9	107.6	109.6	113.6	118.3
Year	1989	1990	1991	1992	1993	1994	1995	1996	1997	1998	1999
CPI	124	130.7	136.2	140.3	144.5	148.2	152.4	157.6	160.3	163	166.6
Year	2000	2001	2002	2003	2004	2005	2006	2007	2008	2009	2010
CPI	172.2	177.1	179.9	184	188.9	195.3	201.6	207.3	215.3	214.5	218.1

Sketch a graph of the CPI over the years given in the table and explain what the graph shows.

Solution

A scatterplot of the CPI over the years 1978 to 2010 appears next. This graph shows that the increase in CPI from 1978 through 2010 was fairly steady, with slightly larger increases around 1980 and 2008.

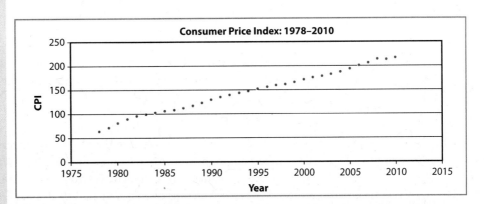

Index numbers such as the CPI allow us to make comparisons. When we look at the cost of something over time, in order to understand the nature of the changes in cost, we might want to look at the increases and prices in dollar units that reflect actual buying power at one particular time. We can reexpress the costs in dollar units, called **constant dollars**, which adjust the monetary units for inflation.

To convert a monetary amount from **current dollars** at *time A* to an amount with the same buying power at another time, say, *time O*, the ratio of the monetary amount to the CPI must be the same for both times; that is,

$$\frac{\text{Dollars at time } A}{\text{CPI at time } A} = \frac{\text{Dollars at time } O}{\text{CPI at time } O}$$

We can write this ratio as

$$\text{Dollars at time } O = \text{Dollars at time } A \times \frac{\text{CPI at time } O}{\text{CPI at time } A}$$

Example 8.3

The following table shows the federal minimum hourly wage rate from 1978 to 2011 and the years in which it increased during that time period.

Year	1978	1979	1980	1981	1990	1991	1996	1997	2007	2008	2009
Wage ($)	2.65	2.90	3.10	3.35	3.80	4.25	4.75	5.15	5.85	6.55	7.25

a. By what percentage did the CPI increase from 1978 to 2009?

b. By what percentage did the minimum wage increase from 1978 to 2009?

c. What do the values calculated in parts (a) and (b) of this example tell you?

d. Convert the 1978 (time *A*) minimum wage into constant 2010 (time *O*) dollars. Repeat for the minimum wage for each of the other years given in the table and record in a new table.

e. Create a graph of the minimum wage in constant 2010 dollars for the years 1978 to 2009 and describe what happened to the minimum wage over the years 1978 to 2009.

Solution

a. We recall how to calculate percent increase. We take the amount of increase in CPI and divide it by the CPI at the time from which we measured the increase, 1978; then we multiply by 100 to convert to percentage. The CPI increased from 65.2 to 214.5, which is an increase of 149.3 units. We then divide this by 65.2 and convert it to a percentage: $\left(\frac{149.3}{65.2}\right) \times 100\% \approx 228.99\%$. So, the CPI increased by approximately 229% from 1978 to 2009.

b. The minimum wage increased from \$2.65 to \$7.25, which is an increase of \$4.60. We divide by 2.65 to get $\left(\frac{4.60}{2.65}\right) \times 100\% \approx 173.58\%$. The minimum wage increased by approximately 174%.

c. These calculations show that the CPI increased by a much higher percentage than the minimum wage did, which means that the minimum wage did not increase as fast as the cost of market basket goods.

d. With time A as 1978 and time O as 2010, we convert the minimum wage dollars at time A of \$2.65 to its dollar value in 2010 (i.e. at time O) using the following formula:

$$\text{Dollars at time } O = \text{Dollars at time } A \times \frac{\text{CPI at time } O}{\text{CPI at time } A}$$

$$1978 \text{ minimum wage in 2010 dollars} = 2.65 \times \left(\frac{218.1}{65.2}\right) \approx 8.86$$

For the 1979 minimum wage in 2010 dollars, we consider time A to be 1979 and use 2.90 as dollars at time A and also use the CPI in 1979 as CPI at time A. We then get 1999 minimum wage in 2010 dollars $= 2.90 \times \left(\frac{218.1}{72.6}\right) \approx 8.71$. Using a similar calculation for each of the years in the table, we get these values:

Year	1978	1979	1980	1981	1990	1991	1996	1997	2007	2008	2009
Minimum Wage	2.65	2.9	3.1	3.35	3.8	4.25	4.75	5.15	5.85	6.55	7.25
Minimum Wage in 2010 \$	8.86	8.71	8.21	8.04	6.34	6.81	6.60	7.00	6.15	6.64	7.37

e. The graph shows that the minimum wage, measured in 2010 constant dollars, was highest in 1978. It declined from 1978 until 1990 and has had slight ups and downs since then. In 2009, it was higher than in earlier years, but still lower than in the years 1978–1981. Even with the increase in the last few years, the minimum wage has not "kept up with inflation."

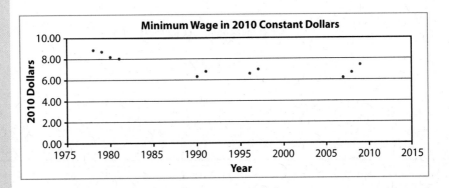

The **Consumer Confidence Index (CCI)** is a measure of consumers' optimism about the state of the economy; it is based on a sample of 50,000 U.S. households and is designed to reflect people's feelings about general business conditions, employment opportunities, and their own income prospects. The next example looks at how this index has changed over time.

Example 8.4

The CCI is set up so that the index number for 1985 is 100, and the other years' CCI values are given relative to the 1985 index of 100. The following table gives CCI values for the years 1978 through 2010:

Year	1978	1979	1980	1981	1982	1983	1984	1985	1986	1987	1988
CCI	106.0	91.9	73.8	77.4	59.0	85.7	102.3	100.0	94.7	102.6	115.2
Year	1989	1990	1991	1992	1993	1994	1995	1996	1997	1998	1999
CCI	116.8	91.5	68.5	61.6	65.9	90.6	100.0	104.6	125.1	131.7	135.3
Year	2000	2001	2002	2003	2004	2005	2006	2007	2008	2009	2010
CCI	139.0	106.6	96.6	79.6	96.0	100.3	105.9	103.4	57.9	45.4	54.5

Source: The Wall Street Journal Almanac 1999, p. 135, and http://future.aae.wisc.edu/data.

a. Sketch a graph of these data and describe what the graph shows.

b. From approximately 1989 to 1992, the graph of CCI is decreasing and concave upward. What do the shape and direction of the graph tell us?

c. Does there appear to be any relationship between the CCI and the CPI?

Solution

a. A graph of the CCI appears next. The CCI goes up and down throughout the time period under consideration, reaching its maximum value in 2000 and its minimum value in 2009. For more than half of this time period, the CCI is below the 1985 level of 100.

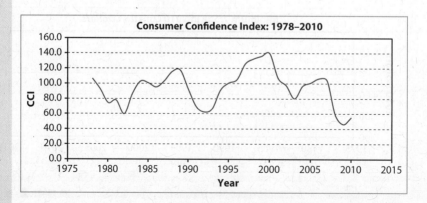

b. Because the graph is decreasing, the CCI is going down; that is, consumers' optimism is decreasing. However, because the graph is concave upward, even though consumer confidence is decreasing, the rate at which it is changing is increasing, so it appears as though the consumers' outlook is improving. A look at the graph for the years after 1992 confirms this.

c. There doesn't appear to be a link between CCI and CPI. The CPI has a clearly increasing trend, while the CCI does not. The following graph shows both CCI and CPI:

The **Fog Index** is one measure of the reading difficulty of a passage of text and is sometimes used by newspaper and magazine editors. (There are other measures of reading difficulty.) The Fog Index takes into account word count, sentence length, and word size. For our purposes, we will define "big" words as words with three or more syllables. We'll also use the following conventions: proper names do not count as big words; compound words that are combinations of easy words, such as *everything*, and words in which one of the three syllables is formed from a suffix such as *-ed* or *-ing* do not count as big words; however, big words that are repeated count each time they are used. The Fog Index (FI) is defined and calculated using the following formula:

$$FI = \left(\frac{\text{number of words}}{\text{number of sentences}} + 100 \times \frac{\text{number of "big" words}}{\text{number of words}} \right) \times 0.4$$

In applying the formula to calculate the Fog Index, we need to be careful and follow the standard order of operations. We need to first calculate the quantity inside the parentheses and multiply this result by 0.4. For example, suppose an article has 300 words, 20 sentences, and 24 "big" words. We first calculate: $\frac{\text{number of words}}{\text{number of sentences}} = \frac{300}{20} = 15$ and $100 \times \frac{\text{number of "big" words}}{\text{number of words}} =$ $100 \times \frac{24}{300} = 100 \times 0.08 = 8$. Then add $\frac{\text{number of words}}{\text{number of sentences}} + 100 \times \frac{\text{number of "big" words}}{\text{number of words}} =$

$15 + 8 = 23$, and finally multiply this number by 0.4; this gives us a Fog Index of $23 \times 0.4 = 9.2$.

A FI value of 9 purportedly indicates a ninth-grade reading level; a value of 12 indicates the twelfth-grade level; 14 is a college-sophomore level.

Example 8.5

Find the Fog Index for the following passage from the article, "Who Will Name the Next Supreme Court Justice?" that appeared in *The Wall Street Journal* on May 21, 2000. Explain how the FI is calculated.

When David M. O'Brien, a government professor at the University of Virginia, took his students on a tour of the Supreme Court recently, they ended up in a private question-and-answer session with a justice who abruptly turned to presidential politics. He blurted out several hot-button issues involving federalism, and said the future composition of the court would dramatically affect the outcome of those cases.

"Vote carefully," the justice, a Republican appointee, implored the students assembled in a stately white oak conference room.

To this justice, and to interest groups on the left and the right, the 2000 presidential campaign is not so much about whether Al Gore or George W. Bush makes it to the White House. It's about whom Mr. Gore or Mr. Bush would put on the Supreme Court, where vacancies are likely, if not in the next four years, then certainly in the next president's potential second term. Three of the nine justices are age 70 or older.

Solution

Before we begin, we need to agree on some criteria for "counting words." We will count a hyphenated word (for example, question-and-answer) as one word and an initial in someone's name or a number will count as a word. Words with an apostrophe will also count as one word. (This follows the conventions used in the word-processing package Microsoft Word.) Thus, the word count for the three paragraph passage is 161. Next, we count "big" words. We do not count as big the words "recently" and "appointee" that reach three syllables because of a suffix. We also don't count as big any words that are proper names. Thus, we count 16 big words and only 6 sentences. (The big words are *government, professor, presidential, politics, several, federalism, composition, dramatically, Republican, assembled, conference, interest, presidential, vacancies, president's, potential.*) Therefore, we calculate the FI as follows:

$$\text{FI} = \left(\frac{161}{6} + 100 \times \frac{16}{161}\right) \times 0.4 \approx (26.83 + 9.94) \times 0.4 \approx 36.77 \times 0.4 \approx 14.7$$

To really understand this index, however, we need to look at the FI for a variety of passages. The first term of the FI involves $\frac{\text{number of words}}{\text{number of sentences}}$, which is the average number of words per sentence. The second term is the percentage of big words in the passage. These two figures are added together and then we multiply that sum by a factor of 0.4.

A rating system can be considered a type of indexing scheme; rating systems are often set up to compare movies, cities, colleges, and other institutions. Sometimes these rating systems organize places or things into categories and other times rating systems rank-order things. Movies are rated by movie industry insiders according to the following categories: G, PG, PG-13, R, or NC-17. Each year, *U.S. News and World Report* releases its ratings of colleges and universities based on diverse criteria. The magazine does not publish an actual index number for each college or university, but they rank-order the colleges based on their system. We encounter rating systems when various organizations rate, for example, the best cities for walking or the best "family-friendly" companies.

To use these rating systems' indexes appropriately, we need to consider the information included in the rating as well as the reliability of that information.

In the next example, we consider what quantities should be included in particular rating systems.

Example 8.6

a. A 268-page report "The State of the Child in Pennsylvania" rates counties on child well-being, using factors such as infant deaths and juvenile delinquency figures. What other quantities would be appropriate to include in a rating system for well-being of children?

b. *U.S. News and World Report* publishes an issue each year that rates the nation's colleges. What variables do you think should be included in such a rating scheme?

c. *Money* magazine rates cities on "livability." What variables do you think should be included in such a rating scheme?

Solution

a. In addition to the variables mentioned, the number of children on welfare, the number of children living in families with incomes below the poverty level, and the proportion of children who finish high school might be included in the rating. (You might think of other variables to also include.)

b. In rating colleges, the proportion of students who graduate in four years, "student satis-faction" with the college's social atmosphere, the proportion of students receiving financial aid, and the proportion of students who pursue graduate or professional school might be taken into consideration. (Note that we could create either a simple or a fairly sophisticated rating system. Also, different individuals might think certain variables are more or less important.)

c. A partial list might include: unemployment rate; tax rate; accessibility to professional sports teams; and distance to and size of shopping malls.

In addition to choosing what to rate, sometimes there is a problem with how to define and measure certain variables. For example, a person's "unemployment status" is one such variable. Is a person who is content in the role of homemaker employed, unemployed, or neither? What about part-time workers looking for full-time employment? What about someone who has no job, is discouraged about looking for a job, and who has given up looking for one? We will consider some of these questions further in the Explorations.

We examine one measure of a country's development in the next example.

Example 8.7

The Human Development Index (HDI) is used by the United Nations to compare countries' development. It combines life expectancy, education attainment, and income indicators in a single number that ranges from 0 to 1. To calculate a country's HDI, we need to know the country's life expectancy at birth (LE), Mean Years of Schooling (MYS), Expected Years of Schooling (EYS), and the Gross National Income per capita (GNIpc) at purchasing power parity (this means it is adjusted for differences in prices across countries) in U.S. dollars. The HDI is calculated using the formula

$$HDI = \left(\frac{LE - 20}{63.2} \times \frac{\sqrt{\frac{MYS}{13.2} \times \frac{EYS}{20.6}}}{0.951} \times \frac{\log(GNIpc) - 2.212}{2.822} \right)^{1/3}$$

The following table gives the four indicators used in calculating the HDI for Costa Rica and Egypt in 2010:

	LE	MYS	EYS	GNIpc (in constant 2008 U.S. Dollars)
Costa Rica	79.1	8.35	11.73	10,870
Egypt	70.54	6.49	11.04	5,889

a. From this information and without making any calculations, which of the two countries do you think has a higher HDI? Explain your answer.

b. Calculate the HDI for each of the two countries.

c. Suppose Costa Rica planned to increase the number of years of schooling to increase its HDI by 0.02. What should Costa Rica's target for MYS be, assuming the other three indicators stay the same?

Solution

a. All four indicators are higher for Costa Rica than for Egypt, and they all are in the numerator of the formula to calculate the HDI, so we would expect Costa Rica to have a higher HDI than Egypt.

b. Substituting Costa Rica's values in the formula, we have

$$HDI_{\text{Costa Rica}} = \left(\frac{79.1 - 20}{63.2} \times \frac{\sqrt{\frac{8.35}{13.2} \times \frac{11.73}{20.6}}}{0.951} \times \frac{\log(10{,}870) - 2.212}{2.822} \right)^{1/3}$$

We use a calculator to perform these operations, approximating to three decimals:

$$HDI_{\text{Costa Rica}} = \left(\frac{59.1}{63.2} \times \frac{\sqrt{0.633 \times 0.569}}{0.951} \times \frac{4.036 - 2.212}{2.822} \right)^{1/3}$$

$$= \left(\frac{59.1}{63.2} \times \frac{0.60}{0.951} \times \frac{1.824}{2.822} \right)^{1/3} = (0.935 \times 0.631 \times 0.646)^{1/3} = (0.381)^{1/3}$$

$$= 0.725$$

In a similar manner, we obtain the HDI value for Egypt: $HDI_{\text{Egypt}} = 0.620$.

c. We found that the HDI for Costa Rica is 0.725, so an increase of 0.02 would give it a value of 0.735. To find the value of MYS needed to achieve this Human Development Index when the other three indicator values are unchanged, we use the equation

$$0.745 = \left(\frac{79.1 - 20}{63.2} \times \frac{\sqrt{\frac{\text{MYS}}{13.2} \times \frac{11.73}{20.6}}}{0.951} \times \frac{\log(10{,}870) - 2.212}{2.822} \right)^{1/3}$$

We obtained this equation from the HDI formula by substituting 0.745 for HDI, and the Costa Rica values for LE, EYS, and GNIpc. We need to solve this equation for the variable MYS.

To simplify the equation, we start by performing the indicated operations inside the parentheses:

$$0.745 = \left(0.935 \times \frac{\sqrt{(0.043)MYS}}{0.951} \times 0.646\right)^{1/3} = \left(\frac{(0.935 \times 0.646)}{0.951}\sqrt{(0.043)MYS}\right)^{1/3}$$

$$= \left(0.635\sqrt{(0.043)MYS}\right)^{1/3}$$

So the equation is $\left(0.635\sqrt{(0.043)MYS}\right)^{1/3} = 0.745$. To solve it, we first raise both sides of the equation to the third power and get $0.635\sqrt{(0.043)MYS} = (0.745)^3$. Next, we divide both sides of the equation by 0.635: $\sqrt{(0.043)MYS} = \frac{(0.745)^3}{0.635} = 0.651$. So, $\sqrt{(0.043)MYS} = 0.651$. Squaring both sides and then dividing by 0.043, we obtain $MYS = \frac{(0.651)^2}{0.043} \approx 9.86$. Thus to increase its HDI by 0.02, Costa Rica needs to raise the mean number of years of schooling to almost 10 years.

Summary

In this topic, we investigated several indexes that are used to show trends over time, like the Dow Jones Industrial Average, the Consumer Confidence Index, and the Consumer Price Index. We also used the Consumer Price Index to look at changes in minimum wage over time. The Fog Index was studied as one way to measure the reading difficulty of a passage of text. Finally, we discussed how rating systems could be viewed as a type of index and looked at what factors should be included in particular rating systems.

Explorations

1. To track increases in cost and to allow for comparisons, the Fan Cost Index (FCI) for various professional sports leagues has been computed for the past several seasons. This index includes the cost of four average-price tickets, four small soft drinks, two small beers, four hot dogs, parking for one car, two game programs, and two souvenir caps. The following tables show the FCI for four professional sports leagues:

MLB Season	FCI ($)	NFL Season	FCI ($)
2000	132.44	2000	279.9
2001	140.63	2001	278.73
2002	145.21	2002	290.41
2003	151.19	2003	301.75
2004	155.52	2004	321.62
2005	164.43	2005	329.91
2006	171.89	2006	346.16
2007	176.84	2007	367.31
2008	191.92	2008	396.36
2009	196.89	2009	412.64
2010	194.98	2010	420.54
2011	197.35	2011	427.21

NHL Season	FCI ($)	NBA Season	FCI ($)
2000–01	264.81	2000–01	281.72
2001–02	239.24	2001–02	277.19
2002–03	240.43	2002–03	254.88
2003–04	253.65	2003–04	261.26
2004–05	season cancelled	2004–05	263.44
2005–06	250.09	2005–06	267.37
2006–07	270.51	2006–07	274.67
2007–08	282.95	2007–08	282.44
2008–09	288.23	2008–09	291.93
2009–10	301	2009–10	289.87
2010–11	313.68	2010–11	289.51

Source: Team Marketing Report, www.teammarketing.com.

a. Discuss the FCI, including if it is a reasonable measure of costs.

b. Look at the pattern of FCI values for the NHL and NBA from 2000–01 to 2003–04 and compare with the values for the four-year period from 2007–08 to 2010–11. What do you observe?

c. Determine whether the cost of attending sporting events has outpaced inflation.

d. Explain the increases or decreases in the FCI for the different sports during the 2000–03 time period.

2. The following table lists first-class postage rates and the years in which they have increased from 1978 to 2009:

Year	1978	1981	1983	1988	1991	1995	1999
Postage in Cents	15	18	22	25	29	32	33
Year	2001	2002	2006	2007	2008	2009	
Postage in Cents	34	37	39	41	42	44	

a. Use the CPI values in Example 8.2 to convert the 1983 postage rate into constant 1981 dollars.

b. Repeat for the other dates in the table from 1981 to 2009. Describe what these values show about postage rates over the years 1981 to 2009.

3. The following graph gives the monthly values of the Consumer Confidence Index (CCI) from January 2010 to July 2011. Describe what the graph shows.

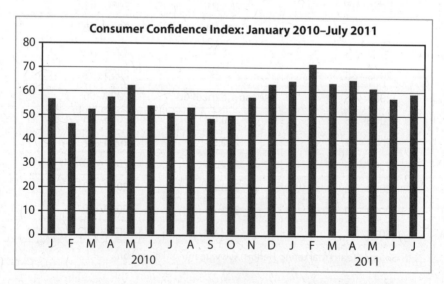

4. The monthly values of the Consumer Price Index (CPI) from January 2010 to July 2011 are given in the following table. Sketch a graph of these data and describe what the graph shows and how the CPI compares with the CCI over the same time period. (See Exploration 3.)

CPI	Jan.	Feb.	Mar.	April	May	June	July	Aug.	Sept.	Oct.	Nov.	Dec.
2010	216.7	216.7	217.6	218.0	218.2	218.0	218.0	218.3	218.4	218.7	218.8	219.2
2011	220.2	221.3	223.5	224.9	226.0	225.7						

5. Describe as precisely as possible how you would measure the following items and indicate any problems that might arise:

   a. A person's employment status

   b. The length of a person's foot

   c. A person's sensitivity toward others

   d. How many calories a person consumes in one day

6. Think about the quantities used for computing the Fog Index of a passage of text. What other quantities might be used to evaluate the reading difficulty of a passage of text?

7. Find an editorial from a recent newspaper.

   a. Compute the Fog Index for the editorial. Indicate any problems or questions that arose in computing the index.

   b. How would the Fog Index for this section change if there were twice as many sentences?

   c. How would it change if there were five more big words? Twice as many big words?

8. Find a short passage from a children's book and compute the Fog Index for the passage. Find a passage from a paper you have written for a class and compute the Fog Index for the passage. Compare these values with the Fog Index computed in Example 8.5.

9. Use the rating systems we looked at in Example 8.6 to answer the following questions:

   a. What additional variables, other than those mentioned in the solution to Example 8.6 (a), could you include in a system that rates counties on child well-being?

   b. What five variables do you think would be most important to include in a system that rates colleges and universities? Give reasons to support your answers.

   c. What additional variables, other than those listed in the solution to Example 8.6(c), should be included in a system that rates cities on "livability"?

10. The following data, gathered by the National Center for Education Statistics, U.S. Department of Education, shows estimated expenditures in unadjusted and constant dollars (in billions) and expenditures per full-time-equivalent student in constant dollars in U.S. public four-year degree-granting institutions of higher education from 1988—89 to 2007—08. Explain in detail what the table shows, including any trends.

Year	Expenditure in Billions: Unadjusted Dollars	Expenditure in Billions: Constant 2002–03 Dollars	Expenditure per Student in Constant 2002–03 Dollars
1988–89	$65.3	$98.2	$21,800
1989–90	70.9	101.7	22,003
1990–91	76.7	104.3	22,011
1991–92	81.3	107.2	22,350
1992–93	86.1	110.0	22,923
1993–94	89.7	111.7	23,435
1994–95	94.9	114.9	24,190
1995–96	97.9	115.4	24,255
1996–97	103.1	118.1	24,778
1997–98	109.2	122.9	25,536
1998–99	115.2	127.4	26,176
1999–2000	124.9	134.3	27,166
2000–01	140.6	146.2	29,092
2001–02	151.1	154.5	29,737
2002–03	159.0	159.0	29,409
2003–04	168.2	165.2	29,500
2004–05	178.0	172.8	30,475
2005–06	187.2	179.1	31,068
2006–07	196.4	184.9	31,434
2007–08	206.7	190.6	31,783

11. A recently released poll rated the top 20 "party" colleges and universities in the country. What variables do you think should be included in such a rating and what importance would you place on each of these variables in such a rating?

12. The following graph shows the value of the Purchasing Managers' Index from August 2008 to July 2011:

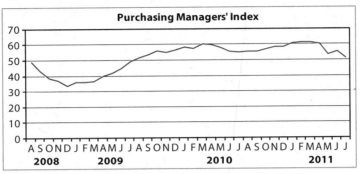

Source: YCharts, http://ycharts.com.

a. Use the Web to find how this index is calculated.

b. Describe what the graph shows in this context.

13. The following information, obtained from the United Nations website, gives the values of the four indicators used in the calculation of the Human Development Index (discussed in Example 8.7 of this topic) for the five most populated countries in Oceania and for the United States.

Country	Life Expectancy			Mean Years of Schooling		
	1990	2000	2010	1990	2000	2010
Australia	76.91	79.76	81.86	11.74	11.86	12.04
Fiji	66.73	67.28	69.22	8.31	10.31	11.04
New Zealand	75.26	78.47	80.60	11.75	12.06	12.52
Papua New Guinea	54.30	57.74	61.59	2.32	3.42	4.33
Solomon Islands	56.66	62.15	67.05	NA	4.5	4.5
United States	76.22	78.01	79.58	12.26	13.22	12.45

Country	Expected Years of Schooling			GNI per capita in 2008 ($)		
	1990	2000	2010	1990	2000	2010
Australia	12.97	20.36	20.46	25,367	32,666	38,692
Fiji	12.67	12.86	13.01	3,557	4,281	4,315
New Zealand	14.63	17.42	19.7	19,438	22,418	25,438
Papua New Guinea	4.79	NA	5.23	1,770	1,978	2,227

Country	Expected Years of Schooling			GNI per capita in 2008 ($)		
	1990	2000	2010	1990	2000	2010
Solomon Islands	NA	6.72	9.66	1,713	2,410	2,172
United States	15.41	15.79	15.75	34,406	43,079	47,094

a. Without performing any calculations, which of the two countries—Australia or the United States—do you think had a higher Human Development Index (HDI) in 2010? Explain your reasoning.

b. Calculate the HDI for the year 2010 for Australia and for the United States.

c. Calculate the HDI for the year 1990 for Australia and for the United States.

d. Find the HDI rate of change per year for each of these two countries from 1990 to 2010.

e. What other indicators do you think would be useful in estimating the degree of human development of a country?

14. Use the information given in the tables in Exploration 13 to answer each of the following questions:

a. What was Papua New Guinea's HDI in 2010?

b. How much higher should Papua New Guinea's Mean Years of Schooling be so that its HDI grows by 0.03 points, assuming no change in the other three indicators?

c. How much higher should Papua New Guinea's GNIpc be so that its HDI increases by 0.1 point, assuming the other three indicators stay the same.

15. Use the data given in Exploration 13 to calculate the HDI for the five most populated countries in Oceania for the years 1990, 2000, and 2010; then make a graph illustrating that information. (Do as many of these calculations as you can with the available data.) Describe what the graph shows about the HDI in those countries and how it has changed in the two decades.

Measurement Difficulties and Indexes

In Topic 8, you read about several indexes that are used to measure complex quantities that are not defined and measured easily. In this activity, you will work with one of those indexes, the Fog Index, which measures reading difficulty. You will analyze different reading passages to get a better sense of how the different components affect the result, and why this index gives a measurement of a text's reading difficulty.

1. Some properties are easier to understand and measure than others. For example, we all understand how to measure the height of a person or the time it takes a particular person to complete an exam. It is not as clear how to measure a person's intelligence or a customer's satisfaction with services delivered to him or her.

   a. List three additional properties of a person that are defined and measured fairly easily.

   b. List three additional properties of a person that are more difficult to define and measure.

2. Consider the property "reading difficulty of a passage of text." For the most part, we can probably agree that such a property exists: Some passages of text are certainly easier to read than others. But how might you measure that property?

a. List three factors that might go into measuring a passage's reading difficulty.

b. Read the following three passages of text:

Passage 1

Everyone agreed to this and off they went, walking briskly and stamping their feet. Lucy proved a good leader. At first, she wondered whether she would be able to find the way, but she recognized an odd-looking tree in one place and a stump in another and brought them on to where the ground became uneven and into the little valley and at last to the very door of Mr. Tumnus' cave. But there a terrible surprise awaited them.

The door had been wrenched off its hinges and broken to bits. Inside, the cave was dark and cold and had the damp feel and smell of a place that had not been lived in for several days. Snow had drifted in from the doorway and was heaped on the floor, mixed with something black, which turned out to be the charred sticks and ashes from the fire. Someone had apparently flung it about the room and then stamped it out. The crockery lay smashed on the floor and the picture of the Faun's father had been slashed into shreds with a knife.[1]

Passage 2

The truck drove to a part of town that George had never seen before. At last it stopped in front of a large building. It was the Museum. George did not know what a Museum was. He was curious. While the guard was busy reading his paper, George slipped inside.

He walked up the steps and into a room full of all sorts of animals. At first George was scared, but then he noticed that they did not move. They were not alive, they were stuffed animals, put into the Museum so that everybody could get a look at them.[2]

Passage 3

To engage in a serious discussion of race in America, we must begin not with the problems of black people but with the flaws of American society, flaws rooted in historic inequalities and longstanding cultural stereotypes. How we set up the terms for discussing racial issues shapes our perception and response to these issues. As long as black people are viewed as a "them," the burden falls on blacks to do all the "cultural" and "moral" work necessary for healthy race relations. The implication is that only certain Americans can define what it means to be American, and the rest must simply "fit in."

The emergence of strong black-nationalist sentiments among blacks, especially among young people, is a revolt against this sense of having to "fit in."[3]

c. Based on your impression from one reading, rank the three passages from least difficult to most difficult to read.

   i. Least difficult

   ii. Next in difficulty

  iii. Most difficult

d. Because longer sentences tend to be more difficult to read than shorter ones, the number of words per sentence in a passage of text would seem to be a reasonable measure of reading difficulty. Set up a table in an Excel spreadsheet, like the one shown here, and record the number of words and the number of sentences for each of the three passages. Then tell Excel to use the appropriate formula to compute the number of words per sentence. Record your answers in this table.

Passage	Number of Words	Number of Sentences	Words per Sentence
1			
2			
3			

e. Based on "words per sentence," rank the three passages from least difficult to most difficult.

   i. Least difficult

   ii. Next in difficulty

   iii. Most difficult

f. Another measure of reading difficulty might be the number of "big" words in a passage of text. Count the number of big words (words having three or more syllables) in each of the previous passages. Note that compound words, such as *everything*, and words in which the third syllable is formed from a suffix such as -ed or -ing should not be counted as big words. Add another column to your spreadsheet, to the right of the **Words per Sentence** column, and label it **Big Words**. Circle all the big words in each passage, count them, and record the number in the appropriate column of your spreadsheet.

g. Add a sixth column to your spreadsheet, and using the appropriate formula, compute the percentage of "big" words in each passage. Write your answers here.

h. Based on "percentage of 'big' words," rank the three passages from least difficult to most difficult.

   i. Least difficult

   ii. Next in difficulty

   iii. Most difficult

i. The Fog Index is a measure of reading difficulty used by newspaper and magazine editors. This measure takes into account both sentence length and word size. Set up another column in your spreadsheet to calculate the Fog Index. Use the following formula, also given in Topic 8, and record the Fog Index for each passage:

$$\text{Fog Index} = 0.4 \times (\text{words per sentence} + \text{percent of "big" words})$$

    i. Passage 1 Fog Index =

    ii. Passage 2 Fog Index =

    iii. Passage 3 Fog Index =

j. A Fog Index value of 9 purportedly indicates a ninth-grade reading level, a value of 12 indicates a twelfth-grade reading level, and a value of 14 indicates a college-sophomore reading level. Based on your calculations, does this seem reasonable? Why or why not?

k. Make the appropriate changes in the spreadsheet you set up to answer the following questions. Be specific.

    i. How does the Fog Index for each of the passages change if the number of "big" words is doubled?

ii. How does the Fog Index for each of the passages change if the number of words and the number of "big" words stays the same, but the number of sentences is doubled?

Summary

In this activity, you examined the Fog Index for several passages of text and looked at whether the Fog Index is a reasonable measure of reading difficulty. You also considered how the Fog Index changes if there are more "big" words or more sentences in the passages.

Notes

1. Lewis, C.S. *The Lion, the Witch and the Wardrobe*. New York: Scholastic Inc., 1981, pp. 53—54.

2. Rey, H. A. *Curious George Gets a Medal*. Boston: Houghton Mifflin Company, 1957, pp. 30—31.

3. West, Cornel. *Race Matters*. Boston: Beacon Press, 1993, p. 3.

Consumer Indexes

In this activity, you will look at how the Consumer Price Index can be used to understand trends involving changes in costs and salaries over time.

Several "Consumer Indexes" are used to measure items related to consumers. In particular, the Consumer Price Index (CPI) is used as a measure of inflation by considering how the price of commonly purchased commodities changes; it measures the price of a market basket of a large number of goods and services purchased by consumers.

1. The following table gives the median weekly earnings, in dollars, of full-time wage and salary workers, 25 years and older, by educational attainment from the years 1980 to 2010 along with the CPI for the same years.

Year	CPI	High School Diploma, No College, Median Weekly Earnings ($)	Some College or Associate's Degree, Median Weekly Earnings ($)	Bachelor's and Higher, Median Weekly Earnings ($)
1980	82.4	266	304	376
1982	96.5	302	351	438
1984	103.9	323	382	486
1986	109.6	344	409	525
1988	118.3	368	430	585
1990	130.7	386	476	638
1992	140.3	403	484	696
1994	148.2	421	499	733
1996	156.9	443	518	758

1998	163.0	479	558	821
2000	172.2	505	596	891
2001	177.1	520	617	921
2002	179.9	535	629	941
2003	184.0	554	639	964
2004	188.9	574	661	986
2005	195.3	583	670	1,013
2006	201.6	595	692	1,039
2007	207.3	604	704	1,072
2008	215.3	618	722	1,115
2009	214.5	626	726	1,137
2010	218.1	626	734	1,144

Source: U.S. Bureau of Labor Statistics, www.bls.gov.

a. By what percentage did the CPI increase from 1980 to 2010?

b. By what percentage did the median weekly salary for workers with a high school diploma and no college increase from 1980 to 2010?

c. By what percentage did the median weekly salary for workers with some college education increase from 1980 to 2010?

d. By what percentage did the median weekly salary for workers with at least a Bachelor's degree increase from 1980 to 2010?

e. Explain what these percentages show about salaries for these groups of workers from 1980 to 2010.

2. Retrieve the file "EA8.2 Median Salaries and CPI.xls" from the text website or WileyPLUS. This file contains the information given in the previous table.

a. You will first create one graph showing median weekly earnings of the three groups of workers. To do this *without* graphing the CPI column, follow these instructions.

Instructions to Create a Scatterplot from Noncontiguous Columns

Highlight the **Year** column (that is, column A) from cell 1 to cell 22. While pressing the **Ctrl** key, highlight the three columns of salary data. When you've highlighted the **Year** column and the three salary columns, release the **Ctrl** key and proceed to create a graph, selecting **Scatter** along with one of the connected-line options.

b. Explain what your graph shows about the salaries of the three groups of workers.

c. You will now add columns that will contain these salary figures converted into constant 2010 dollars. To the right of *each* of the salary columns, create *one* additional column.

Instructions to Insert a New Column

1. Place the cursor anywhere in the column immediately to the right of where you want to insert the new column. From the **Home** tab, click on the arrow to the right of **Insert** in the **Cells** group and select **Insert Sheet Columns**.

2. After you insert three new columns, one to the right of each salary column, enter appropriate headings in cell 1 of each of these three new columns.

d. Next, you'll convert each of the salary values to dollars in year 2010, starting with cell 2 of the "High School Diploma, No College" column. To convert this value to 2010 dollars (interpreted to be dollars at time O), use the ratio

$$\frac{\text{Dollars at time } A}{\text{CPI at time } A} = \frac{\text{Dollars at time } O}{\text{CPI at time } O}$$

Because you want to convert values to time O dollars, when solving for dollars at time O, you get

$$\text{Dollars at time } O = \text{Dollars at time } A \times \frac{\text{CPI at time } O}{\text{CPI at time } A}$$

With time A as 1980 and time O as 2010 (because you're converting to 2010 dollars), you will calculate for the first year in the "High School Diploma, No College" column of the table:

$$\text{1980 salary in 2010 dollars} = 266 \times \frac{\text{CPI in 2010}}{\text{CPI in 1980}}$$

i. Compute this value and write it here: _____

Similarly, for the next year in the "High School Diploma, No College" column, you'll have

$$\text{1982 salary in 2010 dollars} = 302 \times \frac{\text{CPI in 2010}}{\text{CPI in 1982}}$$

ii. Compute this value and write it here: _____

e. In column D (labeled something along the lines of "High School Diploma, No College in 2010$") of your spreadsheet, use the following Excel instructions to compute the "High School" salaries from 1980 to 2010 in 2010 constant dollars.

Instructions to Name a Cell and Drag a Formula with a Value Kept Constant

1. Because you are using CPI in 2010 in all your computations, you will "name" the cell containing this value. Then anytime you enter that name in a formula, it will use that constant value. First, put the cursor in cell B22. You will then see "B22" appear in a white box immediately above column A. This is the **Name box**. Click on the **Name box** and type an appropriate name for the value in the cell, for example, **CPIY2010**. Then press **Enter**.

2. Now you are ready to use this named constant in your formula. Place the cursor in cell D2 and enter the formula: **=C2*CPIY2010/B2**. Note that the 1980 dollar value that you want to convert is in cell C2 and the CPI values for the various years are in column B. Because you need to use the CPI in 2010, which is in cell B22, for all computations in that column, use the name of the constant **CPIY2010** so when you drag down, the cell address doesn't change. (Alternately, you can use an absolute cell reference, **B22**, instead of naming the constant in cell B22, so the value won't change when dragged vertically or horizontally.)

3. Now highlight cell D2 and drag down to fill in the other salaries in column D in 2010 dollars. Check that the two values you computed in Question 2(d) appear in cells 2 and 3 of column D.

   f. Enter the appropriate formulas and convert the "Some College or Associate's Degree" and "Bachelor's and Higher" salary values to 2010 dollars in the columns you added for this purpose.

   g. Now create a scatterplot using the "Year" column and the three columns containing the salaries converted to 2010 dollars. [Remember to use the **Ctrl** key to highlight non-contiguous columns, as you did when making the graph in Question 2(a).] Explain what this graph shows about the salaries of the three groups of workers: "High School Diploma, No College," "Some College or Associate's Degree," and "Bachelor's Degree or Higher."

Summary

By investigating the Consumer Price Index over a 30-year period and using it to compare salaries, in constant dollars, during that period of time, you experienced the effects of looking at values over time in constant dollars. You also learned how to construct a scatterplot using noncontiguous columns, how to add a column, and how to name a cell in Excel.

Personal Finances

After completing this topic, you will be able to:

- Use the terminology associated with personal financial management.

- Analyze relevant formulas to compute simple and compound interest.

- Understand ordinary annuities and use the accumulated savings formula.

- Apply the loan payment formula to understand and analyze installment loans and credit card loans.

How much would you need to invest each month in order to have $1 million when you retire in 40 years? At the end of this topic, you will be able to answer this question and more!

In previous topics and activities, you investigated several problems dealing with personal financial management. For example, you looked at the growth of debt in conjunction with exponential functions and investigated credit card balances in an activity about multiple variable functions. In this topic, we will consider the interest received on savings and the costs and payments associated with various types of loans in more detail.

INTEREST

In order to understand loans in general, we first consider the concept of **interest**. Interest is money that the borrower pays for using the lender's money. If money is lent under a **simple interest** agreement, the loan is for a fixed time period and the borrower repays the money borrowed plus interest at the end of that time period. Simple interest, denoted I, is computed using the formula:

$$I = P \times r \times t$$

where P is the amount of money borrowed, called the **principal**; r is the interest rate expressed as a decimal, and t is the duration of the loan. For example, if the loan is for $5,000, the interest is 12% per year and the loan period is two years, then $P = \$5,000$, $r = 0.12$, and $t = 2$; so the interest is $I = \$5,000 \times 0.12 \times 2 = \$1,200$ for the two-year period. Note that the interest rate and the duration of the loan must be expressed in the same time increment. This means that if the interest rate is given in rate per year, t must be expressed in years; if the interest rate is expressed in rate per month, t must be given in months. For example, if the principal is $5,000, the annual interest rate is 12% per year, which is 1% per month, and the loan period is seven months, then $P = \$5,000$, $r = 0.01$, and $t = 7$. In this case, the interest is $I = \$5,000 \times 0.01 \times 7 = \$350$ for the seven-month period.

Example 9.1

An electronics store is offering a laptop computer for $2,000, with two payment options. Under the first option, there is no payment for the first nine months, but the $2,000 must be repaid at the end of that time, at an annual simple interest rate of 19.8%. Under the second option, the borrower must pay $400 initially and then pay $1,875 at the end of one year.

a. Find the total cost of the first option.

b. Find the annual simple interest rate for the second option.

c. Compare the two options.

Solution

a. To compute the total cost under the first option, we will use time in years. Then the amount of time t is $\frac{9}{12}$ of a year and an annual interest rate of 19.8% means $r = 0.198$. So, the interest paid is $I = P \times r \times t = \$2,000 \times 0.198 \times \frac{9}{12} = \$297$; the total cost of the computer is $2,297. (Alternatively, we could have used time in months, with monthly interest rate of $\frac{19.8\%}{12} = 1.65\%$ and $t = 9$.)

b. To find the interest rate of the second offer, we use the amount of the one-year loan, $1,600, to first find the amount of interest paid: $I = \$1,875 - \$1,600 = \$275$. Then we use the simple interest equation $I = P \times r \times t$, with $I = 275$, $P = 1,600$, and $t = 1$ to solve for r: $275 = 1,600 \times r \times 1$. Thus, $r = \frac{275}{1,600} = 0.171875$ and the annual interest rate is approximately 17.2%.

c. The interest rate under the second option is less than with the first option. The total cost of the computer using the second option is $400 + \$1,875 = \$2,275$, which is $22 less than the cost under the first option.

In finding simple interest, we saw that the interest was computed only on the principal. Interest that is computed on the principal and any accumulated interest is called **compound interest**. Interest paid on savings accounts is normally compound interest. In the next examples, we develop a method for finding compound interest and then compare how interest accrues using different compounding periods.

Example 9.2

A student saves \$2,000 from his summer job and invests it with a single deposit into an account that pays an annual interest of 3%. Assume he does not add or withdraw any money from the account.

a. How much will he have in the account after one year if the interest is compounded annually?

b. How much will he have in the account after nine months if the interest is compounded quarterly?

Solution

a. The annual interest rate is $r = 0.03$, so the interest paid after one year is \$2,000 $\times$ 0.03. The amount in the account after one year is the initial investment plus the interest:

$$\$2,000 + 0.03 \times \$2,000 = (1 + 0.03) \times \$2,000 = 1.03 \times \$2,000 = \$2,060.$$

b. Because interest is paid quarterly, the interest rate paid each quarter is one-fourth of the annual interest rate: $\frac{r}{4} = \frac{0.03}{4} = 0.0075$. At the end of the first quarter, the amount, in dollars, in the account (account balance) is $2,000 + 0.0075 \times 2,000 = (1 + 0.0075) \times 2,000 = 2,015$. At the end of the second quarter, the account balance is the amount at the end of the first quarter, plus the interest on that amount: $2,015 + 0.0075 \times 2,015 = (1 + 0.0075) \times 2,015$. Recall from the previous calculations that $2,015 = (1 + 0.0075) \times 2,000$. So, $(1 + 0.0075) \times 2,015 = (1 + 0.0075) \times (1 + 0.0075) \times 2,000 = (1 + 0.0075)^2 \times 2,000 \approx 2,030.11$. We can use a table to record our findings:

Quarter	Interest Paid This Quarter	Balance End of This Quarter	Formula for Balance End of This Quarter
1	$0.0075 \times 2,000 = 15$	\$2,015	$\left(1 + \frac{r}{4}\right)P = \left(1 + \frac{0.03}{4}\right)(2,000)$
2	$0.0075 \times 2,015 = 15.11$	\$2,030.11	$\left(1 + \frac{r}{4}\right)^2 P = \left(1 + \frac{0.03}{4}\right)^2 (2,000)$
3	$0.0075 \times 2,030.11 = 15.23$	\$2,045.34	$\left(1 + \frac{r}{4}\right)^3 P = \left(1 + \frac{0.03}{4}\right)^3 (2,000)$

The balance at the end of nine months (three quarters) is \$2,045.34.

The table we created in Example 9.2 shows that the balance in the account has exponential growth because the ratio $\frac{\text{Balance at the end of this quarter}}{\text{Balance at the end of last quarter}} = 1 + \frac{0.03}{4} = 1.0075$, a constant. (Exponential growth was discussed in Topic 6.)

The table in Example 9.2 also illustrates the following general **compound interest formula** that gives the amount A in an account after t years, where P denotes the principal or initial deposit, r the annual interest rate, and n the number of compounding periods per year. Note that the number $\dfrac{r}{n}$ in this formula represents the interest rate per period and the exponent nt is the total number of periods in t years:

$$A = \left(1+\frac{r}{n}\right)^{nt} \times P = P\left(1+\frac{r}{n}\right)^{nt}$$

The annual interest rate r is also called the **annual percentage rate** and is often denoted by APR. It must be expressed as a decimal in using this formula.

We now apply the compound interest formula to answer the question in part (b) of Example 9.2. The initial amount invested is $P = \$2,000$, the annual interest rate is $r = 0.03$, the number of compounding periods per year is $n = 4$, and $t = 0.75$ (three-quarters of a year or nine months). So, $A = \$2,000 \times \left(1+\dfrac{0.03}{4}\right)^{(4 \times 0.75)} = \$2,000 \times (1+0.0075)^3 = \$2,000 \times (1.0075)^3 = \$2,045.34$. This agrees with our answer in Example 9.2.

In the next example, we use the compound interest formula to compare the amount of money in an account if the interest is compounded quarterly, monthly, or daily. There are three methods for calculating daily compounding. Historically, accountants used $n = 360$, assuming there are 12 months each year with 30 days. In the modern technological world, many banks now use $n = 365$, and some banks use what's called the "365/360" rule. This rule uses the annual interest rate divided by 360 but then compounds 365 times a year. This gives the lender a slightly higher profit. See Exploration 14 for details. Here we will use $n = 365$ if interest is compounded daily.

Example 9.3

A student has a choice of three banks at which to invest her $2,000 of summer earnings. All three banks offer an annual interest rate of 3%; however, the first bank's interest is compounded quarterly, the second bank's interest is compounded monthly, and the third bank's interest is compounded daily. The student will not deposit or withdraw any money for two years. How much money would she have in the account at the end of two years if she had invested her money at:

a. The first bank?

b. The second bank?

c. The third bank?

Solution

We will use the compound interest formula, $A = P\left(1+\dfrac{r}{n}\right)^{nt}$, to answer the questions. In all three cases, the initial investment is $2,000, the investment period is two years, and the annual interest rate is 3%, so $P = 2,000$, $t = 2$ and $r = 0.03$.

a. At the first bank, the interest is compounded quarterly, so $n = 4$. The amount in the account at the end of two years would be

$$A_{\text{quarterly}} = P\left(1 + \frac{r}{n}\right)^{nt} = 2{,}000 \times \left(1 + \frac{0.03}{4}\right)^{(4)(2)} = 2{,}000 \times (1 + 0.0075)^8 \approx 2{,}123.20.$$

The student will have $2,123.20 in the account after two years, if she invests at the first bank.

b. If the student invests at the second bank, the interest is compounded monthly, so we use $n = 12$:

$$A_{\text{monthly}} = P\left(1 + \frac{r}{n}\right)^{nt} = 2{,}000 \times \left(1 + \frac{0.03}{12}\right)^{(12)(2)} = 2{,}000 \times (1.0025)^{24} \approx 2{,}123.51.$$

In this case, the amount in the account after two years would be $2,123.51.

c. If the interest is compounded daily, $n = 365$ and $A_{\text{daily}} = 2{,}000 \times \left(1 + \frac{0.03}{365}\right)^{(365)(2)} \approx$ 2,123.67. So if the student invests at the third bank, she would have $2,123.67 in the account after 2 years.

Example 9.3 shows that there are only small differences in the total amount in an account after two years, when the interest is compounded annually, quarterly, or monthly. The power of compounding is more evident if we consider a long time period of time, as the next example shows.

Example 9.4

Since the eighteenth century, the U.S. government has had ordinances that established a system for the sale of publicly owned land. A 1785 ordinance allowed the purchase of a "section" of land, with a minimum size of 640 acres, at no less than $1.00 per acre. Suppose, instead of purchasing a 640-acre section of land for $640 in 1785, a family had invested that money at an annual interest rate of 5% and didn't add to or withdraw from the account. How much would the account be worth in 2015, if interest is compounded:

a. Annually?

b. Daily?

Solution

In both scenarios, the money is invested from 1785 to 2015, so it would be invested for $t = 230$ years.

a. If interest is compounded annually, $n = 1$, and we use the formula $A_{\text{annual}} = P\left(1+\frac{r}{n}\right)^{nt} =$ $640\left(1+\frac{0.05}{1}\right)^{(1)(230)} = 640(1.05)^{230} \approx 47{,}832{,}024.19$. So, if interest is compounded annually, the account would be worth \$47,832,024.19.

b. If interest is compounded daily, $n = 365$, so

$$A_{\text{daily}} = P\left(1+\frac{r}{n}\right)^{nt} = 640\left(1+\frac{0.05}{365}\right)^{(365)(230)} = 640(1.000136986)^{83{,}950}$$

$$\approx 63{,}128{,}354.17$$

So, if interest is compounded daily, the account would have \$63,128,354.17, over \$15 million more than if interest is compounded annually!

■ ■ ■ ■ ■ ■ ■ ■ ■ ■

When interest is compounded at periods of time less than one year, the actual gain in the account at the end of one year is not the same as the annual percentage rate times the initial principal, but it is slightly higher, as the following example shows.

Example 9.5   ■ ■ ■ ■ ■ ■ ■ ■

Suppose \$5,000 is invested in an account with an annual interest rate of 6%, and the interest is compounded monthly.

a. What is the account balance after one year?

b. By what percentage does the balance increase in one year?

Solution

a. We use the compound interest formula with $r = 0.06$, $P = 5{,}000$, $t = 1$, and $n = 1$. We find $A = P\left(1+\frac{r}{n}\right)^{nt} = 5{,}000\left(1+\frac{0.06}{12}\right)^{12} = 5{,}000(1.005)^{12} \approx 5{,}308.3891$. So, the account balance after one year is \$5,308.39 (rounded to the nearest cent).

b. The account balance increase in one year is \$5,308.39 − \$5,000 = \$308.39. The percentage increase is $\frac{308.39}{5{,}000} = 0.061678$ or 6.1678%.

■ ■ ■ ■ ■ ■ ■ ■ ■ ■

In Example 9.5 we saw that if we invest $5,000 at an annual interest rate of 6% compounded monthly, the actual interest collected after one year is 6.1678% of the original principal amount. This percentage is called the **annual percentage yield** and is often denoted by *APY*. Banks usually list both the annual percentage rate (*APR*) and the annual percentage yield (*APY*). For the account in Example 9.5, the bank would state $APR = 6\%$ and $APY = 6.1678\%$.

Note that in general, we can calculate the annual percentage yield as the ratio $\frac{A - P}{P}$, where A is the amount in the account after one year, P is the original principal, and $A - P$ is the interest earned on the principal P invested at an annual interest rate of r in an account in which interest is compounded n times per year. Given r and n, we can use any value of P to find A and then use that A and P to find *APY*. We'll see this in the next example.

Example 9.6

A grandfather would like to be able to give his grandson a car when the grandson graduates from college in four years. He decides he will need $25,000 at that time.

a. How much will the grandfather need to invest now in a four-year account that has an annual interest rate of 5.2% compounded monthly in order to have $25,000 in four years?

b. What is the annual percentage yield on this account?

Solution

a. We need to find P, the principal to be invested to yield $A = \$25,000$ at the end of four years. We use the general formula for finding the amount A of money in an account after t years if the annual interest rate is r and n is the number of compounding periods per year:

$A = P\left(1 + \frac{r}{n}\right)^{nt}$, with the principal P as the quantity that we need to determine. In this example, $A = 25,000$, $r = 0.052$, $n = 12$, and $t = 4$. Thus, our equation is $25,000 = P\left(1 + \frac{0.052}{12}\right)^{(12)(4)}$ or $25,000 \approx P(1.0043333)^{48}$. So, $P = \frac{25,000}{1.0043333^{48}} \approx 20,314.33$; the grandfather needs to invest $20,314.33 in the account in order to have $25,000 in four years.

b. With an annual rate of 5.2% compounded monthly, if a principal P is invested for one year, the amount in the account after one year is $A = P\left(1 + \frac{0.052}{12}\right)^{12} \approx P(1.053257)$. Thus, the increase in the account is $P(1.053257) - P = P(1.053257 - 1) = P(0.053257)$, and the *APY* is $\frac{P(0.053257)}{P} = 0.053257$, or approximately 5.326%. Therefore, investing at 5.2% compounded monthly is equivalent to investing at 5.326% compounded annually.

ANNUITIES

Suppose we want to save for something in the future (as the grandfather in Example 9.6 did) but we don't have a lump sum of money to invest initially. Instead, we want to add smaller amounts to our account on a regular basis. There are a variety of savings plans that promote this kind of savings, such as Individual Retirement Accounts (IRAs), Keogh plans, and employee pension plans. A series of fixed, regular payments is called an **annuity**. Leases and rental payments, as well as savings plans in which you deposit a specified amount at fixed intervals of time, are examples of annuities. There are two types of annuities: an ordinary annuity and an annuity due. An **ordinary annuity** is one in which the payments are required at the end of each time period. With an **annuity due**, payments are required at the beginning of each time period. We'll concentrate on ordinary annuities; in the next example, we illustrate how money accumulates with an ordinary annuity.

Example 9.7

A college student is saving to buy a new car and deposits $150 into a savings plan at the end of each month. Suppose that the plan pays interest monthly at an annual interest rate of 6%. Create a table that shows the monthly balance at the end of each month for a six-month period.

Solution

We create a table with columns to keep track of the interest accrued each month as well as the additional deposit each month. We observe that the monthly interest rate is $\frac{r}{n} = \frac{0.06}{12} = 0.005$. The last column of the table keeps track of the new balance in the account each month and represents the sum: previous balance + interest on previous balance + additional deposit. The first column is the new balance from the previous month.

End of Month i	Previous Balance	Interest on Previous Balance	Additional Deposit	New Balance
1	0	0	$150	$150
2	$150	$150 × 0.005 = $0.75	$150	$300.75
3	$300.75	$300.75 × 0.005 = $1.50	$150	$452.25
4	$452.25	$452.25 × 0.005 = $2.26	$150	$604.51
5	$604.51	$604.51 × 0.005 = $3.02	$150	$757.53
6	$757.53	$757.53 × 0.005 = $3.79	$150	$911.32

The table created in Example 9.7 helps us understand how interest and additional deposits accumulate when additional money is deposited into an account. We can also use a formula that gives the relationship between A, the amount in the account at a future date; Q, the payment quantity for each regular payment into the account; r, the annual interest rate in decimal form; n, the number of payment and compounding periods per year (which are assumed to be the same with this formula); and t, the number of years. Here is the **accumulated savings formula**:

$$A = Q \times \frac{\left[\left(1+\frac{r}{n}\right)^{nt} - 1\right]}{\left(\frac{r}{n}\right)}$$

(Sometimes the amount A is called **future value** and denoted by FV.)

In the next example, we'll confirm that this formula gives us the same result we obtained using the table in Example 9.7 for month 6, and find what the account will be worth in the future under the same assumptions.

Example 9.8

Use the accumulated savings formula to find how much money the college student from Example 9.7 (who, at the end of each month, deposits \$150 into a savings plan that pays interest monthly at an *APR* of 6%) has in the account at the end of:

a. Six months

b. One year

c. Four years

Solution

We use the formula

$$A = Q \times \frac{\left[\left(1+\frac{r}{n}\right)^{nt} - 1\right]}{\left(\frac{r}{n}\right)}$$

to find A, with $r = 0.06$, $n = 12$, $Q = \$150$, and different values of t for parts (a–c).

a. Here, $t = \frac{1}{2}$ (years), so

$$A = \$150 \times \frac{\left[\left(1+\frac{0.06}{12}\right)^{(12)(1/2)} - 1\right]}{\left(\frac{0.06}{12}\right)} = \$150 \times \frac{(1.005)^6 - 1}{0.005} \approx \$911.33$$

Note the difference of \$0.01 with the value in the table in Example 9.7, which is due to rounding the calculations to two decimals.

b. With $t = 1$ (year),

$$A = \$150 \times \frac{\left[\left(1+\frac{0.06}{12}\right)^{(12)(1)} - 1\right]}{\left(\frac{0.06}{12}\right)} = \$150 \times \frac{(1.005)^{12} - 1}{0.005} \approx \$1{,}850.33$$

c. Substituting $t = 4$ gives

$$A = \$150 \times \frac{\left[\left(1+\frac{0.06}{12}\right)^{(12)(4)} - 1\right]}{\left(\frac{0.06}{12}\right)} = \$150 \times \frac{(1.005)^{48} - 1}{0.005} \approx \$8{,}114.67$$

We can use the accumulated savings formula to help set a "savings goal" if we want to plan ahead and have money for something particular, as the next example illustrates.

Example 9.9   ■ ■ ■ ■ ■ ■ ■ ■ ■ ■

Suppose a new college graduate wants to accumulate $30,000 for a deposit on a home in eight years by making regular end-of-the-month savings deposits. Assume an annual interest rate of 8% compounded monthly.

a. Find how much the graduate should deposit at the end of each month to reach the goal.

b. Find how much of this accumulated amount comes from the deposits and how much from the interest.

Solution

a. The goal is to accumulate $A = \$30{,}000$ in $t = 8$ years, with $r = 0.08$ and $n = 12$. We want to find Q, so we need to solve the following equation for Q:

$$30{,}000 = Q \times \frac{\left[\left(1+\frac{0.08}{12}\right)^{(12)(8)} - 1\right]}{\left(\frac{0.08}{12}\right)} \approx Q \times \frac{(1.006666667)^{96} - 1}{0.006666667}$$

so

$$Q \approx 30{,}000 \times \frac{0.006666667}{(1.006666667)^{96} - 1} \approx 224.1003737$$

Thus, the graduate needs to make monthly payments of $224.11 to have $30,000 at the end of eight years. Note that to make sure we will reach the goal, we need to round the payment up to the next cent, instead of rounding it to the nearest cent.

b. We assume the annual rate remains at 8% during the eight-year period. At the end of this time period, the graduate has deposited $224.11 per month × 12 months per year × 8 yrs = $21,514.56. So, $30,000 − $21,514.56 = $8,485.44 is the approximate amount of interest earned.

LOANS

We now consider **installment loans** in which the borrower partially repays the loan with equal, regular payments, for example, on a monthly basis, for a fixed amount of time. (Typical installment loans are college tuition loans, home mortgages, and car loans.) In the next example, we calculate how much of the principal (the amount borrowed) is paid off if we make regular monthly payments.

Example 9.10

A student wants to purchase a computer for $2,000 and the store offers an annual interest rate of 18% if the student makes payments of $160.00 each month.

a. How much of the first payment goes toward paying off the principal?

b. How much of the second payment goes toward paying off the principal?

c. Set up a table to determine how much the student will still owe after one year.

Solution

a. When the student makes the initial $160 payment at the end of the first month, the interest for that month is $2,000 × $\frac{0.18}{12}$ = $30. The interest is paid first, leaving $160 − $30 = $130 of the payment to go toward paying off the principal. At the end of the first month, the new principal is $2,000 − $130 = $1870.

b. The interest for the second month is charged on the new principal and is $1,870 × $\frac{0.18}{12}$ = $28.05. After the second month's interest is paid, at the end of the second month, $160 − $28.05 = $131.95 of the payment goes to paying off the principal.

c. The results for the rest of the 12-month period are summarized in the table below. Note that each month the interest is less and the payment toward the principal is greater. At each step, we round all amounts to the nearest cent.

End of Month i	Previous Principal	Payment	Interest Paid (on previous principal)	Payment Toward Principal	New Principal
1	$2,000.00	$160	$30.00	$130.00	$1,870.00
2	$1,870.00	$160	$28.05	$131.95	$1,738.05
3	$1,738.05	$160	$26.07	$133.93	$1,604.12
4	$1,604.12	$160	$24.06	$135.94	$1,468.18
5	$1,468.18	$160	$22.02	$137.98	$1,330.20
6	$1,330.20	$160	$19.95	$140.05	$1,190.15
7	$1,190.15	$160	$17.85	$142.15	$1,048.00
8	$1,048.00	$160	$15.72	$144.28	$903.72
9	$903.72	$160	$13.56	$146.44	$757.28
10	$757.28	$160	$11.36	$148.64	$608.64
11	$608.64	$160	$9.13	$150.87	$457.77
12	$457.77	$160	$6.87	$153.13	$304.64

At the end of the year, the principal remaining is $304.64; one more full payment of $160 plus a smaller payment will exhaust the loan.

If we wanted to pay off the loan in Example 9.10 in exactly 12 equal payments, it's clear that we would need to pay a bit more than $160 each month. We can calculate this amount with the following loan payment formula:

$$Q = P \times \frac{\left(\frac{r}{n}\right)}{\left[1 - \left(1 + \frac{r}{n}\right)^{-nt}\right]}$$

In this formula, Q is the regular payment amount; P is the initial principal, that is, the amount borrowed; r is the annual interest rate; n is the number of regular payment periods per year; and t is the term, in years, of the loan. Note that $\frac{r}{n}$ is the interest rate per payment period and nt is the total number of periods.

Example 9.11 ■ ■ ■ ■ ■ ■ ■ ■ ■

Consider the $2,000 loan to buy a computer at an annual interest rate of 18%.

a. Use the loan payment formula to determine the payment needed to pay off the loan for the computer in 12 regular monthly payments.

b. How would the payments change if we take two years to pay off the loan at the same interest rate?

c. How much interest do we pay on the total loan under the one-year payment plan and how much interest do we pay under the two-year payment plan?

Solution

For this example, $P = 2{,}000$, $n = 12$, and $r = 0.18$.

a. If we pay off the loan in 12 payments, $t = 1$, so

$$Q = P \times \frac{\frac{r}{n}}{\left[1 - \left(1 + \frac{r}{n}\right)^{-nt}\right]} = 2{,}000 \times \frac{\left(\frac{0.18}{12}\right)}{\left[1 - \left(1 + \left(\frac{0.18}{12}\right)^{-(12)(1)}\right)\right]} = \frac{2{,}000 \times 0.015}{1 - (1.015)^{-12}}$$

$$\approx 183.36$$

As expected, this payment of $183.36 per month is greater than the $160 payment in Example 9.10.

b. Now $t = 2$, so

$$Q = P \times \frac{\frac{r}{n}}{\left[1 - \left(1 + \frac{r}{n}\right)^{-nt}\right]} = 2{,}000 \times \frac{\left(\frac{0.18}{12}\right)}{\left[1 - \left(1 + \left(\frac{0.18}{12}\right)^{-(12)(2)}\right)\right]} = \frac{2{,}000 \times 0.015}{1 - (1.015)^{-24}}$$

$$\approx 99.85$$

If we pay off the loan in two years, our monthly payments will be $99.85.

c. With the one-year payment plan, we pay a total of $(12)(\$183.36) = \$2{,}200.32$, so $200.32 is paid in interest. Under the two-year payment plan, we'd pay $(24)(\$99.85) = \$2{,}396.40$, so $396.40 is paid in interest.

Student loans are installment loans. There are different plans for repayment of a federal student loan. The next example considers the standard repayment plan, in which the loan is repaid in fixed monthly payments for up to ten years.

Example 9.12

A recent college graduate owes $24,000 on a federal student loan at an annual interest rate of 6.8% and will make fixed monthly payments for ten years. How much will the graduate pay each month on the student loan?

Solution

To find the amount of the monthly payment we use the loan payment formula where the principal is $P = 24{,}000$. The annual interest rate, written as a decimal, is $r = 0.068$; the number of periods per year is $n = 12$; and the payment period in years is $t = 10$.

$$Q = P \times \frac{\left(\frac{r}{n}\right)}{\left[1 - \left(1 + \frac{r}{n}\right)^{-nt}\right]} = 24{,}000 \times \frac{\left(\frac{0.068}{12}\right)}{\left[1 - \left(1 + \frac{0.068}{12}\right)^{-(12)(10)}\right]}$$

$$\approx 24{,}000 \times \frac{0.0056667}{\left[1 - (1.0056667)^{-120}\right]} \approx 24{,}000 \times \frac{0.0056667}{\left[1 - 0.5075883\right]} \approx 276.193274$$

The graduate could pay the student loan by making monthly payments of $276.19, with one slightly larger payment to account for rounding each payment down.

CREDIT CARDS

Loans accumulated on credit cards differ from installment loans because the credit card company does not require the borrower to pay off the balance in a particular period of time. These types of loans are sometimes referred to as **open-end installment loans** because the borrower can make variable payments each month, with only a minimum payment required by the credit card company. Because credit card interest rates are typically high, if we make only the minimum payment, it takes a long time to pay off the loan, as we saw in Activity 4.1. The following example illustrates this as well.

Example 9.13

Suppose a student has a balance of $1,600 on a credit card; he decides not to make any additional charges and wants to pay a fixed amount every month until the whole balance is paid off. The annual interest rate on the credit card is 18%. How much progress would the student make in six months if he pays $40 each month?

Solution

To find the balance after one month, we need to add to the initial balance of $1,600 the monthly interest on that amount and subtract $40, the monthly payment amount. At an annual rate of 18%, the monthly interest rate is $\frac{0.18}{12} = 0.015$, so we need to calculate $1,600 + 1,600 \times 0.015 - 40$, which we can write in a simpler form as $1,600 \times (1 + 0.015) - 40 = 1,600 \times 1.015 - 40$. So, in this case, the balance after one month is the starting balance multiplied by 1.015, minus 40. If the starting balance is $1,600, the balance next month is $1,600 \times 1.015 - \$40 = \$1,624 - \$40 = \$1,584$.

In the same way, but using $1,584 as the initial balance, we find the balance after two months. This is $1,584 \times 1.015 - \$40 = \$1,567.76$. Similarly, the balance after three months will be $1,567.76 \times 1.015 - \$40 = \$1,551.2764 \cong \$1,551.28$ We continue this way, multiplying the balance one month by 1.015 and subtracting $40 from the result, to find the balance after six months. The following table summarizes the results.

Number of Months Since Balance was $1,600	Balance ($)
1	1,584
2	1,567.76
3	1,551.28
4	1,534.55
5	1,517.57
6	1,500.33

The balance after six months is $1,500.33 so after making six monthly payments, the student has paid $99.67 of the original balance.

There are different ways in which banks calculate the minimum payment, but the minimum payment always depends on the current balance. On a credit card balance of $1,600, the minimum payment might be $40 initially and will get smaller each month as the account balance decreases. Our calculations in Example 9.13 indicate that the progress in paying off a credit card debt making only minimum payments might be quite slow. (See also Exploration 15.) We can use the loan payment formula to calculate the payments needed to pay off a credit card loan in a fixed amount of time.

Example 9.14

A student accumulated a balance of $1,500 on a credit card that has an annual interest rate of 20.99%. Assume that no additional purchases are made using this credit card.

a. How much will the student need to pay each month to pay off the balance over a period of one and a half years?

b. Suppose the credit card company allows the student to make no payments for six months, but charges interest during that time. How much additional interest will the student owe at the end of the six-month period?

c. If after the six-month period of not paying anything to the credit card company, the student wants to pay off the accumulated balance (including the extra interest charged) in one year, how much will the student need to pay each month?

Solution

a. We use the loan payment formula with $P = 1,500$, $r = 0.2099$, $n = 12$, and $t = \frac{3}{2}$. Thus,

$$Q = P \times \frac{\left(\frac{r}{n}\right)}{\left[1 - \left(1 + \frac{r}{n}\right)^{-nt}\right]} = 1,500 \times \frac{\left(\frac{0.2099}{12}\right)}{\left[1 - \left(1 + \frac{0.2099}{12}\right)^{-(12)(3/2)}\right]}$$

$$\approx \frac{1,500 \times 0.017491667}{1 - (1.017491667)^{-18}} \approx 97.8601$$

The student will have to make monthly payments of $97.87 to pay off the loan in one and a half years.

b. The credit card company is earning interest from the student during the first six months; we use the compound interest formula introduced at the beginning of this topic, $A = P\left(1 + \frac{r}{n}\right)^{nt}$, to first find the total amount owed on the credit card at the end of the six months. In this case, $r = 0.2099$, $n = 12$, $t = \frac{1}{2}$, and $P = 1,500$, so

$$A = P\left(1 + \frac{r}{n}\right)^{nt} = 1,500\left(1 + \frac{0.2099}{12}\right)^{(12)(1/2)} \approx 1,664.47.$$ The student will owe $1,664.47, which includes an additional $164.47 in interest.

c. Now we need to calculate the payment needed to pay off the credit card balance of $1,664.47 in one year. So, $P = 1,664.47$ and we use the loan payment formula:

$$Q = P \times \frac{\left(\frac{r}{n}\right)}{\left[1 - \left(1 + \frac{r}{n}\right)^{-nt}\right]} = 1,664.47 \times \frac{\left(\frac{0.2099}{12}\right)}{\left[1 - \left(1 + \frac{0.2099}{12}\right)^{-(12)(1)}\right]} \approx 154.98$$

The student will need to pay $154.98 per month.

Mortgages are installment loans that help people purchase homes. Mortgage interest rates are much lower than credit card rates and other loan interest rates because the home serves as a payment guarantee. Various mortgage options are available from different banks. These include the length of the loan, the down payment required, additional fees to obtain the loan, and whether the loan is a **fixed rate mortgage**, in which the interest rate doesn't change over the term of the loan or an **adjustable rate mortgage**, in which the interest rate changes as the prevailing rates change. (In Activity 9.1 we investigate some of these mortgage option scenarios.)

Summary

In this topic, we investigated simple and compound interest and looked at the effect of compounding using various compounding intervals and lengths of time over which we collect interest. We also determined how much money we should save each month to reach a particular savings goal. Finally, we investigated installment loans and credit card loans, analyzing the cost of borrowing money and evaluating various loan situations.

Explorations

1. Three students each have $1,000 to invest from their summer jobs. Armen invests his money in an account that earns simple interest at an annual interest rate of 5%. Barok invests his money in an account that earns 4.9% interest per year compounded annually. Carrie invests her money in an account that earns 4.8% interest per year compounded monthly. Find how much money each student has in his or her account after:

   a. 2 years

   b. 10 years

   c. 30 years

2. Here's an old story: A man walks into a New York City bank and asks for a $5,000 loan, offering his Ferrari, worth $250,000 as collateral. He tells the loan officer that he needs the money for two weeks for an important venture. The loan officer, having the car as security and after checking references, gives the man the money he requested, with a signed agreement that he will pay the money plus $45 in interest when he returns in two weeks. The bank officer takes the car keys and the car is parked in the bank's underground lot. The man returns in exactly two weeks, pays the loan and interest, and reclaims his car. The bank officer asked him why he was willing to pay such a high interest rate. His reply: Where else can I safely park my car for two weeks in New York City for only $45?

   a. What annual interest rate did the man pay?

   b. How much would the man need to repay at the end of two months if he borrowed $5,000 with the same rules and same annual interest rate?

3. If a Virginia colonist had invested $50 in July 1776 in an account with an *APR* of 5%, how much would that account be worth now if interest is compounded:

    a. Annually?

    b. Quarterly?

    c. Daily?

4. Compare the annual percentage yield (*APY*) for three banks: Bank 1 offers an *APR* of 3.8% compounded daily; Bank 2 offers an *APR* of 4.1% compounded monthly; Bank 3 offers an *APR* of 4.5% compounded quarterly.

5. Some banks advertise that interest is **compounded continuously**. We can think of this as compounding infinitely many times per year (rather than 12 times per year, as with monthly compounding, or 365 times per year, as with daily compounding). With continuous compounding, the compound interest formula has the following form: $A = P \times e^{rt}$, where A is the amount in the account after t years, P is the starting principal, and r is the annual interest rate written as a decimal. The number e that appears in the formula is a special irrational number: $e \approx 2.71828$. [Note that many calculators have a special "e^x" key and Excel uses EXP(x) to evaluate e^x.] Find how much money a student would have in an account if he invested $1,000 with interest compounded continuously, after 2 years, 10 years, and 30 years if the annual interest rate is:

    a. 3%

    b. 5%

    c. 7%

6. A new parent wants to have $80,000 in a college savings account in 18 years. Although it's difficult to predict what the annual interest rate will be over 18 years, we can get a sense of how much the parent should save each month using various assumptions. Calculate the monthly savings needed if the money is invested in an account with monthly compounding and an annual interest rate of:

    a. 4%

    b. 7%

    c. 9%

7. Set up a table like the one given in Example 9.10 and confirm that a monthly payment of $183.36 will result in the loan being paid off in 12 months, as we found using the formula in the solution to Example 9.11(a).

8. Suppose that a friend who spent her junior year abroad has accumulated $2,400 worth of debt on her credit card. The card has an annual interest rate of 19.6%. If she doesn't charge anything additional on the card, how much should she pay each month to pay off the card in:

a.   9 months?

b.   15 months?

c.   3 years?

9.   You have $1,500 in credit card debt on several cards, each of which has an outrageously high interest rate. You want to consolidate your debt and formulate a plan to pay it off in one year, so you are comparing two credit card offers. Card A offers an *APR* of 18.6% while Card B offers no interest for the first six months and then an *APR* of 20.5% after that.

   a.   If you transfer the balance of $1,500 to Card A, find how much you will need to pay each month in order to pay off the whole balance in one year.

   b.   Suppose you pay the amount computed in part (a) of this Exploration but under the scenario offered by Card B. That is, you pay that amount each month, but for the first six months, no interest is being charged, so you are reducing your principal. How much would you need to pay each month for the final six months of the year to pay off the remaining balance on Card B by the end of the year?

   c.   How much interest would you pay under each of the plans?

10. Two car dealers each are offering a $7,000 loan. The first dealer offers an annual interest rate of 7.8% and the loan must be repaid in monthly payments over three years. The second dealer offers an annual interest rate of 8.4%, with monthly payments and a loan term of four years.

   a.   Find the monthly payment under each scenario.

   b.   Find the total amount of interest paid under each scenario.

11. Credit card companies often charge a higher rate of interest for cash advances than they do for purchases. Obtain detailed information from two credit card companies about their interest rates on purchases and cash advances, and analyze these interest rates.

12. Find how much you would need to deposit each month in a retirement account in order to accumulate $1,000,000 in 40 years, assuming:

   a.   The annual interest rate is 3% compounded monthly.

   b.   The annual interest rate is 5% compounded monthly.

   c.   The annual interest rate is 3% compounded quarterly.

13. A student needs to decide between two plans, the standard plan and the extended plan, for repaying his federal student loan. The total amount he will owe on the loan when he graduates is $32,000 and the annual interest rate is 6.8%. Under both plans he will make fixed monthly payments. Under the standard plan, the total amount must be paid in ten years whereas under the extended plan the total amount must paid in 25 years.

   a.   Find the monthly payment under the standard plan.

   b.   Find the monthly payment under the extended plan.

c. Calculate the total interest the student will pay if he chooses the standard plan.

d. Calculate the total interest the student will pay if he chooses the extended plan.

e. Which of the two repayment plans would you choose? Explain the advantages and disadvantages of each.

14. Suppose we borrow $200,000 from a bank at an annual interest rate of 5%, and we are not required to make any payments during the first year. Determine the interest amount that we will owe after the first year in each of the following cases.

a. The interest is compounded daily, using 365 days in a year.

b. The interest is compounded daily, using the 365/360 rule. This rule uses the annual interest rate divided by 360 as the daily rate, and compounds it for 365 days in a year. This means that the formula used to calculate interest is $A = P\left(1 + \dfrac{r}{360}\right)^{365t}$.

(This method of calculating interest has caused some controversy in recent years. See, for example, Edward D. Shapiro's 2010 article, "The 365/360 Method of Calculating Interest: Lenders and Borrowers Square Off," www.muchshelist.com)

c. Explain how the 365/360 rule benefits the lender.

15. In Example 9.13 we saw that a fixed monthly payment of $40 reduced the balance on a credit card debt of $1,600 by about $100 in six months. Assume the student continues to pay $40 each month and does not make any additional charges on the credit card.

a. What will the balance be after six more months?

b. How long will it take for the student to pay off the full balance?

c. What happens if the student doubles the monthly payment amount each month from month 1?

Mortgages

In this activity, you will investigate some considerations that arise when looking for a home mortgage loan. You will determine basic characteristics of several options to compare the consequences of the various choices.

Purchasing a home is a big decision. Often, a down payment of at least 10% of the purchase price of the home is required. A lender will loan you the money for the rest of the purchase price but with additional closing fees. These closing costs include direct fees for things such as a home appraisal and title insurance and also fees for **points** charged; each point is 1% of the loan amount. For example, each point charged on a $150,000 mortgage loan would add $1,500 to the amount borrowed. For a **fixed rate mortgage**, you are guaranteed that the interest rate stated when you sign the loan papers will not change for the term of the loan. With an **adjustable rate mortgage** (ARM), which is more attractive to the lender, your interest rate changes over time. ARMs are generally given as a fixed rate period and then an interval, such as 3/1, which means the rate is fixed for three years and then will change every year based on a rate index. In considering mortgage loan options, you will need to take into account the following:

Interest rate

Amount of the loan

Closing costs, including points

Fixed rate or adjustable rate mortgage

Any prepayment penalties

You have decided to purchase a home and have enough money for a decent down payment as well as the money to cover direct fees for closing costs (which you'll assume do not vary much from one bank to another). You are considering several mortgage options and want to evaluate them. Not counting points, you will need a loan for $130,000. If a mortgage option requires you to pay points, you will need to add that money to your loan amount. Assume the mortgage options you are considering have no prepayment penalties.

1. Lender I offers you a fixed rate 15-year mortgage at an annual interest rate of 5.25%, compounded monthly, with no points.

   a. Find your monthly payments under this option.

   b. Find the total amount of money paid to the lender.

   c. Find the total amount of interest you will pay over the life of the loan.

2. Lender II offers you a fixed rate 30-year mortgage at an annual interest rate of 5.75%, compounded monthly, with one point.

   a. Find your monthly payments under this option. (Don't forget to include the "point" in the loan amount.)

   b. Find the total amount of money paid to the lender.

   c. Find the total amount of interest you will pay over the life of the loan.

3. Excel has a special function that computes the payment required on a loan. Use these instructions to check your computations in Questions 1 and 2.

Instructions to Calculate a Mortgage Payment

1. Open a new sheet in your Excel workbook. In cell A1, enter the label **Loan Amount**; in cell A2, enter the label **Annual Interest Rate**; in cell A3, enter the label **Length of Loan in Yrs**; and in cell A4, enter the label **Monthly Payment**. To check your work from Question 1, enter **$130,000** in cell B1, **5.25%** (do not forget to type the symbol %) in cell B2, and **15** in cell B3.

2. Now go to cell B4 on your worksheet. Go to the **Formulas** tab. Move your cursor to the large symbol *fx* on the left. This is the "Insert Function" symbol; click on it. In the **Insert function** window that appears, select **Financial** for category and **PMT** for function, and then click **OK**.

3. Now you want to fill in the requested values. For **Rate**, the interest rate is in cell B2, so enter **B2/12** (because the mortgage will be paid monthly and this value gives the monthly rate).

4. In **Nper**, enter the total number of payments for the life of the loan. The length of the loan in years is found in cell B3, so in this box enter **B3*12**.

5. **Pv** is the amount of the loan; that amount is found in cell B1, so enter **B1** in this box.

6. You can leave the remaining two boxes blank because the values that are used if left blank are those you want. If you prefer to enter values in these fields, you can enter **0** in the **Fv** box and **0** in the **Type** box. Click **OK**.

7. Change the interest rate to **5.75%** in cell B2 and change the amount and length of the loan to the values used in your calculations for Question 2.

4. Suppose you think that the payments required for the 15-year mortgage are too high for you to afford, but that you'd be able to afford higher payments than the 30-year mortgage requires. You are considering obtaining a 30-year mortgage at the offered rate but paying it off sooner by making larger payments than are required.

　a. Use Excel to find the monthly payments needed to pay the loan off in 20 years and record the monthly payment here.

b. What monthly payments would you need to make in order to pay the loan off in 25 years?

5. You want to compare how much interest is paid during the first two years under each lender's plan. Excel's function to calculate interest can help.

Instructions to Calculate Interest

1. On the same Excel worksheet, in cell D1, enter the label **Month**; in cell E1, enter the label **Interest**; in cell F1, enter the label **Interest Lender I**, and in cell G1, enter **Interest Lender II**. (You'll use column E as a "calculation column" to calculate interest, and then you'll save the values you calculate in columns F and G.) Enter the numbers **1** and **2** in cells D2 and D3, respectively, and drag down to fill in the rest of the months to cover the two-year period.

2. Now you want to find the interest under Lender I's option, so start by entering the appropriate values for "Loan Amount," "Annual Interest Rate," and "Length of Loan" in cells B1, B2, and B3, respectively. Then select cell E2, move your cursor to the symbol *fx*, and click it. For **category**, select **Financial**, and for **function**, select **IPMT** (for interest payment) and then click **OK**.

3. Now you want to fill in the requested values. For **Rate**, the interest rate is found in cell B2, so enter **B$2/12** (since the mortgage will be paid monthly and this value gives the monthly rate). Because you want to drag down to fill in the interest for the remaining months, make this an absolute reference that won't change when you drag down by adding a $ sign between the **B** and the **2**.

4. For **Per**, enter **D2**, because the period is the month.

5. For **Nper**, because the length of the loan in years is found in cell B3, enter **B$3*12**.

6. For **Pv**, which is the amount of the loan found in cell B1, enter **B$1**.

7. You can leave the remaining box blank, or you can enter **0** in the **Fv** box. Then click **OK**.

8. Drag down to fill in the interest for the first two years. Then go to cell E27 and find the sum of all the interest payments under Lender I's option. To do this, type **=SUM(E2:E25)**. Record the total interest in part (a), after these instructions.

9. You want to save these values so that they won't change when you change the interest rate and length of the loan (which you will do to evaluate Lender II's option). To put these interest values in column F, highlight cells E2 through E27 and select **Edit** and

then **Copy**. Then go to cell F2 and select **Edit** and **Paste special**. Choose **Values** by clicking on the white circle to the left of that word and click **OK**. This command pastes the numbers in column F so they are no longer linked to cells B1, B2, and B3.

10. Change the numbers in cells B1, B2, and B3 to reflect Lender II's option. (Remember to include the "points" in the amount borrowed.) The numbers in column E reflect the costs under Lender II's option. Copy and paste (using **Paste special** again) the values in column E to column G. Record the total interest below in part (b).

    a. Total interest paid under Lender I's plan: _____

    b. Total interest paid under Lender II's plan: _____

6. You have another mortgage option—Lender III offers you an adjustable rate 30-year mortgage, with no points, at 4.95% interest for the first year. After that time, the mortgage rate will likely change, but you don't know what the new rate will be. Find your monthly payments for the first year under this option and record that value here.

7. Let's suppose that under Lender III's adjustable rate mortgage option, the interest rate increases by 1% for the second year. Calculating interest paid for the first two years under Lender III's plan is more complicated. Here's one way to do it.

    a. First, use a method similar to what you did for Lenders I and II to find the interest paid for each month of the *first year* under Lender III's plan and record that total here:

    Total interest paid during the first year under Lender III's plan: _____

b. Now use your work in Question 6 to find the total payments made under this plan during the *first year*; then find the amount of principal paid during the first year and the total principal remaining. Record those values here:

Total principal paid during the first year: _____

Total principal remaining: _____

c. Finally, use the IPMT function with the new principal remaining as the amount of the loan, the new interest rate of 5.95%, and 29 years remaining on the loan, to calculate the interest paid during each month of the second year of Lender III's plan. Find the total interest paid during the two years under Lender III's plan and record that total here:

Total interest paid for the first two years under Lender III's plan: _____

8. Consider the monthly payments, the total interest, and any other factors that you think are important and discuss the pros and cons of each of the lender's plans. Which lender would you choose and why?

Summary

In this activity, you learned how to use Excel's loan payment and interest calculation functions. You learned how to copy and paste values so they are no longer linked to cells that may change. Finally, you examined and compared several realistic options for home mortgage loans.

Introduction to Problem Solving

After completing this topic, you will be able to:

- Use a problem-solving process and identify a variety of problem-solving techniques.

- Recognize which problem-solving techniques can be used to solve a particular problem.

- Use a variety of problem-solving techniques to answer questions that concern, environmental, financial, and social matters.

- Recognize when to use various computational methods studied in previous topics.

In this topic, we will outline a four-step problem-solving process as described by George Pólya in 1973 in his book *How to Solve It*. We'll then look more specifically at the first three steps of this process in the context of methods used in Topics 1–9, and we will apply those steps to other problems that relate to environmental, financial, and social issues.

The four steps of Pólya's process are as follows:

A. Understand the problem.

B. Devise a plan.

C. Carry out the plan.

D. Look back.

Each step of the process is important and relies, in some measure, on the other steps. We'll look more closely at the first three steps in this topic and will consider the final step in Topic 15.

UNDERSTAND THE PROBLEM

In order to more fully understand a problem, it is often helpful to use some of the following techniques:

1. Ask questions to clarify the problem. At this stage, we don't want to think about answers, just questions. Asking lots of questions about the problem at hand can help us hone in on the "real" problem. We want to frame the problem and clarify the purpose of the investigation. We might ask questions like the following: What information do we know? What do we want to know? What other information do we need to collect? How can we get the needed information? This technique sometimes involves creative questioning in order to reveal ideas that might otherwise remain hidden.

2. Decide what information is relevant. Sometimes we consider unneeded information or data that confuse a problem or issue; we need to pay attention to what information we really need to use and sort that information from information that is unnecessary.

3. Represent the information in a different form. Looking at our information in a different form will often reveal trends and relationships.
   a. Make a table or chart
   b. Draw a picture, graph, or diagram.
   c. Write an equation or formula.

Example 10.1

You realize that you have been paying increasing amounts of money for the gas you put into your car. You want to find how you can reduce the cost of keeping your gas tank filled. Identify ways to help you understand the problem.

Solution

Here are some questions to help understand the problem; they will help shed light on the high cost of filling the gas tank. How much am I paying per gallon of gas each time I fill the tank? How does this compare with the minimum cost of gas in my region? How many miles do I drive each week and for what purpose? How many miles per gallon does my car get? How does this compare with the average for my make and model of car?

You can create a table to keep information about how often you filled your tank with gas, what you paid per gallon, and how many miles you traveled on that tank of gas before filling.

DEVISE A PLAN

When working to devise a plan, one or both of the following techniques can be useful:

4. Examine a simple case or try several special cases. Looking at special cases may reveal what is going on with the more general case.

5. Break a problem into smaller problems or identify a subgoal or subproblem. By identifying smaller problems and working on them, we can tackle each part and put them together to solve the whole problem.

Example 10.2

For the last several weeks, you have been taking power walks around your town in Maine and have noticed discarded bottles and cans along the sides of the road. The discarded bottles and cans are a problem you want to help solve. Explain how you can understand the problem and devise a plan to solve it.

Solution

Here are some questions you ask to help understand the problem. What can be done with collected cans and bottles? How would I transport the bottles and cans I pick up? Can I pick up bottles and cans and still maintain my walking program? How much of a contribution can I actually make? How will I feel about helping to beautify my community?

The plan to help solve this problem can involve identifying subgoals. First you can collect information about which agencies accept recycled bottles and cans, and what they give for each. You might then brainstorm ideas and test them for carrying the bottles and cans you collect: A plastic bag, a canvas bag, a backpack, an "apron" with large pockets are a few ideas. You can then investigate whether picking up cans and bottles helps or hinders your walking program. Finally, you may try a test week to see how many bottles and cans you can pick up during a week of walks.

CARRY OUT THE PLAN

The following techniques are helpful when carrying out the plan:

6. Work backward. Some problems can be better solved by looking at where we want to go and working backward from that end point.

7. Look for a pattern. Patterns may be revealed by looking at special cases. We then need to make sure that the same patterns hold for the general problem.

Example 10.3

You have been working at a job for a few years and currently live in a rented apartment. You read this recent statistic: "Of the 50 largest cities in the U.S., it is cheaper to purchase a home than to rent in 37 of them." You want to decide whether to rent again or to purchase a home. Explain how you could apply some of the relevant techniques to help understand the problem, devise a plan, and carry out the plan.

Solution

Some questions that can help you understand the problem are: How long do I plan to stay in this city? How much am I paying in rent and utilities each month? What is the cost of a reasonably sized home in my city? Do I have money for a down payment and closing costs? What are the current mortgage rates and how much would my mortgage, taxes, and utilities be each month? What are the projections for housing values in my city over the next several years? What kind of maintenance costs can I expect to incur in a home? How will I feel about owning a home versus renting a home?

Devising a plan will involve collecting information about the home market in your city, individual home costs, mortgages, and closing costs. You would also want to look at the economic projections of home costs for the near future. Finally, you would want to plan visits to prospective homes and collect information to compare the costs of running each potential home.

Carrying out the plan will involve comparing potential homes and the financial information for each. You may be able to identify patterns in mortgage and closing costs for the homes.

In the next example, we consider problem-solving techniques that we used in previous topics.

Example 10.4

Identify one problem-solving technique that was used in each of the following examples from previous topics. Refer to the list of techniques #1−7 above. In each case, write a question related to the content of the example that could be answered using that technique.

a. Example 1.9: We created a stemplot and identified patterns in the data of the acceptance rate for a sample of colleges and universities.

b. Example 2.4: We represented the total number of calories used by a 150-pound person as a function of the number of minutes spent walking.

c. Example 5.4(a): We found the tax an individual should pay for various levels of taxable income.

d. Example 6.4: We developed a formula that gives a family's debt for any month *m*, using data given in a table.

e. Example 6.10: We found the numbers of orders of presentation for taste tests of various numbers of brands of chocolate.

f. Example 8.6(b): We considered what variables should be included in a rating scheme to rate colleges and universities.

Solution

a. In Example 1.9, we used problem-solving technique 3(b). *Draw a picture, graph, or diagram,* by creating a stemplot to represent the given data. We could use this technique to help answer this question: How selective are these colleges, and if I can only apply to five of these colleges, which five should I choose? We also used technique 7. *Look for a pattern,* to identify the patterns revealed by the stemplot.

b. In Example 2.4, we used problem-solving techniques 3(a). *Make a table or chart,* and 3(b). *Draw a picture, graph, or diagram,* by creating a table and a graph from the information given. We could use one of these problem-solving techniques to answer the question: How many minutes of walking should a person add to his or her daily activities to be able to consume an extra piece of fruit or candy bar without gaining weight?

c. In Example 5.4(a), we decided which information line from the tax table instruction was relevant for each of the salaries given. The problem-solving technique represented was 2. *Decide which information is relevant.* In this context, we could use this problem-solving technique to answer the question: How much federal tax corresponds to a specific income?

d. In Example 6.4, we used problem-solving technique 7. *Look for a pattern* to help develop a formula that gave the family's debt as a function of time in months. We could use this formula to find the family's debt after eight months, for example.

e. Example 6.10 provided an example of problem-solving technique 4. *Examine a simple case or try several special cases.* In this example, we looked at several special cases to answer this question: How does the number of presentations grow with the number of brands of chocolate in a taste test? We could use this problem-solving technique to answer the question: If we have *n* brands of chocolate and choose an order of presentation at random, what is the chance that a specific brand is chosen first?

f. In Example 8.6(b), we identified variables that would be useful in ranking colleges. In this way, we broke the problem of rating colleges into several smaller problems, so we were using problem-solving technique 5. *Break a problem into smaller problems or identify a subgoal or subproblem.* In this case, we can use this technique to answer the question: Which college is best for me?

The following example illustrates problem-solving technique 6. *Work backward.*

Example 10.5

Use problem-solving technique 6. *Work backward* to answer the following question. In Topic 4, Exploration 3, we learned that to calculate the body mass index (BMI) of a person, we perform the following steps:

Step 1: Multiply the person's weight by 0.4536 (this gives the weight in kilograms).

Step 2: Divide the weight in kilograms by the square of the height in meters. Height in meters is obtained by multiplying the height in inches by 0.0254.

A person who is 68 inches tall wants to get his or her BMI within the normal range of 20 to 25. What weight range would give this BMI range?

Solution

We first note that for a person who is 68 inches tall, the two previous steps are:

Step 1: Multiply the person's weight by 0.4536.

Step 2: Divide the number obtained by $(68 \times 0.0254)^2 \approx 2.9832$.

So, $BMI = \dfrac{\text{weight} \times (0.4536)}{2.9832}$.

To find the weight of a person with a BMI of 25, we start with BMI = 25 and work backward, reversing the order of the two steps:

Step 1: Multiply the desired BMI of 25 by 2.9832: $25 \times 2.9832 \approx 74.58$.

Step 2: Divide the number obtained in Step 1 by 0.4536: $\dfrac{74.58}{0.4536} \approx 164.4$.

So, the person should weigh approximately 164.4 pounds to have a BMI of 25. Using the same procedure, we can find the weight that corresponds to a BMI of 20:

Step 1: $20 \times 2.9832 \approx 59.664$

Step 2: $\dfrac{59.664}{0.4536} \approx 131.5$

Because BMI increases as weight increases, looking at the two extreme values of BMI is sufficient. Thus, a person who is 68 inches tall should weigh between 131.5 and 164.4 pounds to keep the BMI within the range 20 to 25.

As we have seen in previous topics, drawing a graph can help us identify trends and make estimates and predictions. In the next example, we use problem-solving technique 3(b). *Draw a picture, graph, or diagram* to make an "educated" guess.

Example 10.6

Suppose we own a home and use oil to heat the house. In July 2011, the oil company offers the opportunity to buy, at the current price, the oil needed for the next winter. The following table gives the company's average monthly price per gallon of oil for the last three years. Use problem-solving technique 3(b). *Make a picture, graph, or diagram*, to decide whether or not we should buy the oil in advance.

Year	Jan.	Feb.	Mar.	Apr.	May	June	July	Aug.	Sept.	Oct.	Nov.	Dec.
2008								255.6	400	352	299	266
2009	254	251	234	233	233	249	246	255.6	256	263	283	288
2010	303	298.6	300	304	294	287	284	283.8	283	297	315	334
2011	353	370.5	394	398	389	375	373					

Solution

We graph the function that gives the average monthly price per gallon, with time as the explanatory variable.

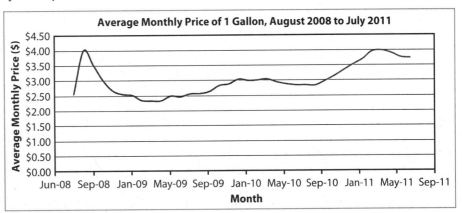

Looking at this graph, we see that the price of heating oil experienced a spike in September 2008 and has had many small variations since that time. Although the price decreased from September 2008 to April 2009, after that time it generally has been increasing. During the past two years, the price was generally lower in the spring and early summer months than in the next winter months. Thus, based on the information we have, it seems reasonable to buy heating oil in advance, if we can afford to pay for it at the current time.

In the next example, we use problem-solving techniques to make an estimate.

Example 10.7

Use the information given in Example 10.6 to estimate the price of heating oil the company might charge in January 2012. Explain how you made that estimate and what problem-solving techniques you used.

Solution

A reasonable way of predicting the price of oil would be to find the regression line for the given data. To do so, we make a scatterplot of the data, using a time scale from 0 to 35 on the horizontal axis. (0 represents August 2008, 1 represents September 2008, and so on, ending with 35 representing July 2011.)

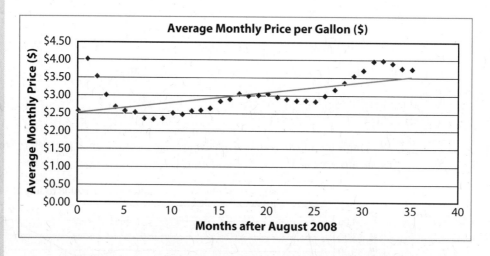

We observe from the scatterplot that there are some cyclic variations over this time period, but we can see the general trend using the regression line. With the help of technology (see Activity 6.2), we obtain an equation for the regression line $y = 0.0293x + 2.5029$. To estimate the price of heating oil in January 2012, we first note that January is the sixth month after July so it corresponds to $x = 41$; then we calculate the corresponding value of y in the equation of the line: $y = (0.0293) \cdot (41) + 2.5029 \approx 3.704$, so we estimate the price of heating oil for January 2012 to be $3.70 per gallon. We applied problem-solving techniques 3(b). *Draw a picture, graph, or diagram* and 3(c). *Use an equation or formula,* to draw the scatterplot and use the equation of the regression line.

Often we need to use more than one technique to solve a problem. In the following example, we consider a problem whose solution might require combining several problem-solving techniques.

Example 10.8

Josh is planning to buy a car that costs $15,000. He plans to make a down payment of $2,000 and to finance the rest with a loan to be paid in five years. Josh needs to decide whether he should get a loan through the car dealership, from the credit union, or if he should accept his uncle's offer of a loan to be paid at the end of five years with a simple interest of 4% per year. Among the three options, Josh wants to choose the loan that will require the smallest monthly payment. Identify which problem-solving technique(s) would help solve Josh's problem and describe how to use them.

Solution

We first need to find out more information about each type of loan and then select the relevant data using problem-solving technique 2. *Decide what information is relevant.* We could also use technique 3(a). *Make a table or chart,* by organizing the relevant loan information in a table. To decide which of the three options is best, we need to find the monthly payment amount under each of the three financing options; thus, we would break the problem into three smaller problems. This is problem-solving technique 5. *Break a problem into smaller problems or identify a subgoal or subproblem.*

In the next example, we will proceed as described in the solution to Example 10.8. The first step is to give specific information about each of the three loans:

1. The car dealer has a special finance offer: a $500 discount on the price of the car and an annual percentage rate (*APR*) of 7.5% on a five-year loan with equal monthly payments.

2. The credit union charges 6% interest per year (that is, *APR* = 6%) on a five-year fixed-installment car loan with monthly payments.

Both car loans are **fixed-installment** (or **amortized**) **loans**, in which the number of payments is determined in advance and the same amount of money is paid each time. In both cases, the interest is compounded monthly and is charged on the principal (the amount of money still owed) at the end of the month. The **loan term** for both loans is five years, and the number of payments is $5 \times 12 = 60$. The **starting loan principal** (initial amount borrowed) is 12,500 if the loan is from the dealership and 13,000 otherwise.

1. If Josh decides to take a loan from his uncle, he will make a payment only at the end of five years, but he will have to set aside the same amount each month during the five-year period of the loan. With simple interest, the interest is calculated on the original principal for each year during the term of the loan.

Example 10.9

Make a table summarizing the information collected that is relevant to Josh's car-financing problem described in Example 10.8.

Solution

The following table summarizes the information on the three possible loans for Josh:

Loan Source	Loan Principal	APR	Loan Term and Type
Uncle	$13,000	4%	5 years—simple interest
Car dealer	$12,500	7.5%	60-month installment loan
Credit union	$13,000	6%	60-month installment loan

We will solve Josh's problem in Activity 10.1.

Credit cards provide an example of a type of installment loan. An open-end installment loan is a loan that requires periodic (usually monthly) payments, but the payments do not have to be of equal amounts. Most credit cards require a minimum payment to cover the interest on the balance owed and just a very small amount of the principal. The longer you take to pay credit card debt, the more interest you pay and the more profit the issuing bank makes. A method often used by banks to calculate finance charges on credit card accounts is the *average daily method*. With this method, the finance charge is the daily interest on the average daily balance on the credit card account for the billing cycle, multiplied by the number of days in the billing cycle. We illustrate this method in the following example.

Example 10.10

Tammy has a balance of $480 on her credit card on January 26. She charges a purchase of $41 on February 3 and a purchase of $63 on February 6, and pays $280 on February 14 (through an electronic check that is posted the same day). There are no other transactions on the account during this billing period, which starts on January 26 and ends on February 24. This credit card charges an interest rate of 16.49% per year on purchases and 21% on cash advances. Calculate the finance charges on Tammy's next credit card statement. Also indicate the problem-solving techniques used to solve this problem.

Solution

To find the finance charges, we need to explore how the average daily balance is computed. We consider two simpler steps (problem-solving technique 5. *Break a problem into smaller problems or identify a subgoal or subproblem*). We first find the average daily balance, and then we find the finance charge.

To find the average daily balance, we create a table (problem-solving technique 3(a)). *Make a table or chart* that shows the date and amount of each transaction made during the billing period and the corresponding balance on each of those days.

Date	Transaction	Balance
1/26		$480
2/3	$41	$521
2/6	$63	$584
2/14	−$280	$304

The average daily balance is the sum of the daily balance for each day of the billing period divided by the number of days in that period. For 8 days (from January 26 through February 2), the daily balance was $480; for 3 days (February 3 through February 5), the daily balance was $521; for 8 days (February 6 to February 13), the balance was $584; and for the last 11 days of the billing period, the balance was $304. We use a formula (problem-solving technique 3(a). *Use an equation or formula*) to find the average daily balance.

The sum of the daily balances is $(480 \times 8) + (521 \times 3) + (584 \times 8) + (304 \times 11) =$ 13,419, and there are 30 days in this billing period, so the average daily balance is $\frac{\$13,419}{30} = \$447.30$.

The daily interest rate is obtained by dividing the annual interest rate by 365, the number of days in a year: $\frac{0.1649}{365} = 0.0004518$. The finance charge on Tammy's credit card account for this billing period is the daily interest on the average daily balance times the number of days in the billing period: $0.0004518 \times \$447.30 \times 30 \approx \$6.06$.

Summary

In this topic, we looked at Polyá's problem-solving process and identified seven problem-solving techniques. We looked at when these were used in previous topics and identified questions that we could answer using these problem-solving techniques. We analyzed problems in which finding the solution required us to use two or more problem-solving techniques. We also explored the average daily method used by some credit card companies to compute interest on credit card charges.

Explorations

1. For each of the problem-solving techniques discussed in Topic 10, find an Example (not mentioned in this topic), an Exploration, or an Activity in which you used the technique.

2. The following table gives the percentage of the U.S. population living below the poverty level for the period from 1997 to 2009:

Year	U.S. Population Living below Poverty Level (%)
1997	13.3
1998	12.7
1999	11.9
2000	11.3
2001	11.7
2002	12.1
2003	12.5
2004	12.7
2005	12.6
2006	12.3
2007	12.5
2008	13.2
2009	14.3

Source: World Almanac and Books of Facts 2011, p. 57.

Do you predict that the poverty rate will increase or decrease in the next several years? Explain which problem-solving technique you used to answer the question.

3. Marian has received an inheritance of $50,000 and would like to invest it in a regular savings account or a money market savings account. The regular savings account pays an interest of 1.49% per year and is compounded quarterly. The money market savings account pays an interest of 1.485% and is compounded daily. Identify which technique (s) from those discussed in this topic would help Marian make a decision and describe how you would use the techniques. (You do not need to solve Marian's problem.)

4. An article in *The New York Times* on August 21, 2011, reported that in 2007, the top 10% of earners received 49.74% of the total income in the United States. Moreover, the top 1% of earners received 23.5% of the total income. This phenomenon is referred to as income inequality. Outline the problem-solving process and techniques you would use to understand the problem and consequences of income inequality for the citizens of the United States.

5. The following table gives U.S. budget amounts spent for the global war on terror operations from 2001 to 2010:

Year	Budget for Global War on Terror (in billions $)
2001–02	33.0
2003	81.2
2004	94.1
2005	107.6
2006	121.4
2007	170.9
2008	185.7
2009	155.1
2010	135.6

Source: U.S. Government Accountability Office, www.gao.gov.

a. What questions would you ask if you were a congressional representative and needed to understand the problem that this money is trying to solve?

b. What steps would you use to devise a plan to solve the problem?

6. Laurie's credit card statement of September 5 showed a debt of $300. On September 15, she pays $50, and on September 20 and September 27, she charges purchases of $21.50 and $35, respectively. She makes no other payments or charges until the end of the billing period, September 6 through October 3 (both days included). Her credit card company charges interest monthly on the average daily balance for purchases at an annual rate of 14.2%.

a. Identify which problem-solving techniques would be useful in answering the question: What will be the balance in Laurie's credit card account at the end of the billing period?

b. Describe how you would use the techniques you identified.

7. Answer the question in Exploration 6 using the techniques you identified.

8. Refer to Exploration 6. Suppose in addition to the payments and purchases described, Laurie takes a cash advance of $100 on September 23. The credit card account charges interest of 21% on cash advances, also computed monthly but only on the daily average of cash advances.

a. What will be the balance on Laurie's credit card account at the end of the billing period?

b. Identify the problem-solving techniques used to answer the question.

9. Your friend needs help with a personal problem. (It may involve a perceived time management problem, problems with snacking on junk food, a "lack of exercise" problem, a

nail-biting problem, or something else—your friend, your choice!) Describe how you would proceed to help your friend understand the problem, devise a plan, and carry out the plan to solve this problem.

10. You are a member of the school board in a district where, at the end of the previous academic year, the standardized test scores of the students at one elementary school in your district were low and resulted in that school being placed on academic probation. The board needs to advise the school's administrators so that the school will be able to get off probation. Describe how you would proceed to understand the problem, devise a plan, and carry out the plan to solve this problem.

10-1

Savings and Loans: Problem Solving and Using Scroll Bars

In the first part of this activity, you will identify problem-solving techniques. In the second part of this activity, you will learn how to create a scroll bar using Excel and then use it to solve problems about savings and loans.

1. Identify all problem-solving techniques used in Question 1 of Activity 6.1 (where you decided which of the magic genie's two offers to choose). For each technique you identify, describe where in the activity it was used.

2. Describe a problem (you can use one solved previously, or you can construct one) that can be solved using the stated problem-solving technique (by itself or in combination with other techniques):

a. Problem-solving technique 3(b). *Draw a picture, graph, or diagram.*

b. Problem-solving technique 4. *Examine a simple case or try several special cases.*

c. Problem-solving technique 6. *Work backward.*

3. In the next part of this activity, you will create an Excel scroll bar to see how interest grows in a savings account.

a. A **scroll bar** is an exciting addition to a spreadsheet that can allow you to vary a value easily. You can vary, for example, the interest rate on an investment in an account over an interval from 1% to 20% to see how the total amount in the account is changing dynamically. Follow the instructions below to investigate a problem using a scroll bar.

Instructions to Create a Scroll Bar

1. You'll first set up a spreadsheet to calculate the interest and total money in the account with an investment of $1,000 over a 15-year period. In cell C1, enter the label **Interest Rate as a Percent** and in cell C2 enter **=1**. In cell B1, enter the label **Interest Rate**

as a Decimal and in cell B2 enter **=C2/100**. (The value in cell C2, which is the interest rate as a percent, will be linked to the scroll bar. The value in cell B2 is the interest rate as a decimal.)

2. In cells A4, B4, and C4, enter the labels **Year, Interest, Total**, respectively.

3. In cell A5, enter **0**; in A6, enter **1**. Highlight both of these cells and drag down to cell A20 to represent 15 years.

4. In cell C5, enter **$1000**, which will be the starting amount in the account. In cell B6, enter **=B2*C5**. (Note that you can type **=B2** and then press the **F4** function key to get the result B2.) In cell C6, enter the formula **=C5+B6**.

5. Highlight cells B6 and C6 and drag down to show the interest and total for each of the 15 years.

6. Next, you will sketch a scatterplot of $x =$ **Year** and $y =$ **Total**. You will need to highlight the label and values in the **Year** column, press **Ctrl**, release the left mouse button, and then go to the **Total** column and highlight the label and values in this column. Now using the **Insert** tab, create the scatterplot, with the points connected by a line. Choose appropriate titles for the axes and chart.

7. To use a scroll bar, you will need to access the **Developer** tab. If the **Developer** tab is not shown on top of the menu ribbon, use the following instructions to load it. Click the **Office Button**. Then click **Excel Options** (at the bottom of the pop-up window). Choose **Popular** on the left menu box, check **Show Developer tab in the Ribbon**, and then click **OK**.

8. Click on the **Developer** tab and then go to the **controls** group and click on the small arrow under **Insert**. On the pop-up menu, under **ActiveX Controls**, find the **Scroll bar** icon. Look for the small picture with the up and down arrows and check that the words **Scroll bar (ActiveX Control)** appear when you place the cursor over it. Click on this icon. The **Design Mode** icon in the **Controls** group will light up and the cursor becomes a thin "plus sign."

9. Move the cursor (the "plus" sign) to the spot where you want the scroll bar. Press and hold the left mouse button. Drag the mouse horizontally to the right to create a horizontal scroll bar, or drag it down to create a vertical scroll bar. When the scroll bar is the size you want, release the mouse button. You should see small open circles, called handles, around the outside of the scroll bar.

10. With the handles still showing, move the cursor to the **Controls** group of the Ribbon and click on **Properties.** This brings up the **Properties** box containing a list of properties in alphabetical order. You will need to fill in three values: **LinkedCell, Max,** and **Min**. Linked cell refers to the cell where you have placed the value you want to change when you scroll. Point to the words **LinkedCell,** click, and type **C2** here. Point to **Min**, click, and enter **1**, and then point to **Max**, click, and enter **20**. Move the cursor to a

cell outside the **Properties** box and click. To close the **Properties** box, click the **x** in the upper-right corner of the box.

11. Click on **Design mode** in the **Controls** group (which should still be highlighted) to exit **Design mode**. You are almost ready to use the scroll bar you designed.

12. You need to change the scale on the y-axis of the scatterplot so the scaling is not automatic, but is fixed. On your finished graph, point to any number on the y-axis and right-click; then click on **Format axis**. In the **Axis Options** window, click on **Fixed** for **Minimum**; then enter **1000** in the **Minimum** box. Also click on **Fixed** for **Maximum and enter 15000** in the box. Click **Close**. (Fixing the scale will allow you to see how the graph changes as the scroll bar changes the value of the interest rate.)

13. You are now ready to move the little box on the scroll bar to increase or decrease the value of the interest rate. This will, in turn, alter the Total column of the spreadsheet and change the graph.

 b. Describe how the graph changes as the scroll bar moves.

4. You can also use a scroll bar to solve Josh's problem (see Example 10.8 in Topic 10). To do this, you will first create a graph of the monthly balance remaining, and then you can change the monthly payment using the scroll bar. Here is how:

 a. Go to a new sheet in your Excel workbook. Set up the new sheet as follows:

  i. In cell A1, enter the label **APR as a Decimal**, and in cell A2, enter **0.06**, the loan's APR (as a decimal) for the loan from the credit union.

  ii. In cell B1, enter the label **Monthly Interest Rate**, and in cell B2, enter **=A2/12**, the monthly interest rate. In cell C1, enter the label **Tentative Payment** and in cell C2, enter **$100**, a tentative payment amount (this will be "attached" to the scroll bar and moved until you find the payment needed to pay off the loan in 60 months).

  iii. In cells A4, B4, and C4, enter **Year**, **Balance Last Month**, and **Balance This Month**, respectively.

  iv. In column A beginning at cell A5, fill in numbers **1** through **60** (for the 60 months in 5 years).

v. In cell B5, enter **$13000** (this is the starting loan principal or balance at the end the month before you start payments).

vi. In cell C5, enter **=B5 + B5*B2 − C2** to calculate the loan balance after one month, that is, at the end of the first month.

vii. In cell B6, enter **=C5**.

viii. Drag to fill in columns B and C up to month 60.

ix. Create a graph that shows "balance this month" as a function of "year." Because you are setting up a scroll bar, you will want to format the y-axis so scaling is fixed.

x. Create a scroll bar with a **LinkedCell** of **C2** (so the payment can be changed by scrolling), with a **Min** of **0** and a **Max** of **1000**.

b. How would you know from the graph that the choice of monthly payment in C5 is the correct one (or a good approximation of it) to pay off the debt in five years? Write the correct amount here.

c. Use the same spreadsheet, changing the interest rate, principal, and compounded period as appropriate, to find all the information needed to solve Josh's problem. Explain how you found the information.

d. Identify which problem-solving techniques (see the list in Topic 10) you used to solve Josh's problem in this activity.

Summary

In this activity, you worked with several problem-solving techniques and looked for situations where specific techniques could be used. You learned how to create a scroll bar in Excel to see how the interest paid and the accumulated total in a savings account change for different annual interest rates. Finally, you used a scroll bar to solve the loan decision problem stated in Example 10.8.

10-2

Asking and Answering a Research Question

In this activity, you will investigate a research question of your choice.

1. Formulate a research question and a hypothesis about the answer to your question. Your question and hypothesis might involve comparing quantitative variables on two different groups, or it may involve looking for a relationship between two quantitative variables. For example, you could look at how much money is spent per student in a collection of southern big-city school districts and compare with per-student spending in a collection of northern big-city school districts. Or, you might look for data that change over time.

2. Collect at least **30** data points to test your hypothesis. Please choose a hypothesis and data that are of interest to you and for which you can find information on the web.

3. Carry out the steps of the problem-solving process, using each of the techniques discussed in this topic, as appropriate.

4. Write an essay that describes the process you used to formulate and answer your research question. In addition to describing the problem-solving process, your essay should also provide an analysis of your research. Your write-up should contain the following:

   - Your question and hypothesis, a table of your data, a description of why you chose the variables you did, and who might be interested in your results

   - A description of how and where you obtained your data, including all book, journal, and website references

   - Several different types of graphs, complete with labels and titles

   - A detailed description of what each of your graphs shows

- Where appropriate, the least-squares regression line and an explanation of how the line can be used

- Concluding paragraphs explaining how one might use the results of your analysis

Summary

In this activity, you used problem-solving techniques to help you formulate and answer a research question.

11

Decision Making

Tanning salons raise skin cancer risk, study indicates

Paul Recer,
Associated Press

Washington—Regularly baking to a golden tan under sun lamps can increase the risk
malignant melanoma, a sometimes fatal skin cancer, and the younger a woman starts
greater the risk, a study says.

P	Q	not P	not Q	P and Q	not (P and Q)	(not P) or (not Q)
T	T	F	F	T	F	F
T	F	F	T	F	T	T
F	T	T	F	F	T	T
F	F	T	T	F	T	T

State	Population	Standard C
Delaware	897,934	0.4084
New Jersey	8,791,894	3.9992
New York	19,378,102	8.8145
Pennsylvania	12,702,379	5.7779
Total	**41,770,309**	

Country	Per Capita Total Spending on Health (in U.S. dollars at average exchange rates)	Per Capita Daily Calorie Supply	Infant Mortality Rate per 1,000 Births
Australia	3,986	3,674	5
Cambodia	36	2,046	62
China	108	2,951	23
France	4,672	3,654	4
Germany	4,209	3,496	4
India	40	2,954	55
Italy	3,136	3,617	4

H ow do you decide what compact disc player or car to buy? Is cost your only consideration? How do you "factor in" reliability or appearance? Should your class trip involve a visit to an amusement park, where inclement weather might close the park on the day of the trip, or should you schedule an indoor concert instead?

Some decisions require us to use information about which there is some uncertainty, such as weather conditions. Other decisions involve information that we already know or that is readily available, such as the price per pound of produce. In Topic 21, we will investigate making decisions that involve uncertainty and probability considerations. In this topic, we look at decisions where the information impacting the decision is assumed to be known for certain or can be obtained. Such decisions are called **decisions under certainty** and include consumer purchase decisions such as buying a car.

After completing this topic, you will be able to:

- Determine when a decision is a decision under certainty and when it involves probability considerations.

- Decide what criteria are important for a variety of decision-making situations.

- Devise a system to rate alternative choices on different criteria.

- Apply two decision-making methods, the cutoff screening method and the weighted sum method, to help make decisions.

Example 11.1

For each of the following decision-making situations, determine if it involves a decision under certainty or not. Give reasons for your answers, but do not actually solve the problems.

a. José is in the fifth grade. His grandfather gave him $1,000 to invest for college, so he will invest it for seven years. He has one opportunity to invest it at a yearly interest rate of 8% compounded annually and, of course, he will put the interest back into the account. He has another opportunity to invest his $1,000 at a yearly interest rate of 7.5%, compounded daily with, again, the interest put back into the account. José must decide where to invest his money.

b. You are in charge of the hot-dog stand for home football games. Because home games are not played every week and because you cannot resell leftover food, you don't want to order too much. On the other hand, you want to service the customers who want refreshments. You need to decide how much food to order.

c. A prominent fast-food company has come up with a new kind of burger that it hopes will be very successful. The company's marketing executives need to decide whether to test-market the new burger in a limited area and then promote it nationally based on its success in the test market, or to promote it nationally without test marketing.

d. Amisha needs to buy a new pair of running shoes. She has found three pairs, each at a different store, that she likes equally well, and all are on sale. The first pair was originally $65. It is on a table marked 25% off, and she has a coupon for an additional 15% off of that sale price. The second pair originally cost $62 and is marked 1/3 off. The final pair was originally $69. It's marked 35% off and today, for one day only, everything at that store is marked down another 10% off of the sale price. Amisha must decide which pair of shoes to buy.

e. The manager of the sanitation crew in a large city must decide what routes to assign to the city's fleet of trash collection trucks in order to collect trash once per week at every home and business in the city.

Solution

The problems described in parts (a) and (d) of this example involve information that is known with certainty and so they involve decisions under certainty. In part (a) of this example, we can figure out exactly how much money José will have after seven years under each scenario because the interest rate is fixed. Amisha knows exactly how much each pair of shoes costs and can determine the final price for each. In part (e) of this example, if the manager assumes that the trucks are all operating so he knows how many trucks there are, it is a decision under certainty. Although this problem is a complicated one, the manager can find out where the streets and alleys and homes and businesses are and can (theoretically, at least) find the most efficient and economical routes to assign the city's trucks. In part (b) of this example, the weather is a factor that impacts sales and involves probability considerations, so the decision in part (b) of this example is not a decision under certainty. Forecasters can track weather patterns over the years and calculate the probabilities associated with various weather conditions on any particular day to help with this decision. Similarly, in part (c) of this example, how the new burger is received involves the uncertainty of consumer demand.

We will investigate some methods for making decisions under certainty. Many of these decisions require comparisons of cost, time, or quality of various alternative choices. The appropriate decision depends on the criteria, and we use various problem-solving techniques in our analysis.

Example 11.2

Assuming that cost is her only criterion (and she wants to minimize it), which pair of shoes should Amisha buy in Example 11.1(d)?

Solution

Because we are dealing with currency, we will round all computed amounts to the nearest cent. The first pair of shoes was $65 but is marked 25% off. Thus, she would pay 75% of $65 or $0.75 \times \$65 = \$48.75$. With her 15% off coupon, she would pay 85% of $48.75 for a final cost of $0.85 \times \$48.75 = \$41.44$. The second pair is marked 1/3 off of $62. Thus, she would only pay 2/3 of $62 for a final cost of $\frac{2}{3} \times \$62 = 41.33$. The third pair was originally $69 but is 35% off. Thus, she would pay 65% of $69 or $0.65 \times \$69 = \$44.85$. The additional 10% off gives the final cost of $\$44.85 - \$4.48 = \$40.37$. So assuming that minimizing cost is her only objective, Amisha should buy the third pair.

We will investigate how to incorporate criteria other than cost into the decision-making process. This process uses some ideas similar to those discussed when we considered ratings in Topic 8, but with a more personal slant.

Example 11.3

Name at least four characteristics or criteria you would want to evaluate or measure in some way before making a decision on each of the following "purchases":

a. A stereo set

b. What college to attend

c. A car

d. An exercise machine

e. A bicycle

Solution

The answers will vary because of individual values and tastes. Some possibilities are given.

a. For a stereo set: cost, reliability, sound quality, repair rate, availability of upgrades

b. For a college: location, size of school, academic reputation, programs available, and attractiveness of the campus

c. For a car: cost, size, gasoline consumption, availability, reliability, and location of dealer

d. For an exercise machine: cost, durability, type, size, and ease of use

e. For a bicycle: weight, comfort, looks, cost, and reliability

We now investigate how to incorporate multiple criteria into making a decision about which digital camera we should purchase.

Example 11.4

Suppose we want to purchase a digital camera and have four brands to consider. We collect information on the price and weight of each of these brands initially. Given the information in the following table, which brand might we choose to purchase and why?

Brand	Price	Weight
Kodak	$860	17 oz
Nikon	$887	14 oz
Olympus	$764	19 oz
Sony	$748	10 oz

Solution

If price and weight were the only criteria, assuming that we want a lightweight camera for ease of carrying, we would likely choose the Sony camera, which is the lightest and least expensive camera.

But what camera would we have decided to buy in Example 11.4 if the least expensive camera had not been the lightest? There are several methods that we can use when we are making a decision based on more than one criterion and the decision is not obvious. In the following examples, we discuss two of these: the cutoff screening method and the weighted sum method.

In the **cutoff screening method**, the decision maker predetermines a cutoff for each criterion. For example, a possible cutoff for the criterion of price is that the price of the camera be no more than $800. With a cutoff determined for each criterion, the decision maker then goes through the criteria one-by-one and eliminates any choices (brands of cameras in the camera example) that do not meet the required cutoff for that criterion. After all criteria have been checked, if one choice remains, then we have a decision. If more than one choice remains, the decision maker can see if he or she wants to make any of the cutoffs more restrictive or if there is another criterion to use to make a final decision. If all choices have been eliminated, then the decision maker must determine if any of the cutoffs can be relaxed or more possible choices included.

Example 11.5

Suppose we include three additional criteria on which to base our decision of which digital camera to buy. The four brands are rated also on print quality, next-shot delay, and flash range. Print quality was rated by comparing the output from the cameras, using a scale from 1 to 10, with 10 being excellent. The ratings and other information from the manufacturer appear in the following table:

Brand	Price	Weight	Print Quality Rating	Next-Shot Delay	Flash Range
Kodak	$860	17 oz	9	6 sec	10 ft
Nikon	$887	14 oz	10	8 sec	18 ft
Olympus	$764	19 oz	10	4 sec	10 ft
Sony	$748	10 oz	6	6 sec	8 ft

a. Using the cutoff screening method with the following cutoffs, decide which camera to buy. Suppose we want a camera that costs less than $900, weighs 1 pound or less, with a print quality of 9 or higher, a next-shot delay of no more than 6 seconds, and a flash range of at least 10 feet. Do any models meet all of these cutoffs? If no models meet the cutoffs, determine what we should do. If one or more models meet the cutoffs, what models are they?

b. Develop a table in which the dollar price of each camera is changed to a "price rating," with the lowest price camera receiving a "high" rating of 10. Set the price ratings of the other cameras relative to the least expensive one. Explain the rating system used.

Solution

a. We will take the criteria in the order listed and eliminate from consideration models that don't meet the cutoffs. The maximum cost of $900 does not help us eliminate any models from consideration. Requiring a weight of 1 pound or less leads us to eliminate the Kodak and Olympus models from consideration.

Brand	Price	Weight	Print Quality Rating	Next-Shot Delay	Flash Range
~~Kodak~~	~~$860~~	~~17 oz~~	~~9~~	~~6 sec~~	~~10 ft~~
Nikon	$887	14 oz	10	8 sec	18 ft
~~Olympus~~	~~$764~~	~~19 oz~~	~~10~~	~~4 sec~~	~~10 ft~~
Sony	$748	10 oz	6	6 sec	8 ft

Having a print quality of 9 or better eliminates the Sony model. Now only the Nikon is under consideration. We check to see if that model meets all the remaining criteria, and we see that it does not. The next-shot delay of the Nikon is 8 seconds, so no model meets all of our criteria. We can decide to look at additional models or relax one of the cutoffs. One possibility is that we might relax the 6 or less second next-shot delay criterion and decide that the Nikon is our best choice.

b. The Sony camera with a price of $748 receives a price rating of 10. Because the camera with the next highest price is within $30, we assign it a rating of 9. There is approximately a $100 price difference between the Olympus and Kodak models, so we assign a price rating of 6 to the Kodak model and a 5 to the Nikon model. Alternatively, we could assign one rating point per $20 (rounding) difference in price. Using that mechanism, we would assign the Kodak a price rating of 4 and the Nikon a price rating of 2. We will use the first approach and use the price ratings given in the following table:

Brand	Price Rating	Weight	Print Quality Rating	Next-Shot Delay	Flash Range
Kodak	6	17 oz	9	6 sec	10 ft
Nikon	5	14 oz	10	8 sec	18 ft
Olympus	9	19 oz	10	4 sec	10 ft
Sony	10	10 oz	6	6 sec	8 ft

In Example 11.5, we assigned a price rating so price is rated using a consistent system in which a higher rating indicates a preferred choice. There were two possible price-rating schemes described in the solution to the example, and there may be additional logical ways to rate the price. What is important is that the rating uses a consistent scale and that we use the same highest rating for all the criteria. We will also adopt a rating system for the weight, next-shot delay time, and flash range, so that a higher rating is the preferred choice and the rating system is consistent with a most preferred rating of 10.

Example 11.6

The following table gives a rating system, based on 10 as the highest, most preferred rating, for the criteria of weight, next-shot delay, and flash range. Explain the logic of the system.

Brand	Price Rating	Weight Rating	Print Quality Rating	Next-Shot Delay Rating	Flash Range Rating
Kodak	6	7	9	9	6
Nikon	5	8	10	8	10
Olympus	9	6	10	10	6
Sony	10	10	6	9	5

Solution

The lightest camera (the Sony) gets a 10 in the weight rating column. We give the next lightest camera (the Nikon) an 8 because there is a jump in weight between the Sony and Nikon cameras. We reduce the rating approximately one point for each 2 ounces of additional weight. For the next-shot delay criterion, the Olympus, with a 4-second delay, is rated 10. The cameras with 6-second delays are rated 9 and the camera with an 8-second delay rates an 8, a one-point rating reduction for each additional 2 seconds of delay. The rating of 10 for flash range is assigned to the Nikon camera, which has the greatest flash range. The Kodak and Olympus cameras are given a flash range rating of 6 (a loss of one rating point per 2 feet of distance in flash range) and the Sony is assigned a 5.

By setting up the ratings within each criterion, we have set the stage for our second decision-making method. We now look at how to use the **weighted sum method** to determine the best camera for our use. Our goal is to get a single numerical rating for each of the cameras, so we can compare them. We want to factor into our numerical rating how important each criterion is to us, relative to the other criteria. To do this, we assign an "importance factor," called a **weight** or **weighting factor**, to each of the following criteria: price, weight rating, print quality, next-shot delay rating, and flash range rating, by giving each a weight between 1 and 10. A weight of 10 is assigned to the criterion that is most important. The weights for the other criteria will be chosen based on how important these criteria are to the decision maker, relative to the most important criterion.

Suppose that print quality is the most important criterion to us as decision makers. Thus, we assign a weight of 10 to print quality. The next most important criterion is price, and it's almost as important as print quality, so we assign it a weight of 9. Next-shot delay is next in importance and gets a weight of 8. Finally, weight rating and flash range are assigned weights of 5, because they are less important to us. (Note that these weights might differ for different decision makers.)

After assigning weights to each of the criteria, we compute a **weighted sum** for each model of camera. To compute the weighted sum for a particular camera model, we multiply the weight for each of the five criteria by the rating of that criterion for that camera. Then we add these terms to get a weighted sum for that model. Because higher ratings and weights indicate preferred alternatives, the preferred choice is the one with the highest weighted sum. The next example illustrates this method.

Example 11.7

Use the weighting factors of 9, 5, 10, 8, and 5 for price, weight, print quality, next-shot delay, and flash range, respectively, to compute a weighted sum for each model of camera.

Solution

We compute the weighted sum for the Kodak model as follows:

$$\text{Sum}_{\text{Kodak}} = (\textbf{wt for price rating}) \times (\text{price rating}) + (\textbf{wt for weight rating})$$
$$\times (\text{weight rating}) + (\textbf{wt for print quality rating}) \times (\text{print quality rating})$$
$$+ (\textbf{wt for next-shot delay rating}) \times (\text{next-shot delay rating})$$
$$+ (\textbf{wt for flash range rating}) \times (\text{flash range rating})$$
$$= \textbf{9} \times 6 + \textbf{5} \times 7 + \textbf{10} \times 9 + \textbf{8} \times 9 + \textbf{5} \times 6 = 281$$

Next we find the weighted sum for the Nikon model:

$$\text{Sum}_{\text{Nikon}} = \textbf{9} \times 5 + \textbf{5} \times 8 + \textbf{10} \times 10 + \textbf{8} \times 8 + \textbf{5} \times 10 = 299$$

The weighted sums for the other models are computed in a similar manner and are displayed in the table below:

Brand	Price Rating	Weight Rating	Print Quality Rating	Next-Shot Delay Rating	Flash Range Rating	Weighted Sum
Weight for Weighted Sum	9	5	10	8	5	
Kodak	6	7	9	9	6	281
Nikon	5	8	10	8	10	299
Olympus	9	6	10	10	6	321
Sony	10	10	6	9	5	297

Using the weighted sum method, we choose the model with the highest weighted sum. For our choice of weighting factors, this method would lead us to choose the Olympus camera.

We could use the weighted sum method to rank the four cameras we chose by putting them in order according to the weighted sum. So using this system, the Olympus is ranked #1, followed by the Nikon, the Sony, and the Kodak, as #2, #3, and #4, respectively. In the next example, we see how the rankings might change using different weighting factors for the weighted sum.

Example 11.8

Use the ratings of the cameras as determined for Example 11.7, but change the weighting factors used for the criteria to 9, 10, 7, 7, and 5 for price, weight, print quality, next-shot delay, and flash range, respectively. Compute the weighted sum for each camera with these weighting factors and determine the ranking of the cameras from most preferred to least preferred.

Solution

We use the new weighting factors to compute the weighted sums as follows:

$$\text{Sum}_{\text{Kodak}} = \mathbf{9} \times 6 + \mathbf{10} \times 7 + \mathbf{7} \times 9 + \mathbf{7} \times 9 + \mathbf{5} \times 6 = 280$$
$$\text{Sum}_{\text{Nikon}} = \mathbf{9} \times 5 + \mathbf{10} \times 8 + \mathbf{7} \times 10 + \mathbf{7} \times 8 + \mathbf{5} \times 10 = 301$$
$$\text{Sum}_{\text{Olympus}} = \mathbf{9} \times 9 + \mathbf{10} \times 6 + \mathbf{7} \times 10 + \mathbf{7} \times 10 + \mathbf{5} \times 6 = 311$$
$$\text{Sum}_{\text{Sony}} = \mathbf{9} \times 10 + \mathbf{10} \times 10 + \mathbf{7} \times 6 + \mathbf{7} \times 9 + \mathbf{5} \times 5 = 320$$

Using these weights, the Sony is ranked #1, the Olympus is #2, followed by the Nikon and Kodak as #3 and #4, respectively.

Examples 11.7 and 11.8 show that using different weights can change the weighted sum and thus change the ranking. When using the weighted sum method or when using a weighted sum to rank different choices, it is important to have a clear rationale for why we choose the weights we use. When considering rankings and ratings that have been set up by someone else, it is also important to understand how they were determined.

Summary

In this topic, we studied two methods we can use to help us make decisions that involve "certain" information—that is, information that is known or that we assume to be known. These methods are the cutoff screening method and the weighted sum method. We discussed various criteria that might be important to consider for making different decisions, and we also investigated how to create consistent ratings of possible choices, relative to each criterion. Finally, we looked at the link between the weighted sum method and rankings of various alternatives.

Explorations

1. For each situation described below, determine if it involves a decision under certainty or not.

    a. Tyrell has a new job and has to decide between two medical insurance plans. If he chooses the first plan, the employer pays for the full cost of insurance, but Tyrell will have a copayment of $20 for every visit to the doctor's office and a copayment of $50 for each emergency room visit. If he chooses plan B, then he will need to pay $100 per month, but will have no copayment when he visits a doctor or the emergency room.

b. Michelle wants to celebrate her son's tenth birthday and asks him if he prefers to go to an amusement park for a day and take one friend, or have a sleep-over party with six friends. Michelle's son does not have a special preference, so Michelle will choose the option that is least expensive.

c. You have a list of groceries you need to buy this week and have access online to brands and prices at each of three stores in your area. You need to decide where to shop this week for your groceries.

2. Name four characteristics or criteria you would want to evaluate or measure in some way before making a decision to purchase each of the following:

a. A bike helmet

b. A fax machine

c. A microwave oven

d. A car

3. You are considering the purchase of a computer and have consulted several computer magazines that contain ratings and prices for the five models you are investigating. You decide to include price, speed, and expansion capabilities as three criteria on which you will judge the models under consideration. The ratings for speed and expansion (with 10 being the highest rating) and the prices are summarized in the following table:

Brand	Price	Speed Rating	Expansion Rating
Dell	$949	8	10
NEC	$859	8	8
Gateway	$729	6	6
Hewlett-Packard	$1,129	10	8
Sony	$1,200	8	6

a. One model **dominates** a second model if it is better in all criteria than the second model. Do any of the models given in the table dominate another model? If so, which one dominates and which one is dominated? Explain how you would use this information.

b. Change each value in the price column to a relative rating based on a highest rating of 10. Explain the system used for this rating.

c. Assign a weighting factor to each of the three criteria of price, speed, and expansion to use the weighted sum method for making a decision. Give a justification for your choices.

d. Use your weighting factors to find the weighted sum for each of the five models. Which model would you choose to buy, based on your weighted sum?

e. What other criteria might you want to include before you decide which computer to buy?

4. Use weights of 5, 10, and 7 for price, speed, and expansion, respectively, in Exploration 3 to find the weighted sum of each model and determine which model is preferred using the weighted sum method.

5. An election for one of your state senators is approaching and you need to choose one of three candidates: Candidate A, Candidate B, or Candidate C. You decide to rate the candidates on three criteria based on the candidate's views on social issues, job creation, and entitlement reform. The following table contains your ratings on a scale from 1 to 10:

	Social Issues	Job Creation	Entitlement Reform
Candidate A	9	6	4
Candidate B	8	7	7
Candidate C	3	9	8

a. You want a candidate with a rating of at least 5 on social issues, and at least 6 on each of the other two criteria. Using the cutoff screening method with those cutoffs, decide which candidate will get your vote.

b. Assign a weighting factor to each of the three criteria and use the weighted sum method to choose who will receive your vote. Explain why you chose these weighting factors.

c. What additional criteria would you use to make your voting decision?

d. Use weights of 4 for social issues, 6 for job creation, and 10 for entitlement reform to find the weighted sum for each candidate and determine which candidate is preferred.

6. Consider the following table that lists the price for each of four models of clothes dryers and rates drying performance and ease of use based on a high score of 10:

Model	Maytag	Frigidaire	Hotpoint	General Electric
Price	$629	$494	$319	$539
Performance rating	10	8	9	8
Ease-of-use rating	8	8	6	9

a. If you were to use the cutoff screening method to make a decision about which dryer to choose, what cutoffs would you choose for each of the criteria?

b. Choose the preferred model based on the cutoffs you identified in part (a) of this Exploration. Explain how you arrived at your choice.

c. Explain why it is not appropriate to use the price values as given in the table in a weighted sum.

d. Assign a weighting factor to each of the three criteria of price, performance, and ease of use, to use for the weighted sum method for making a decision. Give a justification for your choices.

e. Use your weighting factors from part (d) to find the weighted sum for each of the four models. Which model would you choose to buy, based on the weighted sums?

7. Consider the values given in the following table for a selection of countries:

Country	Per Capita Total Spending on Health (in U.S. dollars at average exchange rates)	Per Capita Daily Calorie Supply	Infant Mortality Rate per 1,000 Births
Australia	3,986	3,674	5
Cambodia	36	2,046	62
China	108	2,951	23
France	4,627	3,654	4
Germany	4,209	3,496	4
India	40	2,459	55
Italy	3,136	3,671	4
Kenya	34	2,090	64
Mexico	564	3,145	17
Poland	716	3,375	7
South Korea	1,362	3,058	4
Spain	2,712	3,371	4
Turkey	465	3,357	28
United States	7,285	3,774	6

Sources: The New York Times Almanac 2011, pp. 493–495, and World Resources Institute, 2007, EarthTrends: Environmental Information, http://earthtrends.wri.org.

a. Assign weights to the three given social indicators (per capita total spending on health, daily calorie supply, and infant mortality rate) to specify their importance in contributing to a social index for a country. Justify your choices.

b. Set up and describe a way to use the given social indicators and your weights from part (a) of this Exploration to give a single social index for each of these countries. Justify your choices and explain what the rating shows.

c. What other variables might you include in such an index?

8. Consider the social index for countries presented in Exploration 7.

a. How sensitive is your social index to the choice of weights you assigned for each of the three social indicators? (Pick a different set of weights and compute the social index to help answer this question.)

b. How sensitive is your social index to the choice of relative ratings you assigned within each of the social indicators? How did you determine this?

9. Suppose you are considering a job offer in each of four cities. The cities and the salary offered for each job are given in the following table. You don't want to make your decision based solely on salary, but want to consider the "livability" of the city as well. Climate is one factor to consider.

City	Salary	Climate		
Seattle	$46,000			
Orlando	$41,000			
Chicago	$49,000			
Philadelphia	$44,000			

a. Identify two other criteria on which you will base your decision and fill in the table, assigning a relative rating to each city for each criterion.

b. Use the weighted sum method to make a decision about which job to choose.

10. The entertainment industry uses rating schemes and critics' judgments to boost sales and create advertisements. For each of the following, decide on at least four criteria you would use to set up a rating scheme. Then pick three specific examples for each and rate them on the four criteria. Finally, use one of the methods discussed to decide which gets the highest overall rating.

a. Newly released movies

b. First-run television shows

      c. Music videos

      d. Compact disc recordings of music

      e. Fast-food restaurants

11.   Suppose you used a scale of 1 to 100 to assign the weights to rank each of the criteria important to a decision (instead of the 1-to-10 system used previously). Do you think the results of the decision might change? Experiment with this scale using the data in Example 11.6 to help answer the question.

11-1

Ranking Cities: Ratings and Decisions

You will practice using several methods for making decisions that involve analyzing characteristics of possible alternative choices. These methods can be used for a variety of decisions and ratings, from consumer purchases to rating cities or films.

The following table gives information about twelve U.S. cities. Suppose you are considering job offers in each of these cities and want to decide, on the basis of the characteristics presented in the table (population, average January daily temperature, serious crimes per 100,000 population, percent unemployed, per capita income, and geographic region), which city you would most prefer. Assume that the jobs are fairly similar, so only the characteristics of the cities will influence your decision.

City	Population	Average Daily Temperature, January (°F)	Crimes per 100K	Unemployed (%)	Per Capita Income ($)	Region
Atlanta	536,472	41	6,812.8	9.4	35,453	Southeast
Boston	644,064	28.6	4,106.3	6.2	31,856	Northeast
Charlotte	797,733	39.3	4,963.1	10.3	30,984	Southeast
Dallas	1,306,775	44.6	5,608.2	7.7	26,716	South
Honolulu	950,268	71.4	4,498.2	5.3	38,770	South Pacific
Minneapolis	385,704	11.8	5,798.0	5.5	29,551	North Central

New York City	8,336,002	31.5	2,256.5	8.2	30,498	East
Omaha	464,628	21.1	4,220.2	4.7	26,123	Central
San Francisco	818,594	51.1	4,655.8	8.5	45,478	West
Seattle	620,195	40.1	5,917.7	7.8	40,868	Northwest
Phoenix	1,544,427	53	4,492.7	7.9	24,460	Southwest
Washington, DC	601,723	34.6	5,751.2	5.5	42,078	East

Sources: The Federal Bureau of Investigation, www.fbi.gov; U.S. Census Bureau, www.census.gov; and U.S. Bureau of Labor Statistics, www.bls.gov.

1. Without examining the data in the table closely, based on your impression of the cities, which one do you think you'd choose and why?

2. Suppose you decide on the following cutoffs for the characteristics: population, no more than 600,000; average January temperature, at least 25°; crimes per 100,000, less than 8,000; unemployed, 4.5% or less. Use the cutoff screening method with these cutoffs, and additional characteristics and cutoffs if you need any, to determine which city you would choose. Describe your reasoning.

3. Retrieve the data set containing the previous table, "EA11.1 Cities.xls," from the text website or WileyPLUS. In the next several steps, you'll look at the given characteristics one-by-one, and set up a rating system for each, based on your preferences.

a. For the characteristic of population, set up a rating system in which *your* most preferred rating receives a 10 and the least preferred gets a 1. (Your ratings are based on your personal preferences, which may be different from your neighbor's or instructor's preferences.) Record these ratings for each city in the corresponding row of column H of the spreadsheet and enter an appropriate heading for column H. Explain the rationale for your rating system.

b. For the characteristic of average January temperature, set up a rating system based on a system in which your most preferred rating receives a 10 and the least preferred gets a 1. Record these ratings in column I of the spreadsheet and enter an appropriate heading for column I. Explain the rationale for your rating system.

c. Choose three of the four remaining characteristics that appear in the table and that you feel could influence your decision of which city to choose. List these characteristics here.

d. For each of the characteristics you just listed in part (c), set up a rating system in which the most preferred rating receives a 10 and the least preferred gets a 1. Record these ratings in columns J, K, and L of the spreadsheet. For each of these columns, include an appropriate column heading. Explain the rationale for each of your rating systems.

e. Now look at the characteristics on which you will base your decision. These characteristics include population, average January temperature, and the three additional characteristics you chose. Assign weights to these characteristics, using a scale of 1 to 10, with 10 being the most important characteristic for you. Record your five characteristics and corresponding weights here, and explain why you picked the weights you did.

    i.

    ii.

    iii.

    iv.

    v.

f. You are now ready to calculate the weighted sum of the ratings for each city. For each city, you will calculate: (weight) × (assigned rating) for each characteristic and then add these. Do this using the following instructions, and then write your result for Atlanta's weighted sum. _____

Instructions to Calculate the Weighted Sum

1. In your spreadsheet, enter the weights you picked for each characteristic in part (e) in row 15 of the corresponding columns that contain the ratings, that is, columns H, I, J, K, and L. In cell G15, enter the label **weights=** .

2. Name each of the cells containing the weights using an appropriate name. (Activity 8.2 described how to do this by using the **Name box** below the **Clipboard** group on the spreadsheet.) For example, you could name the value in cell H15 **popwt**, with no spaces in the name.

3. In cell M2, enter a formula such as the following, depending on the names you chose for the named cells, for the weighted sum ranking for Atlanta:
= popwt*H2+tempwt*I2+crimewt*J2+unempwt*K2+incomewt*L2

4. Drag the formula down to find the total weighted sum for each city in the list. (Note that the named cells for the weights do not change when you drag down the formula.)

g. List the city with the highest weighted sum and the city with the lowest weighted sum. (Before you answer, you may want to sort the data by "weighted sum." To do this, highlight all columns, then click on the **Home** tab, go to the **Editing** group, and click on the down arrow next to "Sort & Filter." Click on **Custom Sort** and Sort by **Column M.**) Are you surprised by these rankings?

h. List at least three additional characteristics you might want to include in a ranking of desirable places to live.

4. For what other types of decisions might you use an analysis such as the one you carried out for the cities?

5. Discuss the advantages and disadvantages of the weighted sum method for helping you make decisions.

Summary

You investigated two methods for making decisions and also explored how to rate cities on various criteria and how to rank the cities using those ratings. You also learned how to decide on weighting factors for the criteria and used Excel to compute the weighted sum. Finally, you practiced naming a cell and entering a formula in Excel.

Inductive Reasoning

Tanning salons raise skin cancer risk, study indicates

Paul Recer,
Associated Press

Washington—Regularly baking to a golden tan under sun lamps can increase the risk o
malignant melanoma, a sometimes fatal skin cancer, and the younger a woman starts th
greater the risk, a study says.

P	Q	not P	not Q	P and Q	not (P and Q)	(not P) or (not Q)
T	T	F	F	T	F	F
T	F	F	T	F	T	T
F	T	T	F	F	T	T
F	F	T	T	F	T	T

State	Population	Standard Qu
Delaware	897,934	0.4084
New Jersey	8,791,894	3.9992
New York	19,378,102	8.8145
Pennsylvania	12,702,379	5.7779
Total	**41,770,309**	

Country	Per Capita Total Spending on Health (in U.S. dollars at average exchange rates)	Per Capita Daily Calorie Supply	Infant Mortality Rate per 1,000 Births
Australia	3,986	3,674	5
Cambodia	36	2,046	62
China	108	2,951	23
France	4,672	3,654	4
Germany	4,209	3,496	4
India	40	2,954	55
Italy	3,136	3,617	4

Reasoning is an essential activity of the human brain. We use reasoning to draw conclusions daily. A conclusion is useful as part of human knowledge only when it has been obtained from valid reasoning. In this topic and the next, we discuss the two major forms of reasoning: inductive reasoning and deductive reasoning.

Inductive reasoning argues from particular cases to a general rule; **deductive reasoning** argues from general cases to particular cases. We use inductive reasoning when we draw a general conclusion from experiments or particular observations. The truth of a conclusion obtained through an inductive argument from valid premises is, at best, highly likely to be true, but not necessarily true. Deductive reasoning is used when conclusions are made through logical inference from the premises. A conclusion obtained through valid deductive reasoning from true premises is necessarily true.

After completing this topic, you will be able to:

- Distinguish between inductive and deductive reasoning.

- Recognize and use different forms of inductive reasoning.

- Decide whether or not a conclusion reached through inductive reasoning is valid.

- Identify assumptions made when using inductive reasoning and identify when a conclusion reached through valid inductive reasoning might be false.

Example 12.1 ■ ■ ■ ■ ■ ■ ■ ■ ■

Decide whether each of the following situations describes inductive or deductive reasoning—that is, decide if the conclusion must follow from the premises or not, assuming the premises are true. In each case, decide if the conclusion is reasonable from the information given and state what assumptions are made to draw the conclusion.

a. My fourteen-year-old brother likes to play the guitar. My cousin, who is fifteen, also likes to play the guitar. Most teenagers like to play the guitar.

b. Thirty-four million Americans wear contact lenses and 85% of them choose soft contacts. Because 85% of 34 million is 28,900,000, we can conclude that almost 29 million Americans wear soft contact lenses.

c. A British study followed 9,000 women through pregnancy and after childbirth. Researchers recorded symptoms of depression at 18 and 32 weeks of pregnancy and at 8 weeks and 8 months after childbirth. These depression symptom scores were compared with depression scores for women at other times of their lives. The researchers concluded that symptoms of depression were not more common during pregnancy and after childbirth (postpartum) than at other times in a woman's life. (*Source:* BMJ Group, www.bmj.com)

Solution

a. This is an example of inductive reasoning. We drew a conclusion after observing two particular cases. This conclusion does not seem to be valid, however, because two teenagers from the same family can hardly be representative of all teenagers. The (faulty) assumption here is that these two cases are sufficient to make a general statement about the whole group.

b. This paragraph shows a case of deductive reasoning. From the fact that 85% of the 34 million Americans who wear contact lenses wear soft contact lenses, it follows that there are $0.85 \cdot 34,000,000 = 28,900,000$ Americans who wear soft contact lenses. In turn, this means that almost 29 million Americans wear soft lenses. The conclusion is not only reasonable, it is definitely true. We are assuming that the information given is true—that 85% of those who wear contact lenses wear soft lenses and that 34 million Americans wear contact lenses.

c. The researchers used inductive reasoning to conclude that there is no such thing as "postpartum depression," but that depression occurs among women who have given birth recently just as frequently as among those who have not given birth recently. The conclusion is a reasonable one to make, assuming that the data collected are accurate, there are no biases in the selection of the 9,000 women, and that eight weeks after delivery is close enough to the delivery date so that any case of depression after delivery would still be present then. (Not everyone agrees with this last assumption. In fact, some experts say the number of women suffering from depression might be higher because some cases might have been missed in the study where the first measure was done after eight weeks.)

Forms of Inductive Reasoning

There are four main types of inductive reasoning: prediction, generalization, causal inference, and analogy. A **prediction** is a form of inductive argument that concludes with a claim about what will happen in the future, based on past or present observations. Financial analysts and weather forecasters often use this form of reasoning. The following, from the article "NOAA Hurricane Outlook Indicates an Above-Normal Atlantic Season" issued from the National Oceanic and Atmospheric Administration on May 19, 2011, is an example of a prediction:

> Across the entire Atlantic Basin for the six-month season, which begins June 1, NOAA is predicting the following ranges this year:
>
> - 12 to 18 named storms (winds of 39 mph or higher), of which:
> - 6 to 10 could become hurricanes (winds of 74 mph or higher), including:
> - 3 to 6 major hurricanes (Category 3, 4, or 5; winds of 111 mph or higher)
>
> Each of these ranges has a 70 percent likelihood; these indicate that activity will exceed the seasonal average of 11 named storms, six hurricanes and two major hurricanes.

As with any prediction, there is no absolute certainty that there will be more storms than last year. But weather experts are coming to this conclusion based on their experience and their interpretations of the computer models.

A second form of inductive reasoning is **generalization**. This occurs when a conclusion is drawn about a whole class or group based on the knowledge of some cases from that group. Studies that make conclusions from sampling methods use generalization. (Sampling is discussed in more detail in Topic 17.) A 2009 nationwide survey by the National Highway Traffic Safety Administration (NHTSA) estimates that at any given time of the day, 9% of drivers are using a handheld phone. This means that approximately 672,000 vehicles on the road at any one time are driven by someone on a handheld phone. The survey also revealed that 10% of drivers in the 16- to 24-year-old-group were observed visibly manipulating handheld electronic devices while driving. (*Source:* National Highway Traffic Safety Administration, www.nhtsa.dot.gov.)

Here the conclusion that, at any given time, 9% of those who are driving are at the same time using a handheld phone, is based on the observation of 43,000 drivers. This is a sample that does not include all American drivers. As with most cases of inductive reasoning, the conclusion is not necessarily true, but is highly probable, assuming the study was well designed.

Another form of inductive reasoning is **causal inference**, in which a conclusion is made about the cause of some situation when only the result is known. This form of reasoning is often used in daily life. The following example is excerpted from the story "Echo Guilt," from *Tell Me Everything and Other Stories*, by Joyce Hinnefeld:

> One day in the summer my husband, Jack, found a frog at the edge of our vegetable garden, in the back yard behind this house that we've rented, three

hours north of New York City on a steep slope up from the Hudson River. Convinced that this was an animal that had somehow strayed too far from its native environment—surely the river—Jack teased the frog into a paper bag.

In this passage, Jack's conclusion that the frog has come from the river is very likely and perhaps the most reasonable conclusion to make. Of course, this is highly probable, but it is not necessarily the case. It could be, for example, that someone who lives nearby bought it at a pet store and somehow the frog escaped.

A fourth form of inductive reasoning occurs when we make a conclusion about something (events, people, objects) because of its similarity with other things. This is called reasoning by **analogy**. We use this form of reasoning when we interpret something unknown by relating it with something we know with similar characteristics. In September 2007, a local newspaper reported that the U.S. Coast Guard received dozens of calls from people from the New Jersey shore to South Carolina describing a burning aircraft crash along the coast. Because the Federal Aviation Administration had no reports of missing aircraft and searches of the coast showed no evidence of a crash, experts concluded that what the residents saw was most likely part of a meteor shower.

Those who thought they saw a burning airplane used analogy to arrive at that conclusion. They saw a burning ball of fire falling from the sky, similar to the ones they have probably seen in the movies many times that represent falling airplanes. Because of the similarity, they concluded it must have been a falling airplane. In this case the analogy did not work; that is, the conclusion was not correct.

Example 12.2   ■ ■ ■ ■ ■ ■ ■ ■ ■

Each of the following scenarios contains inductive reasoning. Decide what type of inductive reasoning each uses and explain how you arrived at that decision.

a. Here are excerpts from a December 2008 report issued by the National Center for Education Statistics of the U.S. Department of Education (www.nces.ed.gov). It summarizes results from a survey on homeschooling.

> Since 1999, the National Household Education Surveys Program (NHES), conducted by the U.S. Department of Education's National Center for Education Statistics (NCES) in the Institute of Education Sciences, has collected nationally representative data that can be used to estimate the number of homeschooled students in the United States.

> Data were collected for students ages 5 through 17 with a grade equivalent of kindergarten through 12th grade. Interviews were conducted with the parents of 10,681 students, including 290 homeschooled students.

Data from the 2007 NHES survey show an estimated 1.5 million students (1,508,000) were homeschooled in the United States in the spring of 2007. This represents an increase from the estimated 1.1 million students who were homeschooled in the spring of 2003. In addition, the percentage of the school-age population that was homeschooled increased from 2.2 percent in 2003 to 2.9 percent in 2007. Data from the 1999 NHES showed an estimated 850,000 homeschooled students in the United States—about 1.7 percent of the school-age population. The increase in the homeschooling rate (from 1.7 percent in 1999 to 2.2 percent in 2003 to 2.9 percent in 2007) represents a 74 percent relative increase over the 8-year period and a 36 percent relative increase since 2003.

As with the results from any sample survey, the numbers and percentages discussed in this Issue Brief are estimates of the actual numbers and percentages of homeschooled students in the population. Although 1.5 million is the best estimate of the number of homeschoolers from the 2007 NHES, another similar sample survey might produce a different estimate.

b. The following excerpt is from a report released in December 2009: (*Source:* U.S. Bureau of Labor Statistics, www.bls.gov)

EMPLOYMENT PROJECTIONS: 2008−18
Total employment is projected to increase by 15.3 million, or 10.1 percent, during the 2008−18 period, the U.S. Bureau of Labor Statistics reported today. The projections show an aging and more racially and ethnically diverse labor force, and employment growth in service-providing industries. More than half of the new jobs will be in professional and related occupations and service occupations. In addition, occupations where a postsecondary degree or award is usually required are expected to account for one-third of total job openings during the projection period. Job openings from replacement needs—those which occur when workers who retire or otherwise leave their occupations need to be replaced—are projected to be more than double the number of openings due to economic growth.
The projected growth for the 2008—18 period is larger than the increase of 10.4 million over the 1998−2008 period, or 7.4 percent. The relatively slow growth rate for the earlier 10-year period was affected by the recession which began in December 2007, and the projected growth rate is higher than would otherwise be expected because the 2008 starting point is a recession year.

c. A genetics researcher at the National Cancer Institute, Dr. Amar Klar, studied handedness in humans. He noted that in most mice, the heart is on the left side, but in some mutant mice, the heart is on the right side. In work with these mutant mice, scientists have found that normal mice have a particular gene that is absent in the mutant mice. If two mutant mice have offspring, about 50% of them have hearts on the right while the rest have hearts on the left. Dr. Klar theorized that a similar mechanism could work in humans to explain why identical twins can have different hand preferences. "Dr. Klar's explanation is that these twins lack the right-handed gene, and each of them has an equal chance of being right-handed or left-handed." (*Source:* "On Left-Handedness, Its causes and Costs," *New York Times,* May 16, 2000)

d. From the story "The Speckled Band," in *The Adventures of Sherlock Holmes,* by A. Conan Doyle:

> "You have come in by train this morning, I see."
>
> "You know me then?"
>
> "No, but I observe the second half of a return ticket in the palm of your left glove. You must have started early, and yet you had a good drive in a dog-cart, along heavy roads, before you reached the station."
>
> The lady gave a violent start, and stared in bewilderment at my companion.
>
> "There is no mystery, my dear madam" said he, smiling. "The left arm of your jacket is spattered with mud in no less than seven places. The marks are perfectly fresh. There is no vehicle save a dog-cart which throws up mud in that way, and then only when you sit on the left-hand side of the driver."

Solution

a. The conclusion about the number of children being homeschooled is obtained from a sample survey of 10,681 students. One type of induction used is generalization, which allows the surveyors to draw a conclusion about the total number of homeschooled children in the general population. (Deductive reasoning was used to draw conclusions about the increases in the home schooling rate.)

b. The reasoning used here is a prediction. Based on an analysis of the job market, economic conditions, and population demographics, analysts at the U.S. Bureau of Labor Statistics predict how many new jobs there will be and in what general occupation classes.

c. From the observation that the heart is on the right side in mice who are missing a certain gene, Dr. Klar concludes that there is a specific gene present in all right-handed persons. He is using analogy to draw this conclusion.

d. Here, Sherlock Holmes uses inductive reasoning in the form of causal inference to conclude that his visitor has traveled by dog-cart and train and has arrived recently. He arrives at this conclusion after observing a return train ticket in the lady's hand and fresh mud stains on her clothes.

■ ■ ■ ■ ■ ■ ■ ■ ■ ■ ■

Inductive reasoning leads to conclusions that are likely to be true, but might not be true. In Example 12.3, we will consider why specific conclusions might not be true.

Example 12.3

For each of the scenarios described in Example 12.2, identify what premises are used and explain why it is possible that the conclusion might be false.

Solution

a. The premises used here are that the households surveyed are a good sample of the general population and therefore the information obtained from them can be generalized. The conclusion could be false if the households surveyed contained, for example, a disproportionately large number of households with college-educated adults. This might result in too high an estimate of homeschooled children. (The report acknowledged that another similar sample survey might produce a different estimate.)

b. One premise is that the analysis is accurate and that conditions will remain generally as they are in the analysts' assumptions. The prediction might not be true if a drastic, unforeseen change occurs in these conditions.

c. One premise is that there is a gene in humans similar to the one found in the normal mice. As with any analogy, the conclusion might be false. Although there may be some similarities between two objects or situations, it does not necessarily follow that they are identical in all aspects. In this case, there are similarities in the basic biology of mice and humans, but we know also that many differences exist. For this reason, the researcher can be sure of his theory only if he identifies the human gene he suspects exists.

d. Two premises are that the ticket is current and that the mud was not planted as a diversion. This conclusion could be false if, for example, the lady's ticket was an old one that she just found in her purse or the mud stains were perhaps purposely made to give the impression she had recently traveled by dog-cart.

In the previous example, we saw that even with good inductive reasoning the conclusion may not be true. Bad inductive reasoning often leads to false conclusions. A generalization based on two specific people from the same family, as in Example 12.1(a), does not result in a valid conclusion. In the next example, we will look at other cases of inductive reasoning in which the conclusion is not based on sound reasoning.

Example 12.4

For each of the following scenarios, describe the fallacies in the reasoning:

a. Since it has rained for two full days, I predict that tomorrow will be a beautiful, sunny day.

b. Smoking can't be as bad as they say it is. My aunt Margaret, who smokes a pack of cigarettes a day, is 80 years old and still in good health.

c. As reported in *The New York Times*, July 29, 2001, the number of head injuries among bicycle riders has increased 10% since 1991, while the use of helmets has increased considerably; therefore, I will not use my helmet anymore because it seems that using helmets increases the chances of experiencing a head injury.

Solution

a. This seems more like a wish than a prediction. A prediction must be based on some relevant evidence or study. To predict whether it will be sunny or not, we must analyze the atmospheric conditions. It is quite possible to have three or more consecutive rainy days, so the fact that it has already rained for two days is not a reason to think that it will be sunny tomorrow.

b. The reasoning is invalid in this case because we are using one individual case to draw a general conclusion about the effects of smoking in the human body. In this case, we do know that this conclusion is wrong because there is much scientific evidence, based on valid inductive reasoning, of the health risks of smoking.

c. This is a misuse of the causal inference form of reasoning. Although the two facts are occurring together, there is no reason to believe that one is causing the other. (According to the newspaper article, some experts believe people are assuming riskier behaviors because of the sense of security a helmet provides.)

As we saw in several examples, inductive reasoning is sometimes done by sampling a population, analyzing the results from the sample, and then concluding that these results will hold for the whole population. This can be done through an observational study or by an experiment. In an **observational study**, individuals are observed and some variable or variables of interest are measured; the researcher does not attempt to influence the responses. An **experiment** is carried out when the researcher deliberately imposes some treatment on individuals in order to observe their responses. Note that the term *study* is a general term used to describe either an observational study or an experiment.

Example 12.5

For each scenario, determine if the study described is an observational study or an experiment. Identify the explanatory and response variables in the study and determine if it is possible to conclude that there is a cause and effect relationship between the explanatory and response variables.

a. An article released by WebMD Health News (www.webmd.com) on August 10, 2010, reports on a Danish study that was published in the *American Journal of Clinical Nutrition* in September 2010. The study showed that pregnant women who consume artificially sweetened soft drinks may be at an increased risk for an early preterm delivery. "We observed a positive association between the intake of artificially sweetened soft drinks and the risk of preterm delivery, [but] no association was observed for sugar-sweetened soft drinks," conclude study researchers. The researchers followed 59,334 pregnancies for a six-year period of time. They found that women who drank at least one diet soda per day were 38% more likely to have their baby early compared with women who drank no diet sodas. Women who drank at least four diet sodas each day were 78% more likely to have their baby early. (Preterm delivery is defined as giving birth before 37 weeks of pregnancy.)

b. An article in the *British Journal of Sports Medicine*, and accessed on the National Institutes of Health website (www.ncbi.nlm.nih.gov), reported on a study to test how well a tart cherry juice blend performed in preventing the symptoms of exercise-induced muscle damage. During the study, fourteen male college students drank 12 ounces of a cherry juice blend or a placebo, twice per day for eight consecutive days. On day 4, the study participants performed a series of elbow contraction exercises. Strength, pain, muscle tenderness, and relaxed elbow angle were recorded before and for four days after the exercises. The experiment was repeated two weeks later with subjects who took the placebo initially now taking the cherry juice, and subjects who took the juice taking the placebo. They used the opposite arm for the exercise for the second round of exercises. Results showed that strength loss and pain were significantly lower when cherry juice was consumed compared to the placebo. Relaxed elbow angle and muscle tenderness were not significantly different.

Solution

a. The Danish study is an observational study. The researchers observed how many diet sodas the women consumed but did not impose a treatment. The explanatory variable is number of diet sodas and the response variable is the number of weeks of gestation before delivery. Although the researchers concluded there was a positive association between intake of diet sodas and risk of preterm delivery, they could not conclude cause and effect. There may be other reasons for such an association. For example, it could be that more of the diet soda drinkers were smokers, which is also associated with a higher incidence of preterm delivery, or the diet soda drinkers could have been deficient in some critical vitamin, like vitamin D.

b. This study is an experiment in which half of the participants were given fake cherry juice (a placebo) and half were given real cherry juice, and then the experiment was repeated.

The explanatory variable is the cherry juice or placebo, and the response variables include elbow strength, pain level, muscle tenderness, and relaxed elbow angle. Although the number of study participants was small, the study appears to be well planned and executed because both treatments were given to each subject. It is reasonable to conclude that the cherry juice caused less strength loss and less pain in the exercisers. It is still possible that there are other influential factors and the conclusion might be wrong. This is why experiments like these are often followed by a similar study that includes a larger number of subjects.

Summary

In this topic, we discussed the differences between inductive and deductive reasoning and noted that correct deductive reasoning always leads to a true conclusion, but a conclusion reached through inductive reasoning is not necessarily true. We also analyzed different forms of inductive reasoning—prediction, generalization, causal inference, and analogy—and investigated fallacies in inductive reasoning.

Explorations

1. The following passage is taken from *L is for Lawless* (New York: Henry Holt, 2005, p. 237), one of the books in a mystery series by Sue Grafton. In this scene, two characters are discussing something that had happened years ago. A man had pulled a bank robbery and was arrested some time later, but without the cash and the jewels he stole. In this passage, identify where deductive reasoning is used and where inductive reasoning is used and explain how they are different.

   "Unless he had time to go to some other town and come back," I said. "It's like saying you always find something the last place you look. I mean, it's self-evident. Once you find what you are looking for, you don't look any place else. The last you saw him, he had the sacks full of cash. By the time he was arrested, they were gone.

   Therefore, the money had to have been hidden some time in that period. By the way, you never said how long it was."

   "Half a day."

   "So he probably didn't have time to get far."

   "Yeah, that's true. . . . "

2. Consider each of the following reasoning scenarios. For each, identify if inductive or deductive reasoning is used, determine if the conclusion is reasonable from the information given, and state what assumptions are made to draw the conclusion.

a. Arrest rates in large cities in the United States for certain offenses such as disorderly conduct, drunkenness, and vagrancy have been declining. This shows that America's large cities are becoming more peaceful.

b. An advertisement for a particular brand of fruit bar claims that mothers will feel good about feeding it to their children because it is made from "real fruit juice."

c. I go to the store with $5.00 in my wallet. I want to buy 7 pounds of bananas that cost $0.49 per pound. I will have enough money also to buy a small candy bar.

d. The majority of Americans prefer football over other sports. This conclusion was based on a survey taken by middle school students of their peers. Their survey showed that approximately 65% preferred football as their favorite sport.

e. Reports of the price of crude oil indicate that the price has fallen by $30 a barrel from April 2011 to August 2011, closing on August 8, 2011, at $81.31 a barrel. Economists say that drivers could see regular gasoline prices of $3.25 a gallon next month, which is $0.40 lower than August 8 levels.

f. My airplane from San Francisco to Allentown, Pennsylvania, stops in Chicago. I have stopped in Illinois.

3. Each of the following scenarios contains inductive reasoning. Decide what type of inductive reasoning each contains.

a. "The men, Dorothy thought, were about as old as Uncle Henry, for two of them had beards." (*Source*: L. Frank Baum, *The Wonderful Wizard of Oz*, New York: Books of Wonder, 1987, p. 21.)

b. "From the limited polls that have been taken on this issue (stem cell research), we know that Americans wanted Mr. Bush to advance research with astonishing potential and to reassure them that, in so doing, the United States would not cross some terrible line into unthinkable evil." (*Source:* "Bush's Gift to America's Extremists," *The New York Times*, August 19, 2001.)

c. "The global sell-off began the moment investors were able to weigh in on Standard & Poor's historic decision to revoke its AAA rating from U.S. debt and the rout continued in a day that echoed the market chaos of the financial crisis three years ago." As a result of this event, many investors were fearful that the financial situation of 2008 would be repeated. (*Source:* "Frightened Market Plummets," *The Morning Call*, August 9, 2011.)

d. "'Ayi!' she gasped. On the side of a ground swell lay Jello, his body torn in bloody shreds, his face contorted. Beside him lay her backpack!
Instantly she knew what had happened; Amaroq had turned on him. Once Kapugen had told her that some wolves had tolerated a lone wolf until the day he stole meat from the pups. With that, the leader gave a signal and his pack turned, struck, and tore the lone wolf to pieces. 'There is no room in the wolf society for an animal who cannot contribute,' he had said.

Jello had been so cowed he was useless. And now he was dead." (*Source:* Jean Craighead George, *Julie of the Wolves,* New York: Harper & Row, 1972, p. 121.)

4. For each of the scenarios detailed in Exploration 3, describe what premises are used and why the conclusion might be false.

5. For each study, determine if it is an observational study or an experiment. For each, identify the explanatory and response variables.

   a. In a report of a study on teens and Facebook, psychologist Larry Rosen reported at a meeting of the American Psychological Association in August 2011 that teens who use more technology, such as video games or the Internet, tended to have more stomachaches and sleeping problems and missed more school than students who used those technologies less. In observations of middle school, high school, and college students, the researchers found that students who checked social networks or text messages every few minutes had lower test performance than students who focused for longer periods of time.

   b. A study reported in a 2011 issue of the *American Journal of Medicine* (lead author Dr. William Shrank) reported that patients were less likely to fill prescriptions when their doctors specify that brand name drugs can't be substituted with generics. In the study, the researchers analyzed all prescriptions that were filled by CVS Caremark at CVS stores and online over the course of one month. In total, 5.6 million prescriptions were filled by 2 million patients. When there was no request for a brand name prescription (that is, no "dispense as written" label) on a new prescription for patients with chronic disease, about 8% went unfilled, compared to close to 12% when patients themselves said they didn't want a generic.

   c. An article published in 2011 in the journal *Psychological Science* reports on a study by researchers Deanna Kuhn and Amanda Crowell. They created a new curriculum for teaching reasoning skills to middle school students that emphasizes discussion. A group of 48 sixth-graders was taught over a three-year period using the new curriculum, while a group of 23 students in a separate class was taught using more traditional methods of reasoning with techniques such as essay-writing. After each year, all students wrote essays on entirely new topics. The researchers analyzed these essays for evidence of reasoning skills. Students who participated in the new teaching method fared better on all measures of reason-based skills.

6. The word *proof* is used in everyday language as well as in more technical conversations and writing.

   a. Describe the ways in which the word *proof* is used in everyday language.

   b. Write several paragraphs to describe what would be convincing proof for an argument using inductive reasoning

   c. Write several paragraphs to describe what would be convincing proof for an argument using deductive reasoning.

7. Find newspaper or magazine articles that illustrate the four types of inductive reasoning: prediction, generalization, causal inference, and analogy. Explain how each article exhibits the particular type of reasoning.

Analyzing Studies: Inductive Reasoning

In this activity, you will investigate possible flaws when using inductive reasoning. First, you will read newspaper articles that summarize situations in which inductive reasoning is used. You will then decide if the study described is an observational study or an experiment. Finally, you will analyze each situation and decide which inductive reasoning flaws might be present.

Inductive reasoning often involves a general conclusion or relation that results from specific examples or experiences. Four types of inductive reasoning are prediction, generalization, analogy, and causal inference. Such reasoning is sometimes done by sampling a population, obtaining results from the sample, and inferring that these results will hold for the whole population. One way to do this is through an **observational study** where individuals are observed and some variable or variables of interest are measured, but the researcher does not attempt to influence the responses. On the other hand, in an **experiment**, the researcher deliberately imposes some treatment on individuals in order to observe their responses.

Because of the nature of inductive reasoning, the conclusions resulting from an inductive argument are not guaranteed to hold. Studies and experiments must be carefully planned and carried out so errors are minimized and the results are likely to hold (even though still not guaranteed). The following are some of the most commonly occurring inductive reasoning errors in quantitative research.

Correlation and Causation

If a group of people or objects is measured with respect to two variables, and if neither of the variables is experimentally manipulated, then the simple finding of a relationship between the two variables does not necessarily mean that one variable has a causal influence on the other. It is quite possible that both of the variables are linked to some unmeasured third variable.

Experimenter Effect

If there are two or more conditions of a manipulated treatment variable, and if the treatments are administered to the subjects by someone familiar with the researcher's hoped-for results, then it is possible that the comparison groups will be treated differently, biasing the results. This unconscious biasing of the results is sometimes called **expectancy**.

Sampling Bias

If subjects are sampled in such a way as to make them unrepresentative of the total group from which they are drawn, conclusions about the total group are not valid. Particular care must be taken when interpreting the results of questionnaires returned anonymously. If the return rate is 30%, for example, the researcher cannot assume the 30% responding is representative of the total population.

Subject Selection

When two or more groups of subjects receive different treatments, usually with one of the groups serving as a control group, only random selection of subjects to the groups will protect against the possibility that some extraneous variable (age, sex, education level, for example) is having an unwanted effect on the results.

Subject Effect

The subjects who receive a particular treatment might figure out, even if not told directly, what treatment they have been given, what treatment the other group(s) in the study received, and the intent of the research. This information may cause the groups to perform differently on the response variable(s) simply because of the way the subjects expect their treatments to affect their behavior, attitudes, or knowledge.

Valid Data/Self-Report

When subjects self-report data, they might be consciously or unconsciously motivated to withhold their honest thoughts. Or the data collected might constitute quantitative information on the wrong response variable. Additionally, sometimes subjects simply make mistakes when responding to a questionnaire.

On the following pages, you will find news articles containing inductive reasoning. For each article, answer these questions:

a. Write out, in question form, a question that the research and reasoning was designed to answer. (Note that the article's headline might guide you but might also be misleading.)

b. Determine if the research was an observational study or an experiment. (Be careful about assuming it is a study just because the word *study* is used in the headline or article.) Explain how you arrived at your answer.

c. Identify any inductive reasoning flaws you think might be present in the study or experiment and what the researchers might have done to avoid the flaws.

d. Think about the researcher's conclusion. If possible, suggest a plausible explanation for the outcome of the study that is different from the explanation that the researcher has offered.

1. Health News

Air cleaners help kids with asthma

-Reuters

Indoor air cleaners may help asthmatic children who live with a smoker breathe a little easier, a study finds.

The devices are no cure-all and parents of children with asthma should strive for a smoke-free home, researchers said. But in cases where that's not yet a reality, they say, air cleaners may lessen the ill effects of secondhand smoke.

The study looked at the effects of HEPA air cleaners among 115 inner-city Baltimore children who had asthma and lived with a smoker.

The researchers gave one-third of the families an air cleaner for their living room and one for the child's bedroom. Another third received the devices plus education sessions with a nurse on the dangers of second-hand smoke. The final third served as the control group.

Over six months, the study found, children in the air-cleaner groups showed improvements in wheezing, coughing and other symptoms, based on surveys of their parents.

Source: The Morning Call, August 7, 2011.

a. Research question:

b. Observational study or experiment and how you determined this:

c. Possible inductive reasoning flaws and remedies:

d. Possible alternative explanation for conclusion:

2. Tanning Dangers

Tanning salons raise skin cancer risk, study indicates

Paul Recer,
Associated Press

Washington—Regularly baking to a golden tan under sun lamps can increase the risk of malignant melanoma, a sometimes fatal skin cancer, and the younger a woman starts the greater the risk, a study says.

The study, which analyzed the lifestyles and melanoma risks for women between the ages of 30 and 50, found what the researchers said was the strongest evidence yet that artificial sun tanning can be dangerous to healthy skin.

Melanoma risk is highest among fair-skinned people in Australia, New Zealand, Europe and North America. Since the 1950s, the rate of the skin cancer has tripled in Norway and

Sweden, where light skin is common. About 50,000 cases of melanoma are diagnosed annually in the United States and about 7,500 people die of the disease each year, according to American Academy of Dermatology officials.

In the study, appearing this week in the *Journal of the National Cancer Institute*, an international group of researchers analyzed data from the Women's Lifestyle and Health Cohort Study in Norway and Sweden. In 1991 and 1992, 106,379 women completed extensive questionnaires about their exposure to sunlight and to artificial tanning. In 1999, the researchers rechecked the women's cancer status using the national health registries in Norway and Sweden.

The researchers found 187 cases of malignant melanoma diagnosed among the study group during the eight-year follow-up period.

They found that women of any age or hair color who regularly visited tanning salons once or more per month increased their chance of developing melanoma by 55 percent.

The risk was highest for young adults. Compared with women who never used a solarium, women who reported using artificial tanning systems once or more per month when they were between the ages of 20 and 29 increased their risk of melanoma by about 150 percent.

"Our results provide stronger evidence than those of other studies that solarium use is associated with an increased risk of melanoma," the authors of the study wrote.

"This is just one of many papers that have suggested a link between indoor tanning and the development of melanoma skin cancer," said Dr. James M. Spencer, vice chairman of dermatology at the Mount Sinai School of Medicine in New York. He said studies have also linked artificial tanning to basal cell and squamous cell carcinoma, two common types of skin cancer.

Spencer said it is well known that ultraviolet light causes skin cancer.

"Whether you get it at the indoor tanning parlor or at the beach, (UV light) is a carcinogen," he said.

People may have good reasons to work and play in natural sunlight, where they can protect themselves with sun block, Spencer acknowledged. However, "There is no compelling reason to go to a tanning salon," he said. "It is just for a cosmetic tan that fades in a couple of weeks and can cause you a lifetime of trouble."

Source: The Morning Call, October 15, 2003.

a. Research question:

b. Observational study or experiment and how you determined this:

c. Possible inductive reasoning flaws and remedies:

d. Possible alternative explanation for conclusion:

3. Television Time

Study: Lots of television time can hurt children's reading ability

PARENTS MUST UNDERSTAND PITFALLS OF MEDIA, RESEARCHER SAYS

Siobhan McDonough,
Associated Press

WASHINGTON—Children who live in homes where the television is on most of the time might have more trouble learning to read than other children, a study says.

Tuesday's report, based on a survey of parents, also found that children 6 months to 6 years spend about two hours a day watching television, playing video games or using computers. That's roughly the same amount of time they spend playing outdoors and three times as long as they spend reading or being read to.

The study, by the Kaiser Family Foundation and Children's Digital Media Centers, found about one-third of children 6 and younger live in homes where a television is on most or all the time. In those "heavy TV households," 34 percent of children ages 4 to 6 can read, compared with 56 percent in homes where the TV is on less often.

CHILDREN AND THE MEDIA

Children's exposure to electronic media, according to a study:

- Children 6 and younger spend about two hours a day with a TV, video games or a computer.
- 48 percent of children 6 and under have used a computer; 30 percent have played video games.
- 90 percent of parents set rules about what their children can watch and 69 percent control how much time their children can watch TV.
- Children with time-related rules spend about a half-hour less per day in front of the TV than other children.
- 68 percent of children under 2 will be in front of a screen for an average of just over two hours a day.
- 36 percent of children 6 and younger have TVs in their bedrooms.
- 27 percent of 4- to 6-year olds use a computer each day, and those who do spend an average of about an hour at the keyboard.
- Those who live in households where the TV is on always or most of the time are less likely to read every day.
- 72 percent of parents say computers mostly help with children's learning.

Source: The Morning Call, October 29, 2003.

   a.   Research question:

   b.   Observational study or experiment and how you determined this:

   c.   Possible inductive reasoning flaws and remedies:

d. Possible alternative explanation for conclusion:

4. Hypnosis and Surgery

Hypnosis a help in surgery, study says

Associated Press

LONDON—People who were hypnotized while undergoing surgery without a general anesthetic needed less pain medication, left the operating room sooner and had more stable vital signs than those who were not, according to a study in this week's issue of *The Lancet* medical journal.

Trance states have been used for hundreds of years by both witch doctors and modern surgeons to help sick people.

But there had been little scientific evidence that hypnosis really works to reduce pain during surgery.

"Despite how long hypnosis has been around and the dramatic effects it has shown, there are very few properly designed clinical studies that demonstrate that it is more than a placebo," said David R. Patterson, professor of rehabilitation medicine, surgery and psychology at the University of Washington in Seattle, who was not connected with the study.

"This really solidifies the evidence. For acute pain . . . it's not arguable anymore."

The study, led by Dr. Elivira Lang of Beth Israel Deaconess Medical Center in Boston, involved 241 people of similar health and age who had operations to open clogged arteries and veins, relieve blockages in the kidney drainage system or block blood vessels feeding tumors.

The patients were divided into three groups—one that experienced normal interactions with doctors and nurses, another that received extra attention from an additional person in the operating room who made sure nobody said anything negative, and a third who were helped to hypnotize themselves.

All the patients were able to give themselves as much pain medication as they wanted through an intravenous tube.

The hypnosis group—who were guided through visualizations of scenarios they found pleasant—fared best, but the patients receiving extra attention also benefited.

About half the patients in those two groups needed no drugs, while the rest gave themselves only half the amount of medication as those undergoing the operation with no special attention.

The hypnotized patients were the only ones who said the pain did not get worse as the surgery progressed. They also had fewer problems with blood pressure and heart rate during the operation.

Their operations also finished 17 minutes earlier than those conducted without the special attention.

Lang suggested that time was saved in those operations because the surgeon's attention was diverted less often by events such as the patient's being oversedated, unstable blood pressure or vomiting.

The study was funded by the National Institutes of Health.

Source: The Morning Call, April 28, 2000.

a. Research question:

b. Observational study or experiment and how you determined this:

c. Possible inductive reasoning flaws and remedies:

d. Possible alternative explanation for conclusion:

Summary

In this activity, you investigated possible flaws in inductive reasoning by analyzing news reports of studies.

13

Deductive Reasoning

Tanning salons raise skin cancer risk, study indicates

Paul Recer,
Associated Press

Washington—Regularly baking to a golden tan under sun lamps can increase the risk
malignant melanoma, a sometimes fatal skin cancer, and the younger a woman starts
greater the risk, a study says.

P	Q	not P	not Q	P and Q	not (P and Q)	(not P) or (not Q)
T	T	F	F	T	F	F
T	F	F	T	F	T	T
F	T	T	F	F	T	T
F	F	T	T	F	T	T

State	Population	Standard C
Delaware	897,934	0.4084
New Jersey	8,791,894	3.9992
New York	19,378,102	8.8145
Pennsylvania	12,702,379	5.7779
Total	**41,770,309**	

Country	Per Capita Total Spending on Health (in U.S. dollars at average exchange rates)	Per Capita Daily Calorie Supply	Infant Mortality Rate per 1,000 Births
Australia	3,986	3,674	5
Cambodia	36	2,046	62
China	108	2,951	23
France	4,672	3,654	4
Germany	4,209	3,496	4
India	40	2,954	55
Italy	3,136	3,617	4

After completing this topic, you will be able to:

- Relate logical statements given in words to symbolic logical statements, and use truth tables to assess logical statements.

- Identify logical statements and formulate their negations.

- Recognize and use the three different types of compound logical statements.

- Formulate and use the contrapositive and converse of an if-then statement.

- Form simple deductive arguments and analyze correct and incorrect deductive reasoning.

Deductive reasoning is the form of reasoning that we use to derive logical consequences from given true statements. We often use deductive reasoning when we want to prove a point, whether it is a mathematical theorem, a legal argument, or a scientific theory. Deductive reasoning always leads to true statements, provided the premises are true.

The following excerpt from an address to the House of Representatives by Representative David N. Cicilline (Democrat, Rhode Island) gives examples of logical statements and deductive reasoning (*Source: Congressional Record*, Government Printing Office, www.gpo.gov, November 15, 2011):

Mr. Speaker, last month the Congressional Budget Office released a report that examined household income distribution between 1979 and 2007. The most disturbing figure to me in this report is that the top 1 percent of income earners have seen their average real after-tax household income grow by 275 percent. Middle-income Americans saw an increase of 40 percent over the same period of time.

This report illuminates a sad fact: Income inequality in our country is growing at a staggering pace. The report is pointing out what many of my constituents tell me as I travel around my district from Cumberland to Pawtucket to Newport, from community dinners and talking to business owners: This economy is not working for the majority of middle class families. In fact, the hardworking middle class of our country is being hollowed out, a middle class made up of people that are just trying to provide a good life for themselves and their families. My real fear is that if we let that happen, we'll never get it back.

In these comments, the representative uses deductive reasoning when he draws the conclusion that income inequality is growing at a high rate. His conclusion follows from data released by the Congressional Budget Office.

It is important to be able to tell whether or not a specific form of deductive reasoning is valid. When the reasoning is valid, then the truth of the premises necessarily implies the truth of the conclusion. The validity of reasoning is governed by the rules of logic. We will explore the kinds of statements used in reasoning and the basic logic principles that lead to valid reasoning.

A **statement** is a sentence that is either true or false, but not both. We explore this definition of a statement in the first example.

Example 13.1

For each of the following sentences, decide whether it is a statement or not, and explain why you gave the answer you did.

a. "I am the man who accompanied Jacqueline Kennedy to Paris, and I have enjoyed it." (*Source:* President John F. Kennedy.)

b. The population of Pennsylvania is 12,281,054, and the population of the United States is 150,287,967.

c. "What does it mean to live in a diverse nation, where not everybody looks like you do, or thinks like you do, or comes from the same neighborhood as you do?" (*Source:* President Barack Obama, "Back to School" speech, September 2011, www.educationnews.org.)

d. "From global airlines and shipping giants to small manufacturers, all kinds of companies are feeling the strain as European banks pull back on lending in an effort to hoard capital and shore up their balance sheets." (*Source:* "Crisis in Europe Tightens Credit Across the Globe," *The New York Times*, November 28, 2011.)

e. Please do not go out in this storm.

f. "This economy is not working for the majority of middle class families." (*Source:* Representative David Cicillini, *Congressional Record*, Government Printing Office, www.gpo.gov, November 15, 2011.)

Solution

The sentences in parts (a), (b), (d), and (f) are all statements because they are true or false. We might or might not know whether they are true or false, but we know that one or the other must be the case. (Note that the truth value of the statement in part (f) depends on how the speaker has defined "is working.") The sentence in part (c) of this example contains a question, which is neither true nor false, so it is not a statement. The sentence in part (e) of this example is not a statement because it is a request, which is neither true nor false.

NEGATION OF A STATEMENT

The **negation** of a statement is the statement obtained by negating the original statement. If we represent the original statement by P, its negation is **not P**. If P is true, then "not P" is false; and if P is false, then "not P" is true; that is, P and "not P" have opposite truth values. We can summarize the relationship between a statement P and its negation using a **truth table**. A truth table gives all possible truth values for the statements under consideration. Here is a truth table for a statement P and its negation:

P	not P
T	F
F	T

The table shows that P can either be true (T) or false (F). If P is true, then "not P" is false. If P is false, then "not P" is true, by the definition of "not P." For example, the negation of the statement "The door is open" is "The door is closed" because each of these statements is true exactly when the other one is false.

We must be careful about the language we use when forming the negation of a statement. For example, the statement "There are less than 15 students in the classroom" is not the negation of "There are 15 students in the classroom," because these two statements could have the same truth value. An instance of this would be if the actual number of students in the classroom were 16. In such a case, both statements would be false. The negation of a statement must have the opposite truth value in all instances.

Example 13.2

Based on the results of a study published in a scientific journal, we can make the following statement: "Using appropriate technology to control greenhouse gases in four highly polluted major cities—São Paulo, Brazil; Mexico City; Santiago, Chile; and New York City—would save 64,000 lives over the next 20 years." Find the negation of this statement.

Solution

The negation of this statement is "Using appropriate technology to control greenhouse gases in four highly polluted major cities—São Paulo, Brazil; Mexico City; Santiago, Chile; and New York City—would not save 64,000 lives over the next 20 years." This new statement is true if the original statement is false, and it is false if the original statement is true. (We could also negate the original statement by saying, "It is not the case that using appropriate technology to control greenhouse gases in four highly polluted major cities— São Paulo, Brazil; Mexico City; Santiago, Chile; and New York City—would save 64,000 lives over the next 20 years," but that is an awkward statement and would not be a particularly useful negation.)

COMPOUND STATEMENTS

Simple statements are often combined into more complex statements, called **compound** statements. To form a compound statement, we connect simple statements through **logical connectors**. The truth value of a compound statement depends on the truth value of each of the components and on the connector used. The three connectors that are used most often are **and**, **or**, and **if-then**. If we represent the original statements by P and Q, then using these connectors, we obtain three compound statements: **P and Q**; **P or Q**; and **if P then Q**.

A statement of the form **P and Q** is called the **conjunction** of P and Q. For example, the statement (from an address to the Senate by Senator Chuck Grassley (R-IA), *Congressional Records*, www.congress.gov, July 26, 2001):

> "These two efforts will provide complete elimination of the marriage penalty for low- and many middle-income working families **and** will also benefit married couples with higher incomes."

is the conjunction of the statement P: "These two efforts will provide complete elimination of the marriage penalty for low- and many middle-income working families," and the statement Q: "These two efforts will benefit married couples with higher incomes."

The conjunction "P and Q" is true when both of the statements, P and Q, are true, and it is false if either one of P or Q is false or both P and Q are false. This relationship is summarized in the truth table given next. Note that we need to include four rows in the table to cover all possible combinations of truth values for both P and Q. When P is true, there are two possibilities for Q, true or false. Similarly, when P is false, Q can be true or false.

P	Q	P and Q
T	T	T
T	F	F
F	T	F
F	F	F

The conjunction can also be expressed by connecting words other than the word *and*, such as the words *while*, *but*, and *yet*. For example, the statement "P and Q" given previously could be expressed as

These two efforts will provide complete elimination of the marriage penalty for low- and many middle-income working families **but** will also benefit married couples with higher incomes.

A compound statement of the form **P or Q** is called a **disjunction**. It is true when *at least* one of P or Q is true, that is, when P is true or Q is true or both are true. The following statement is an example of a disjunction:

This evening, I will study for the history test **or** I will complete the math project.

The compound statement is composed of the statement P: "This evening, I will study for the history test," and the statement Q: "This evening, I will complete the math project." The compound statement will be true in the case where the speaker only studies for the history exam that evening, or the speaker only completes the math project, or the speaker does both—studies history and completes the math project—that evening. The truth table for the statement "P or Q" covers all possible combinations of truth values for the two simpler statements P and Q:

P	Q	P or Q
T	T	T
T	F	T
F	T	T
F	F	F

Note that the word *or* in English may also mean that only one and not both of the two connected statements is true. In standard logic and in our work, we will assume "P or Q" is true when P, or Q, or both are true, as the table indicates.

A compound statement of the form **if P, then Q** is called a **conditional** statement. In this situation, the statement P is called the **antecedent** and Q is called the **consequent** of the conditional statement. For example, the statement "If you pass the final exam, then you will pass the course" is a conditional statement. Here, the antecedent is "You pass the final exam," and the consequent is "You will pass the course."

The conditional statement "if P, then Q" is true when Q is true (and P is either true or false), or when P is false (and Q is either true or false). That is, it is true in all cases, except the case when the antecedent (P) is true and the consequent (Q) is false.

The truth table for "if P, then Q" is as follows:

P	Q	if P, then Q
T	T	T
T	F	F
F	T	T
F	F	T

The conditional statement "if P, then Q" does not assert that Q is true in all cases; it just says that Q is true when P is true. For example, the conditional statement "If you pass the final exam, you will pass the course" is not saying that you will pass the course. The conditional statement would still be true in the case where you fail the final, whether you fail the course or not.

We analyze another conditional in the following example.

Example 13.3

Consider the following conditional statement (also made by Senator Chuck Grassley in his 2001 address before Congress):

> "If the first $6,000 of a single individual is taxed at 10%, then the first $12,000 of a married couple filing jointly will be taxed at 10%."

a. Give the antecedent and the consequent of the conditional statement.

b. In each of the following cases, decide whether the conditional statement is true or false and explain why:

i. Suppose the tax rate on the first $6,000 for a single individual is 10% and the tax rate on the first $12,000 for a married couple filing jointly is 10%.

ii. Suppose the tax rate on the first $6,000 for a single individual is 10% and the tax rate on the first $12,000 for a married couple filing jointly is 12%.

iii. Suppose the tax rate on the first $6,000 for a single individual is 8% and the tax rate on the first $12,000 for a married couple filing jointly is 12%.

Solution

a. The given statement is of the form "if P, then Q." The antecedent, P, is the statement "The first $6,000 of a single individual is taxed at 10%." The consequent Q is the statement "The first $12,000 of a married couple filing jointly will be taxed at 10%."

b. **i.** In this case, the conditional statement is true because it is of the form "if P, then Q," where both, the antecedent P and the consequent Q, are true.

ii. Under the given assumptions, the "condition" that the rate on the first $6,000 for a single individual is 10%, has been established, but the "consequence" that the rate for the first $12,000 for married couples is also 10% has not happened. So, our conditional statement is false because it is of the form "if P, then Q," where P is true but Q is false.

iii. Because the tax rate on the first $6,000 for a single individual is 8%, the "condition" that the rate on the first $6,000 for a single individual is 10% is not true. So, the conditional statement is true whether or not the rate for married couples is 10%. In this case, our conditional statement is of the form "if P, then Q," where P is false and Q is false, which makes it a true statement.

There are other ways of expressing the conditional. Sometimes the word *then* is omitted, resulting in a statement of the form "if P, Q." For example, the conditional statement in Example 13.3 can be expressed as

If the first $6,000 of a single individual is taxed at 10%, the first $12,000 of a married couple filing jointly will be taxed at 10%.

Other ways of expressing the same statement are to say "Q, if P" or "Q whenever P," as the following rewordings of the previous example show:

The first $12,000 of a married couple filing jointly will be taxed at 10%, if the first $6,000 of a single individual is taxed at 10%.

The first $12,000 of a married couple filing jointly will be taxed at 10% whenever the first $6,000 of a single individual is taxed at 10%.

In the following example we identify different types of simple and compound statements.

Example 13.4

For each of the following statements, decide whether the statement is simple or compound. If the statement is compound, decide whether it is of the form "P and Q," "P or Q," "not P," or "if P, then Q," and give P and Q. (These statements are all taken from two speeches published in *Vital Speeches of the Day*: Murray Weidenbaum, "Breaking the Deadlock in U.S. Trade Policy," June 2001, and William Brody, "The Intellectual Climate of the U.S.," July 2001.)

a. ". . . Globalization is responsible for abuses of labor rights and of the environment and it reduces the sovereignty of individual nations."

b. "Eventually, of course, some deals will have to be made if any legislative action is to occur at all."

c. "For a variety of reasons, including disagreements between the developed and developing nations, the World Trade Organization meetings in Seattle concluded in failure."

d. "If these issues are not managed intelligently and creatively, the domestic consensus in favor of open markets may ultimately erode."

e. "The people hurt by globalization are being ignored while the winners are enjoying all the benefits."

f. "We may have a faculty member teaching microeconomics at Harvard in the fall, in Singapore in the winter, and at Hopkins in the summer, or we may have faculty members doing collaborative research across institutions."

Solution

a. This statement is a compound statement of the form "P and Q," which is a conjunction. Here, P is "Globalization is responsible for abuses of labor rights and of the environment," and Q is "It [globalization] reduces the sovereignty of individual nations."

b. This is a conditional statement, that is, a compound statement of the type "if P, then Q." In this case, the statement P (the antecedent) is "Any legislative action is to occur," and Q (the consequent) is "Some deals will have to be made."

c. This statement is a simple statement.

d. This statement is a compound statement of the form "if P, then Q." P, the antecedent, is "These issues are not managed intelligently and creatively," and Q, the consequent, is "The domestic consensus in favor of open markets may ultimately erode."

e. This is a compound statement. It is a conjunction that can be expressed in the form "P and Q," where P is the statement, "The people hurt by globalization are being ignored," and Q is the statement "The winners are enjoying all the benefits."

f. This is a disjunction, that is, a compound statement of the form "P or Q." Here, P is "We may have a faculty member teaching microeconomics at Harvard in the fall, in Singapore in the winter, and at Hopkins in the summer"; Q is "We may have faculty members doing collaborative research across institutions."

NEGATION OF A COMPOUND STATEMENT

The negation of a compound statement has a truth value opposite that of the statement. Sometimes it is easier to determine whether a statement is true or false by looking at its negation. The negation of any statement P can be expressed as "It is not the case that P." Usually, however, this is not a useful form of the negation. For example, the negation of the statement (from Senator Chuck Grassley's previously mentioned address)

> "These two efforts will provide complete elimination of the marriage penalty for low- and many middle-income working families **and** will also benefit married couples with higher incomes."

could be stated in a quite non-useful form as

> "It is not the case that these two efforts will provide complete elimination of the marriage penalty for low- and many middle-income working families and will also benefit married couples with higher incomes."

A more useful form of the statement's negation is

> "These two efforts will **not** provide complete elimination of the marriage penalty for low- and many middle-income working families **or** will **not** benefit married couples with higher incomes."

Note that this last statement is of the form "(not P) or (not Q)."

The disjunction "(not P) or (not Q)" is the logical **equivalent** to the statement "not (P and Q)," and therefore it is the negation of "P and Q" because the statements "(not P) or (not Q)" and "not (P and Q)" are both true or they are both false. So,

"not (P and Q)" is equivalent to "(not P) or (not Q)"

This is the case because the statement "(not P) or (not Q)" has the opposite truth value of "P and Q". To see this, note that "(not P) or (not Q)" is false only when both "not P" and "not Q" are false. This occurs when both P and Q are true. This is precisely the only case when "P and Q" is true. A truth table summarizes this discussion:

P	Q	not P	not Q	P and Q	not (P and Q)	(not P) or (not Q)
T	T	F	F	T	F	F
T	F	F	T	F	T	T
F	T	T	F	F	T	T
F	F	T	T	F	T	T

In the table, the last two columns have exactly the same truth values, which means the statements are equivalent. To find the truth values for the last column, we use the "not P" column and the "not Q" column. The *or* statement is false only when both "not P" and "not Q" are false, which is when P and Q are both true.

In a similar manner, we can see that the negation of a disjunction "P or Q" is the conjunction "(not P) and (not Q)." That is,

"not (P or Q)" is equivalent to "(not P) and (not Q)"

Here is a truth table that shows this relationship:

P	Q	not P	not Q	P or Q	not (P or Q)	(not P) and (not Q)
T	T	F	F	T	F	F
T	F	F	T	T	F	F
F	T	T	F	T	F	F
F	F	T	T	F	T	T

The negation of a conditional statement "if P, then Q" is a statement logically equivalent to "not (if P, then Q)." We look for a statement that is true exactly when "if P, then Q" is false. This is the case when P is true and Q is false. The statement "P and (not Q)" is true precisely when P is true and (not Q) is true, that is, when P is true and Q is false. We can then say that

"not (if P, then Q)" is equivalent to "P and (not Q)"

A truth table confirms this relationship:

P	Q	not Q	if P, then Q	not (if P, then Q)	P and (not Q)
T	T	F	T	F	F
T	F	T	F	T	T
F	T	F	T	F	F
F	F	T	T	F	F

In the following example, we use the equivalent statements shown in the truth tables to compose useful negations of various compound statements.

Example 13.5

Give a useful negation of each of the following statements (from Example 13.4):

a. "... Globalization is responsible for abuses of labor rights and of the environment and it reduces the sovereignty of individual nations."

b. "Eventually, of course, some deals will have to be made if any legislative action is to occur at all."

c. "... The World Trade Organization meetings in Seattle concluded in failure."

d. "If these issues are not managed intelligently and creatively, the domestic consensus in favor of open markets may ultimately erode."

e. "The people hurt by globalization are being ignored while the winners are enjoying all the benefits."

f. "We may have a faculty member teaching microeconomics at Harvard in the fall, in Singapore in the winter, and at Hopkins in the summer, or we may have faculty members doing collaborative research across institutions."

Solution

a. Because this is a compound statement of the form "P and Q," its negation is of the form "(not P) or (not Q)." So, a useful negation of the given statement is "Globalization is not responsible for abuses of labor rights and of the environment, or it does not reduce the sovereignty of individual nations."

b. The negation of the conditional "if P, then Q" is a statement of the form "P and (not Q)." So, a useful negation of the given statement is "Some legislative action will occur and (but) no deals will be made."

c. This is a simple statement. A useful negation is "The World Trade Organization meetings in Seattle did not conclude in failure" or "The World Trade Organization meetings in Seattle were successful."

d. A useful negation of this conditional statement is "These issues are not managed intelligently and creatively and the domestic consensus in favor of open markets may not erode."

e. The negation of this statement is of the form "(not P) or (not Q)." It can be expressed as "The people hurt by globalization are not being ignored or the winners are not enjoying all the benefits."

f. This is a disjunction of the form "P or Q," so its negation is the conjunction "(not P) and (not Q)." A useful negation of the given statement is "We may not have a faculty member teaching microeconomics at Harvard in the fall, in Singapore in the winter, and at Hopkins in the summer, and we may not have faculty members doing collaborative research across institutions."

CONTRAPOSITIVE AND CONVERSE OF A CONDITIONAL STATEMENT

The conditional statement "if P, then Q" is central to deductive reasoning. In such a statement, the roles of P and Q are very different. Reversing their roles or using their negations results in new conditional statements. Two often used conditional statements related to "if P, then Q" are the **contrapositive** and the **converse.** The **contrapositive** of "if P, then Q" is the conditional statement **if (not Q), then (not P).** Note that to state the contrapositive of the statement "if P, then Q," we switch the positions of P and Q and negate both. The contrapositive of a conditional statement is equivalent to the original conditional. (You are asked to use a truth table to verify this in Activity 13.1.) This means that the contrapositive is true when the original statement is true, and it is false when the original statement is false. For example, consider again the following statement:

> If the first $6,000 of a single individual is taxed at 10%, **then** the first $12,000 of a married couple filing jointly will be taxed at 10%.

Its contrapositive is

> If the first $12,000 of a married couple filing jointly is **not** taxed at 10%, **then** the first $6,000 of a single individual is **not** taxed at 10%.

The **converse** of the conditional statement "if P, then Q" is the conditional statement **if Q, then P**; that is, the converse is the conditional statement obtained by interchanging the consequent and antecedent of the original statement. The converse is not equivalent to the original statement. It is possible for the converse to be false when the original statement is true or the converse might be true and the original statement might be false. It is also possible that a statement and its converse are both true or are both false. Consider once more the statement:

> If the first $6,000 of a single individual is taxed at 10%, **then** the first $12,000 of a married couple filing jointly will be taxed at 10%.

Its converse is

> If the first $12,000 of a married couple filing jointly is taxed at 10%, **then** the first $6,000 of a single individual is taxed at 10%.

Example 13.6　■ ■ ■ ■ ■ ■ ■ ■ ■

Consider the conditional statement "If the Democratic candidate from New York is elected to the House of Representatives, then the Democrats will have a majority in the House."

a. Give the contrapositive of the conditional statement.

b. Give the converse of the conditional statement.

Solution

a. We first identify P and Q in the given conditional statement; P is the statement "The Democratic candidate from New York is elected to the House of Representatives," and Q is the statement "The Democrats have a majority in the House." The contrapositive is "if not Q, then not P," which is "If the Democrats do not have a majority in the House of Representatives, then the Democratic candidate from New York was not elected to the House."

b. The converse of the given statement is "If the Democrats have a majority in the House of Representatives, then the Democratic candidate from New York was elected."

QUANTIFIED STATEMENTS

We often use statements that assert a truth about some or all elements of a set. These statements are called **quantified statements**, and contain words such as *all, every, no, there is,* or *there exists* called **quantifiers**. These are examples of quantified statements:

"All citizens can vote."

"Some dogs are dangerous."

"No candidate is sufficiently ahead in the polls."

"There is a country that has a name starting with the letter *C*."

Quantified statements can be grouped into two general classes: those equivalent to statements containing the **universal quantifier**, *all*, and those equivalent to statements containing the **existential quantifier**, *there exists*. For example, the statement "Some dogs are dangerous" contains an existential quantifier since it can be rephrased as "There exists a (or there exists at least one) dog that is dangerous." Note that the quantifier "some" does not necessarily mean that there is more than one.

Sometimes we will need to reword statements to identify the quantifier. For example, the statement "No candidate is sufficiently ahead in the polls" contains a universal quantifier because it is equivalent to "All candidates are not sufficiently ahead in the polls." We look at additional statements involving quantifiers in the next example.

Example 13.7

For each of the following statements, identify the type of quantifier it contains, and write an equivalent statement using one of the quantifiers *all* or *there exists*.

a. "Every team here is exceptional."

b. ". . . Some of the material was accessed in 2004."

c. "No known human has ever received an injection of embryonic stem cells. . . ."

d. " . . . Not all of the files from the computer have been examined."

e. "No one is missed more than the Americans." (This refers to the fact that the American baseball team was eliminated before arriving at the 2004 Olympic Games in Greece.)

[Statements in part (a) and (e) of this example are from the article "Olympic Baseball Pallid Without U.S-Cuba Clash," *The Providence Sunday Journal*, August 13, 2004; statements in parts (b) and (d) of this example are from the article "Seized Terror-Target Files Were Accessed in Spring," *The Wall Street Journal*, August 13, 2004; and the statement in part (c) of this example is from the article "Trace of Human Stem Cells Put in Unborn Mice Brains," *The New York Times*, December 13, 2005.]

Solution

a. The statement is equivalent to the statement "All teams here are exceptional." It involves a universal quantifier.

b. The statement is equivalent to "There exists some material that was accessed in 2004." It involves an existential quantifier.

c. The statement is equivalent to "Every human has not received an injection of embryonic stem cells. . . ." It contains a universal quantifier.

d. The statement is equivalent to the statement "Some files from the computer have not been examined," or "There exists a file that has not been examined." It involves the existential quantifier.

e. An equivalent statement is "Every other team is not missed as much as the Americans," or "All other teams are not missed as much as the Americans." This statement involves a universal quantifier.

Special care is needed when stating the negation of a quantified statement because the type of quantifier in the statement is different from the type of quantifier in its negation. For example, the negation of "Some dogs are dangerous" is "All dogs are not dangerous," or equivalently, "No dog is dangerous." (Note that the statement "Some dogs are not dangerous" is **not** a negation of "Some dogs are dangerous," because both statements are true.)

The general form of the negation of a statement involving the universal quantifier, such as "all A's are B," is "some A's are not B" (or equivalently, "There exists an A that is not B"), because if it is not true that all A's are B, then there must be at least one A that is not B. Similarly, the general form of the negation of a statement involving the existential quantifier, such as "some A's are B," is "all A's are not B." The next example illustrates how to formulate useful negations of quantified statements.

Example 13.8

Write a useful negation of each of the statements in Example 13.7.

Solution

a. The given statement ("Every team here is exceptional") contains a universal quantifier, so its negation contains an existential quantifier. The negation is "There is a team here that is not exceptional," or "Some teams here are not exceptional."

b. Because this statement ("Some of the material was accessed in 2004") uses the existential quantifier, its negation will have the universal quantifier: "All of the material was not accessed in 2004," which is best expressed as "None of the material was accessed in 2004."

c. The negation of the statement "No known human has ever received an injection of embryonic stem cells. . . ." involves an existential quantifier. It can be expressed as "Some humans have received an injection of embryonic cells. . . ."

d. The original statement "Not all of the files from the computer have been examined" is equivalent to "Some files from the computer have not been examined." Its negation is "All files from the computer have been examined."

e. The negation of the statement "No one is missed more than the Americans" involves an existential quantifier, "There is another team that is missed as much as the Americans."

DEDUCTIVE ARGUMENTS

We conclude this topic with a brief discussion of what makes a deductive argument a good deductive argument. A deductive argument consists of premises (or hypotheses or assumptions) and conclusions that follow logically from those premises. There are two key elements for a good deductive argument: (1) The premises are true, and (2) the reasoning is valid. When these two elements are present, the conclusion is unquestionably true. The two forms of valid deductive reasoning are **direct reasoning** or **Modus Ponens**, and **indirect reasoning** or **Modus Tollens**.

Direct reasoning or **Modus Ponens**, is stated as follows:

If the statement "if P, then Q" is true, and P is also true, then Q must be true.

An example of a good deductive argument using direct reasoning is the following: "If two quantities x and y are related through an equation of the form $y = kx$ (where k is a constant), then y is directly proportional to x. The number of calories c that a 150-pound person uses when walking for m minutes is $c = 5.4m$. Then, the number of calories is directly proportional to the number of minutes the person walks."

Indirect reasoning or **Modus Tollens** is stated as follows:

If the statement "if P, then Q" is true and Q is false, then P must be false.

An example of an indirect deductive argument is the following: "If the response variable is directly proportional to the explanatory variable, then the response variable is a linear function of the explanatory variable. The relative energy released by an earthquake is not a linear function of the earthquake's magnitude on the Richter scale. Conclusion: The relative energy released by an earthquake is not directly proportional to its magnitude on the Richter scale." We look at additional samples of valid and invalid reasoning in the next example.

Example 13.9

For each of the following arguments, decide whether the reasoning is valid or not. If it is valid, decide which form of deductive reasoning is used. If it is not valid, explain why not.

a. If the Flyers win today's game, then they advance to the Stanley Cup playoffs. They did not advance to the Stanley Cup playoffs. Therefore, they did not win today's game.

b. If the price of electricity goes up, our family will pay over $95 a month in electricity. The price of electricity is the same as last month. Then our family will not pay more than $95 this month.

c. When you are caught driving over the speed limit, you get a ticket. You were given a ticket this morning. So, you must have been driving over the speed limit.

d. If the interest rates go down, then more houses are sold. The interest rates have gone down. Then the number of houses sold has increased.

Solution

a. Let P be the statement "The Flyers win today's game," and let Q be the statement "They advance to the Stanley Cup playoffs." The given argument has the form "if P, then Q"; "not Q." Then, "not P." Assuming the statements "if P, then Q" and "not Q" are true, this is a valid argument. The form of reasoning used is indirect reasoning.

b. This is not a valid argument. Let P be the statement "The price of electricity goes up," and let Q be the statement "Our family will pay over $95 a month in electricity." The given argument has the form "if P, then Q"; "not P." Then "not Q." This reasoning is not correct because assuming that the statements "if P, then Q" and "not P" are true does not mean that "not Q" must be true. The truth of the statement "if P, then Q" guarantees that when P is true, then Q must be true, but says nothing about the truth of Q when P is false. In the case of the given argument, P is false ("not P" is true), so we cannot conclude that Q is false (or "not Q" is true). Note that the electricity bill may be larger this month if, for example, the air conditioner was used more often.

c. This is not a valid argument. Let P be the statement "You are caught driving over the speed limit" and Q be the statement "You get a ticket." The given argument has the form "if P, then Q"; "Q." Therefore, "P." Even when the statement "if P, then Q" is true, and Q is true, the truth of P does not follow. In this case, you might have been given a ticket for driving with bad tires, for example, or for not stopping at a stop sign.

d. This is a valid argument. Let P be the statement "The interest rates go down," and let Q be the statement "More houses are sold." The given reasoning has the form "if P, then Q"; "P." Therefore, "Q." The reasoning form is direct reasoning.

Summary

In this topic, we investigated statements used in deductive reasoning and considered three types of compound statements: conjunctions, disjunctions, and conditional statements. We explored how to negate simple and compound statements and considered how to formulate the contrapositive and the converse of a conditional statement. We also studied quantified statements, using the universal and existential quantifiers, and we practiced valid deductive reasoning. We used truth tables to help understand compound statements and logical reasoning.

Explorations

1. Identify which of the following are statements, and for those that are statements, say whether they are true or false:

   a. Be quiet.

   b. George W. Bush was president during the years 2000 and 2008.

   c. There are more than 2,000 students currently enrolled at Pennsylvania State University.

   d. All college students attend parties at least once a week.

   e. Would you like to go to the movies?

2. For each of the following, determine if the expression is a statement or not, and explain why you gave the answer you did:

a. "The human fascination with whales has led to a new counterweight to the pro-whaling forces—the hundreds of companies running whale-watching operations in 87 countries, including those seeking an end to the ban on commercial hunting." (*Source:* "Save the Whales! Then What?" *The New York Times*, August 17, 2004.)

b. "His conducting has gained technical assurance over the years."(*Source:* "Hearing Echoes of Yesterday," *The New York Times*, August 17, 2004.)

c. "A musical by Shostakovich? The colossal and inscrutable 20th century composer who has come to epitomize the tragic plight of the artist compelled to play a public role in a totalitarian state?"(*Source:* "Hearing Echoes of Yesterday," *The New York Times*, August 17, 2004.)

d. "Conformity is the jailer of freedom and the enemy of growth."(*Source:* President John F. Kennedy.)

e. "And we all have seen the Pew Report which shows that white wealth is 20 times more than African American wealth, 18 times more than Hispanic wealth, and that more African Americans live in extreme poverty." (*Source:* Representative Donna Christensen, *Congressional Record*, Government Printing Office, www.gpo.gov, November 14, 2011.)

f. "So, again, we call on the leadership of this body to enact a jobs agenda." (*Source:* Representative Donna Christensen, *Congressional Record*, Government Printing Office, www.gpo.gov, November 14, 2011.)

3. For each of the following statements, decide whether the statement is simple or compound. If the statement is compound, decide whether it is of the form "P and Q," "P or Q," or "if P, then Q," and identify each of the statements P and Q.

a. My GPA is over 3.0.

b. Some courses at Kansas State University meet four times a week, while other courses meet three times a week.

c. If the homework is not handed in on time, I will lose points on the grade.

d. All students take Writing 100 or mathematics in their first term at college.

e. She likes to watch soccer and she plays the clarinet.

f. He will go to the play or he will come to the party.

g. If Huntington-Hill is used to apportion representatives, then Delaware gets 1 seat and Pennsylvania gets 19 seats.

h. If the graph of a company's profit over the years 2004 through 2012 is increasing and concave downward, then the rate at which the company's profit increases is decreasing.

4. Give a useful negation of each of the statements in Exploration 2. Make sure your statement is clear and understandable.

5. Give a useful negation of each of the statements in Exploration 3. Make sure your statement is clear and understandable.

6. The truth table given in the text showed that the statement "not (P or Q)" is logically equivalent to "(not P) and (not Q)." Give examples to show how these two statements are equivalent.

7. Use truth tables to show that the negation of "if P, then Q" is not equivalent to "if P, then (not Q)." Explain how your truth table shows that these statements are not equivalent.

8. For each of the following statements,

   i. Identify whether the statement is simple or compound. If the statement is compound, decide whether it is a conjunction, disjunction, or a conditional statement.

   ii. Give a useful negation.

   [Statements (a) through (d) are from a congressional address by Representative Donna Christensen, *Congressional Record*, Government Printing Office, www.gpo.gov, November 14, 2011; statements (e) through (i) are from *Remarks by the President on Economic Growth and Deficit Reduction*, www.whitehouse.gov, September 19, 2011.]

   a. "The number of people in high-poverty neighborhoods increased by nearly 5 million people since 2000, when 18.4 million metropolitan residents, 7.9% of the total, lived in high-poverty neighborhoods."

   b. "The number of people in high-poverty neighborhoods stabilized in the 1990s and the concentrated poverty rate fell, fueling optimism that faith-based initiatives and rising prosperity were reversing a crisis that had grown dire in the 1980s."

   c. "If this trend continues, it is a very bad prognosis for the economic health of our Nation."

   d. "America is the land of opportunity and all of us, not just the 43 members of the Congressional Black Caucus but all 441 or, really, all 541, need to be working together to make sure that it is for all and not just for some."

   e. "During this past decade, profligate spending in Washington, tax cuts for multimillionaires and billionaires, and two wars have turned a record surplus into a massive deficit, and that left us with a big pile of IOUs."

   f. "If we don't act, the debt will eventually crowd out everything else, eventually affecting us from investing in things like education and Medicaid."

   g. "If we're not willing to ask those who've done extraordinarily well to help America close the deficit and we are trying to reach that same target of $4 trillion, then the logic, the math says everybody else has to do a whole lot more: We've got to put the entire burden on the middle class and the poor."

   h. "Either we gut education and medical research, or we've got to reform the tax code so that the most profitable corporations have to give up tax loopholes that other companies don't get."

   i. "Social Security is not the primary cause of our deficits, but it does face long-term challenges as our country grows older."

9. For each of the if-then statements (conditionals) given next, state the converse and the contrapositive:

   a. If he is guilty, he will be convicted.

   b. If he is convicted, then he must be guilty.

   c. If I study for the exam, then I will get a good grade.

   d. If you want your clothes to be really clean, then use TIDE.

   e. If I run this red light, I will get to my class on time.

10. In each of the following situations, if the hypotheses allow a valid reasoning process, state the conclusion, and describe why it is a valid conclusion. If the statements do not fit any valid reasoning process, write "no valid conclusion" and explain why there is no valid conclusion.

   a. In order to drink legally in Pennsylvania, you must be 21 years of age or older. You are not yet 21.

   b. If you use Brand H laundry detergent, your clothes will be "whiter than white." You do not use Brand H laundry detergent.

   c. If you study for at least two hours, you will pass the test. You fell asleep and did not study at all.

   d. If a student plays football at this school, he cannot play soccer. John does not play soccer.

   e. If I can save enough money to afford the trip, I will go to Aruba on spring break. I did not go to Aruba on spring break.

   f. If a student plays football at this school, he cannot play soccer. Eric plays football.

11. For each of the following quantified statements, (i) decide whether the statement involves a universal or an existential quantifier, (ii) write the statement using *all* or *there exists*, and (iii) decide whether the statement is true or false (give a reason for your answer).

   a. There is a country that has a name starting with the letter *C*.

   b. In the United States, every citizen votes in the presidential elections.

   c. No European country has a larger population than the United States.

   d. Some college students have full-time jobs outside the college.

12. Write the negation of each of the quantified statements in Exploration 11.

13. Explain, using the context of the statement, why the contrapositive statement given in the solution of Example 13.6(a) must have the same truth value as the original statement, but the converse given in the solution of Example 13.6(b) might not have the same truth value as the original statement.

Code-Breaking and Deductive Reasoning

In the first part of this activity, you will exercise your deductive reasoning skills in the same way you do when you play some board games. In the second part of this activity, you will work with truth tables and the contrapositive and converse of a conditional statement.

Breaking the Code

Many code-breaking schemes use mathematics and deductive reasoning to decipher a code. The following setup is based on the game *Mastermind* by Invictus.

The object is to decipher a secret three-letter code word consisting of the letters A, E, I, O, and U. Letters may be repeated one, two, or three times in the word. For example, the following could be possible secret code words: EEE, OEO, IOU.

The code-breaker makes a series of three-letter guesses, and each one is followed by a reply containing information about whether any of the letters is correct and in the correct position, or if any letter is correct but not in the correct position. Each reply of **x** means that one letter is correct and is in the correct position, although there is no information about which letter is the correct one. Each reply of **o** means that one letter is correct but is not in the correct position. Again, there is no information about which letter is the correct one.

For example, suppose the code-breaker guesses I O U and the reply is **x o**. This might mean that the I is correctly placed in the first position and U is in the code but is not in the third position. If that's the case, then the U would have to be in the second position, and you also know that there is no O in the secret code word. (How do you know this?)

However, there are other possibilities: The O could be the correct letter in the correct position; the I might be in the third position in the secret code word; and there might not be a U in the word.

1. Here is a series of guesses and replies to help decipher a secret code word:

Guess Number	Code-Word Guess	Reply
1	A I I	x
2	O I U	x o
3	E I O	o

Answer the following questions about the secret code word that you are trying to find:

a. If there is an I in the secret code word, where is it? How do you know this?

b. If there is an I in the secret code word, can there be an A? Why or why not?

c. If there is an I in the secret code word, can there be an O? Why or why not?

d. If there is an I in the secret code word, can there be an E? Why or why not?

e. Can there be two I's? Why or why not?

f. If there is an I, can there be a U? Why or why not?

g. What do the answers to parts (a)–(f) lead you to conclude about the existence of an I in the secret code word?

h. If there is an A, where is it?

i. If there is an A, what are the other two letters?

j. If there is an A, what is the code?

k. Is this the only possible code that fits the previous clues?

2. Here is another series of guesses and replies for a new secret code word:

Guess Number	Code-Word Guess	Reply
1	A E O	o o
2	E I A	x
3	U O A	x

Fill in a "useful" consequent in each of the blank spaces, and answer the questions.

a. If U is the correct letter in guess #3, then _____.

b. Can U be the correct letter in guess #3?

c. If O is the correct letter in guess #3, then _____.

d. Can O be the correct letter in guess #3?

e. If A is the correct letter in guess #3, then _____.

f. What is the secret code word? Is this the only possible code word?

If-then Statements

3. Write the converse and the contrapositive of each of the following statements:

a. "If there is an I in the code word, then there is not an A in the code word."

i. Converse:

ii. Contrapositive:

b. "If teenagers are caught in a car with alcohol, then they will lose their license."

    i. Converse:

    ii. Contrapositive:

4. Look at the examples of if-then statements and their converses and contrapositives in Question 3. Using these examples, explain why the contrapositive says the same thing as the original statement, but the converse does not.

5. In the following exercises, you will compare the truth values of the conditional statement "if P, then Q" with the truth values of its contrapositive and converse.

    a. Fill in the following truth table:

P	Q	not P	not Q	if P, then Q	if (not Q), then (not P)
T	T				
T	F				
F	T				
F	F				

    b. Explain why the truth table you completed in part (a) shows that the contrapositive of the statement "if P, then Q" is equivalent to the original conditional statement.

c. Fill in the following truth table:

P	Q	if P, then Q	if Q, then P
T	T		
T	F		
F	T		
F	F		

d. Explain why the truth table you completed in part (c) shows that the converse of the statement "if P, then Q" is not equivalent to the original conditional statement.

Summary

In this activity, you practiced using logic by playing a code-breaking game and formulating conditional statements to help you break the code. You also composed truth tables for the contrapositive and converse of the conditional statement "if P, then Q" and constructed the contrapositive and converse of several conditional statements.

13-2

Compound Statements Used in Reasoning

In this activity, you will analyze compound statements found in news articles and quotes from famous people. You will formulate and analyze the negation of these statements. You will also investigate the equivalence of statements using truth tables.

A compound statement is formed from two other statements using logical connectors. You'll work with three different connectors: *and*, *or*, and *if-then*.

1. For each of the following compound statements, identify the two statements P and Q that are joined to make the compound statement. Then give the form of the compound statement ("P and Q," "P or Q," "if P, then Q"), and identify the type of compound statement (conjunction, disjunction, conditional).

   a. "If the freedom of speech is taken away then dumb and silent we may be led, like sheep to the slaughter." (*Source:* George Washington.)

      i. P:

      ii. Q:

    iii. Form:

    iv. Type:

b. "You can keep Johnson in Washington, D.C., or you can send him back to his Texas cotton patch." (*Source:* Malcolm X, "The Ballot or the Bullet Speech," April 12, 1964.)

    i. P:

    ii. Q:

    iii. Form:

    iv. Type:

c. "This is a world of compensations, and he who would *be* no slave, must consent to *have* no slave." (*Source:* Abraham Lincoln, Letter to Henry Pierce, April 6, 1859.)

    i. P:

    ii. Q:

    iii. Form:

    iv. Type:

d. "[Senator Kerry said Monday that] he would have voted to give the president the authority to invade Iraq even if he had known all he does now about the apparent dearth of unconventional weapons or a close connection with Al Qaeda." (*Source:* "Kerry Says His Vote on Iraq Would Be the Same Today," *The New York Times,* August 10, 2004.)

   i. P:

   ii. Q:

   iii. Form:

   iv. Type:

e. "The Cubans have lost some prominent players to defection and retirement, but still have more than enough to win another gold." (*Source:* "Olympic Baseball Pallid without U.S-Cuba Clash," *The Providence Sunday Journal,* August 15, 2004.)

   i. P:

   ii. Q:

   iii. Form:

   iv. Type:

f. "It's going to be easier for us to hire our vets if the overall economy is going strong." (*Source:* Remarks by President Barack Obama, www.whitehouse.gov/the-press-office/ 2011/11/21/remarks-president-and-first-lady-bill-signing.)

  i. P:

  ii. Q:

  iii. Form:

  iv. Type:

The negation of a statement is a statement that is false if the original statement is true and is true if the original statement is false. One way to negate a statement P is to say, "It is not the case that P." However, there are generally more useful ways to negate a statement, especially a compound statement.

2. For a statement of the type "P and Q" to be true, both statement P and statement Q must be true. So, the negation of the statement "P and Q" is that at least one of the statements P, Q is not true; that is, "(not P) or (not Q)."

  a. Complete the following truth table that summarizes the relationship between the compound statements "P and Q" and "(not P) or (not Q):"

P	Q	P and Q	not P	not Q	(not P) or (not Q)
T	T				
T	F				
F	T				
F	F				

b. Explain how the truth table shows that the negation of the statement "P and Q" is equivalent to the statement "(not P) or (not Q)."

3. Complete the following truth table and explain how it shows that the negation of the statement "P or Q" is equivalent to the statement "(not P) and (not Q):"

P	Q	P or Q	not P	not Q	(not P) and (not Q)
T	T				
T	F				
F	T				
F	F				

4. Complete the following truth table and explain how it shows that the negation of the statement "if P, then Q" is equivalent to the statement "P and (not Q):"

P	Q	if P, then Q	not Q	P and (not Q)
T	T			
T	F			
F	T			
F	F			

5. Give a useful negation of the following statements relating to the code-breaking activity (Activity 13.1):

    a. A is in the code or I is in the code.

    b. O is in the code and U is not in the code.

    c. If E is in the code, then it is in the first position.

    d. If U is in the code, then A is not in the code.

6. Give a useful negation of the following statements:

    a. Children caught with cigarettes or other tobacco products on public school property in Pennsylvania often pay larger fines than stores that are cited for selling tobacco to minors.

    b. If no one gets more than half the vote on November 2, a runoff will be held in December.

c. If teenagers are caught in a car with alcohol, then they will lose their license.

d. "You can keep Johnson in Washington, D.C., or you can send him back to his Texas cotton patch."

e. House prices are down but mortgage loans are hard to obtain.

Summary

In this activity, you practiced identifying P and Q in the three types of compound statements. You used truth tables to confirm the relationship between the negation of a compound statement and an equivalent compound statement. You also formulated the negation of a variety of compound statements.

13-3

Quantified Statements and Deductive Reasoning: Direct and Indirect Reasoning

In the first part of this activity, you will identify the type of quantified statement and will formulate negations of quantified statements. In the second part of this activity, you will use the two forms of deductive reasoning to draw conclusions from given statements; you will also identify the type of deductive reasoning used to draw a conclusion.

Quantified statements can be expressed using one of the two quantifiers: the universal quantifier *all*, or the existential quantifier *there exists*.

1. For each of the following statements,

    i. Identify the type of quantifier it contains, and write an equivalent statement using the quantifier *all*, or *there exists*.

    ii. Write the negation of the statement.

a. "Some astronomers hope that some of these functions can be performed by the James Webb Space Telescope—NASA's Hubble successor, overdue and over budget, now scheduled for a launch in 2018." (*Source:* "Science Hot on Trail of Livable Planets, but Tenants May Be Slime," *The New York Times*, December 3, 2011.)

    i.

    ii.

b. "Each period of our national history has had its special challenges." (*Source:* Harry S. Truman, Inaugural Address, January 20, 1949.)

    i.

    ii.

c. "No army can withstand the strength of an idea whose time has come." (*Source:* Victor Hugo, www.quotes.net.)

    i.

    ii.

d. "Several other Republicans and a chorus of Democrats have also questioned the change, with some proposing legislation to prohibit it." (*Source:* "Senators Criticize Decision to Allow Scissors on Planes," *The New York Times*, December 13, 2005.)

    i.

    ii.

A deductive argument consists of premises (or hypotheses or assumptions) and conclusions that follow logically from those premises. There are two key elements for good deductive reasoning: (1) The premises are true and (2) the reasoning is valid.

The two forms of valid deductive reasoning are summarized here.

Direct Reasoning (Modus Ponens)

If the statement "if P, then Q" is true, and the statement P is true, then the statement Q is true.

Indirect Reasoning (Modus Tollens)

If the statement "if P, then Q" is true, and the statement Q is false, then the statement P is false.

2. In each of the following cases, use direct or indirect reasoning to draw a conclusion from the statements given, and identify the type of reasoning used.

    a. If I drink coffee between 7 and 8 PM, I cannot fall asleep until midnight.
        This evening I drank coffee at 7:15 PM.

        i. Conclusion:

        ii. Type of reasoning used:

    b. If you are younger than 21, you cannot drink alcohol legally.
        You can legally drink alcohol.

        i. Conclusion:

        ii. Type of reasoning used:

3. In each of the following situations, assume that the premises are true. If the premises allow a conclusion through valid reasoning, state the conclusion and describe why it is valid. If a conclusion is not possible, write "no valid conclusion" and explain why there is no valid conclusion.

    a. If a salesperson is rude with the customers, then the salesperson will be fired.
        Shanon, a salesperson, is rude with the customers.

b. If a salesperson is rude with the customers, then the salesperson will be fired.
John, a salesperson, was fired.

c. If a salesperson is rude with the customers, then the salesperson will be fired.
Bob, a salesperson, is not rude with the customers.

d. If you liked the play, then you'd like the movie.
You did not like the movie.

e. If you liked the play, then you'd like the movie.
You liked the movie.

f. If you liked the play, then you'd like the movie.
You liked the play.

g. All American presidents have been men. (Note this is equivalent to stating, "If a person has been an American president, then the person has been a man.")
John Kennedy was a man.

h. All American presidents have been men.
Dana Brown has not been an American president.

i. All American presidents have been men.
T. Roosevelt was president.

Summary

In this activity, you practiced working with quantified statements and their negations. You also used the two basic forms of deductive reasoning: direct reasoning (Modus Ponens) and indirect reasoning (Modus Tollens), and identified invalid reasoning.

Apportionment

Tanning salons raise skin cancer risk, study indicates

Paul Recer,
Associated Press

Washington—Regularly baking to a golden tan under sun lamps can increase the risk of malignant melanoma, a sometimes fatal skin cancer, and the younger a woman starts the greater the risk, a study says.

P	Q	not P	not Q	P and Q	not (P and Q)	(not P) or (not Q)
T	T	F	F	T	F	F
T	F	F	T	F	T	T
F	T	T	F	F	T	T
F	F	T	T	F	T	T

State	Population	Standard Q
Delaware	897,934	0.4084
New Jersey	8,791,894	3.9992
New York	19,378,102	8.8145
Pennsylvania	12,702,379	5.7779
Total	**41,770,309**	

Country	Per Capita Total Spending on Health (in U.S. dollars at average exchange rates)	Per Capita Daily Calorie Supply	Infant Mortality Rate per 1,000 Births
Australia	3,986	3,674	5
Cambodia	36	2,046	62
China	108	2,951	23
France	4,672	3,654	4
Germany	4,209	3,496	4
India	40	2,954	55
Italy	3,136	3,617	4

After completing this topic, you will be able to:

- Use and understand the terminology of apportionment.

- Recognize several methods of apportionment.

- Use several methods of apportionment to decide how many representatives each subgroup of a larger group should have, given a fixed total number of representatives.

- Calculate geometric means.

The number of representatives from each state in the House of Representatives is proportional to the state's population. This number is updated every 10 years when the national census takes place. Throughout the years, several different methods, called methods of apportionment, have been used or proposed to decide how many representatives each state should have. Apportionment methods are useful in any situation in which a small number of delegates or representatives is to be chosen to represent a larger population that consists of separate groups, in a manner consistent with the size of these groups. For example, we could use an apportionment method to decide how many student council members each dorm would elect, in such a way that the number of members is proportional to the size of the dorm. In this topic, we will discuss several apportionment methods.

The American Congress consists of the Senate and the House of Representatives. The Senate is composed of two senators from each state. The House, on the other hand, has representatives from all states, but the number of representatives per state varies according to the state's population. This structure was established by the United States Constitution, which includes the following paragraph from Article I, Section 2:

> Representatives and direct taxes shall be apportioned among the several states which may be included within this Union, according to their respective numbers, which shall be determined by adding to the whole number of free persons, including those bound to service

for a term of years, and excluding Indians not taxed, three fifth of all other persons. The actual enumeration shall be made within three years after the first meeting of the Congress of the United States, and within every subsequent term of ten years, in such manner as they shall by law direct. The number of representatives shall not exceed one for every thirty thousand, but each state shall have at least one representative. . . .

Since 1790 a national census has been conducted every 10 years to determine each state's population. The actual population considered has changed with time. Originally, some sectors of the population were not counted, but since the Fourteenth Amendment in 1868 the total emancipated slave population was included, and in 1940 it was determined that no Native American would be classified as "not taxed."

Apportionment is the process of distributing, according to some plan, the number of seats to which each state is entitled in the U.S. House of Representatives, in proportion to that state's population based on the 10-year census. It is also used for other processes of distribution proportional to population. The result of such a process is also called apportionment.

The total number of seats in the House is determined by Congress and has changed from the 65 set by the constitution to the current number of 435. It is not possible to have an exact proportional distribution of seats as the following example shows.

Example 14.1

According to the 2010 census, the population of New York State is 19,378,102 and the total population of the United States is 308,745,538. Assuming that each state is allotted a number of seats in the House of Representatives in exactly equal proportion to its population, how many of the total 435 seats would correspond to New York? Explain why this is not a practical answer.

Solution

Because the total number of seats is 435, and the total population is 308,745,538, each seat corresponds to $\frac{308,745,538}{435} \approx 709,759.858$ people. To find the number of representatives for New York, we then divide the state's population by the number of people per seat: $\frac{19,378,102}{709,759.858} \approx 27.302$.

The number of seats that corresponds to New York is 27.302. Because the number of seats per state has to be a whole number, the number of seats for New York cannot be given by this answer. We will need to round this number up or down to obtain either 27 or 28 seats.

The fact that the number of seats allotted to each state must be an integer number makes apportionment a difficult task. Congress has used various methods of apportionment at different times, each method designed to correct problems that the previously used method had, but introducing a new bias. Mathematically, it is not possible to find a "perfect" method, so the choice of method is a combination of mathematics and politics. We will describe six apportionment methods after we introduce some appropriate terminology.

The **standard divisor** is the nation's total population divided by the total number of seats: standard divisor $= \dfrac{\text{total population}}{\text{total number of seats}}$. It can be interpreted as the average number of people represented by any one House member. This is the number we calculated in Example 14.1 for the population given by the 2010 census and for the current total number of seats of 435.

A state's **standard quota** (also called the **exact quota**) is the number obtained by dividing the state's population by the standard divisor:

$$\text{standard quota} = \frac{\text{state's population}}{\text{standard divisor}}$$

This is the fraction of the total number of seats that the state is entitled to have. If the standard quota for each state were always an integer, then this would be the allotted number of seats for each state and a fair apportionment would be guaranteed. Because the standard quota is generally a noninteger, it is natural to select the integer just below this number or the integer just above it. However, ordinary round-off does not quite work as the following example shows.

Example 14.2

Suppose the states of Delaware, New Jersey, New York, and Pennsylvania want to form a Middle Atlantic States Council with 15 members, in such a way that representation is proportional to each state's population and each state has at least one representative. The following table gives the population for each of these four states, from data collected in the 2010 census.

State	Population
Delaware	897,934
New Jersey	8,791,894
New York	19,378,102
Pennsylvania	12,702,379

a. Find the standard divisor and the standard quota for each of the given states. Keep at least three decimal places.

b. Round off each standard quota using the usual round-off method (that is, if its decimal part is 0.5 or larger, the standard quota is rounded up to the next integer; otherwise, it is rounded down). Determine if these quotas can be used to apportion seats for the 15-member council, so that each state has at least one representative.

Solution

a. The total population to be represented on this council is the sum of the populations of the four states: $897{,}934 + 8{,}791{,}894 + 19{,}378{,}102 + 12{,}702{,}379 = 41{,}770{,}309$. Using this number as the total population and 15 as the total number of seats, we compute the standard divisor:

$$\text{standard divisor} = \frac{41{,}770{,}309}{15} \approx 2{,}784{,}687.267$$

Now we compute each state's standard quota:

$$\text{Delaware's standard quota} = \frac{897{,}934}{2{,}784{,}687.267} \approx 0.3225$$

$$\text{New Jersey's standard quota} = \frac{8{,}791{,}894}{2{,}784{,}687.267} \approx 3.1572$$

$$\text{New York's standard quota} = \frac{19{,}378{,}102}{2{,}784{,}687.267} \approx 6.9588$$

$$\text{Pennsylvania's standard quota} = \frac{12{,}702{,}379}{2{,}784{,}687.267} \approx 4.5615$$

b. Rounding off to the nearest integer, we get 0, 3, 7, and 5. Because every state must have at least one representative, we cannot assign 0 to Delaware, so we assign 1 seat to Delaware. The following table summarizes our result:

State	Population	Standard Quota	Rounded-Off Quota	Apportionment
Delaware	897,934	0.3225	0	1
New Jersey	8,791,894	3.1572	3	3
New York	19,378,102	6.9588	7	7
Pennsylvania	12,702,379	4.5615	5	5
Total			**15**	**16**

This method would result in one more seat than the 15 seats the council will have. We cannot use this method to apportion the 15 seats in a manner proportional to the population of each state and under the condition that all four states are represented.

QUOTA METHODS

The methods of apportionment used by Congress since 1850 can be classified into two groups: **quota methods** and **divisor methods**. Quota methods work with the standard divisor and round off the standard quotas, according to some rules, while divisor methods work with divisors other than the standard divisor.

Quota methods consist of assigning seats in such a way that the final number corresponding to any given state is either the largest integer below its standard quota, called the **lower quota**, or the smallest integer above its standard quota, called the **upper quota**. Thus, if a state's standard quota is, for example, 6.356, then a quota method will assign to this state either 6 or 7 seats. Because the total number of seats is fixed, any apportionment method must allow for some states to receive their lower quota, while other states will receive their upper quota as the final number of seats. Two quota methods are described next.

Hamilton's Method

Hamilton's method, also called the Hamilton/Vinton method, consists of assigning the lower quota (except when it is 0) to each state at first and then assigning any remaining seats to the states whose standard quotas have the largest fractional (decimal) part. To implement this method, we follow these steps:

1. Find the standard divisor.

2. Find each state's standard quota.

3. Assign to each state the integer part of its standard quota (this is the lower quota), unless this is 0. If the integer part of the standard quota for any state is 0, assign 1 seat to that state.

4. If there are seats remaining, then, eliminating any state that had 0 as its lower quota, find the state that has the largest fractional part and assign one more seat to this state. If there are still remaining seats, assign the next seat to the state with the next largest fractional part. Continue in this manner until all seats are assigned.

Example 14.3

Suppose that the Middle Atlantic States Council discussed in Example 14.2 will now have 19 members. Use Hamilton's method to apportion the number of seats for each of the four states.

Solution

We will follow the four steps described previously:

a. The standard divisor is the total population of the four states divided by the total number of seats:

$$\text{standard divisor} = \frac{41,770,309}{19} \approx 2,198,437.316$$

b. To find each state's standard quota, we divide the state's population by the standard divisor and obtain these numbers (rounded to four decimals):

$$\text{Delaware's standard quota} = \frac{897,934}{2,198,437.316} \approx 0.4084$$

$$\text{New Jersey's standard quota} = \frac{8,791,894}{2,198,437.316} \approx 3.9992$$

$$\text{New York's standard quota} = \frac{19,378,102}{2,198,437.316} \approx 8.8145$$

$$\text{Pennsylvania's standard quota} = \frac{12,702,379}{2,198,437.316} \approx 5.7779$$

c. Choosing the lower quota when the lower quota is not 0, and 1 if the lower quota is 0, we first assign Delaware 1, New Jersey 3, New York 8, and Pennsylvania 5.

d. The total number of seats assigned so far is $1 + 3 + 8 + 5 = 17$. We have two remaining seats, so we look at the fractional parts of those states with nonzero lower quotas. These fractional parts are as follows: New Jersey 0.9992, New York 0.8145, and Pennsylvania 0.7779. Thus, New Jersey, with the largest fractional part, gets the first remaining seat. Because there is another seat remaining, we look for the state with second largest fractional part. The second remaining seat goes to New York.

We summarize the results in the following table:

State	Standard Quota	Integer Part or 1	Fractional Part	Apportionment by Hamilton's Method
Delaware	0.4084	1		1
New Jersey	3.9992	3	0.9992	4
New York	8.8145	8	0.8145	9
Pennsylvania	5.7779	5	0.7779	5
Total		**17**		**19**

Lowndes' Method

The only difference between Lowndes' method and Hamilton's method is that Lowndes' method uses the **relative fractional part** to decide which states get the extra seats. The relative fractional part of a number is the ratio of its fractional part divided by its integer part. For example, the relative fractional part of 3.79 is $\frac{0.79}{3} \approx 0.2633$. To use Lowndes' method, we follow Steps 1 through 3 of Hamilton's method and then replace Step 4 with the following:

4. If there are any remaining seats, compute the relative fractional parts of the standard quotas (with the exception of those that had a lower quota of 0) and assign one of the remaining seats to the state with the largest relative fractional part. If seats still remain, we assign the next seat to the state with the second largest relative fractional part. We continue in this manner until all seats are assigned.

Example 14.4

Use Lowndes' method to apportion the number of representatives in the 19-member council discussed in Example 14.3.

Solution

We have already computed each state's standard quota and their integer and fractional parts (see the table at the end of Example 14.3). Now we calculate each state's relative fractional part for all states other than Delaware, which has the lower quota of 0 and automatically gets 1 representative. The relative fractional parts are as follows:

$$\text{New Jersey's relative fractional part} = \frac{0.9992}{3} \approx 0.3331$$

$$\text{New York's relative fractional part} = \frac{0.8145}{8} \approx 0.1018$$

$$\text{Pennsylvania's relative fractional part} = \frac{0.7779}{5} \approx 0.1556$$

The state with the largest relative fractional part is New Jersey, and Pennsylvania has the second largest relative fractional part, so they each get one additional representative. The apportionment is given in the last column of the following table:

State	Standard Quota	Integer Part or 1	Relative Fractional Part	Apportionment by Lowndes' Method
Delaware	0.4084	1		1
New Jersey	3.9992	3	0.3331	4
New York	8.8145	8	0.1018	8
Pennsylvania	5.7779	5	0.1556	6
Total		**17**		**19**

The use of the relative fractional part instead of the fractional part gives smaller states a slight advantage over larger states. This is so because if two states have equal fractional parts, the relative fractional part of the state with the smaller standard quota will be larger. (Consider two states with standard quotas of 3.6781 and 7.6781. Their relative fractional parts are $\frac{0.6781}{3} \approx 0.2260$ and $\frac{0.6781}{7} \approx 0.0969$, respectively.)

DIVISOR METHODS

Recall that the quota methods, Hamilton's and Lowndes', use the standard divisor $\frac{\text{total population}}{\text{total number of seats}}$. Divisor methods use **modified divisors** instead of the standard divisor. The quotient of a state's population divided by a modified divisor is called the state's **modified quota**. Each divisor method uses a different procedure to round a state's modified quota, and the modified divisor is chosen in such a way that when the modified quotas are computed and rounded off to an integer value, the total number of resulting seats is exactly the number of seats available.

We describe four divisor methods: Jefferson's method, Adam's method, Webster's method, and the Huntington−Hill method.

Jefferson's Method

Jefferson's method is also known as the method of greatest divisors. The method consists of finding a (modified) divisor D so that when each state's modified quota is *rounded down* and all are added, then the total number is exactly the number of seats to be allotted. For example, if a state's population is 12,005,678 and the modified divisor is $D = 1,789,000$, then this state's modified quota is $\frac{12,005,678}{1,789,000} \approx 6.71$, so the number of seats allotted to this state would be 6, which is 6.71 rounded down.

Because we do not know in advance which divisor D will work, we might have to try a few values of D until we find the right one. (We can estimate whether D must be larger or smaller than the standard divisor, but we do not know how much larger or smaller.) We illustrate this method with an example.

Example 14.5

Use Jefferson's method to apportion the 19 seats of the Middle Atlantic Council discussed in Examples 14.3 and 14.4.

Solution

In Example 14.3, we computed each state's standard quota using the standard divisor $= \frac{41,770,309}{19} \approx 2,198,437.316$ and obtained the values in the following table:

State	Population	Standard Quota
Delaware	897,934	0.4084
New Jersey	8,791,894	3.9992
New York	19,378,102	8.8145
Pennsylvania	12,702,379	5.7779
Total	**41,770,309**	

If we round down each standard quota, with the exception of Delaware's, which we need to set as 1, we have Delaware 1, New Jersey 3, New York 8, and Pennsylvania 5. This gives us a total of 17, while the total number of seats is 19. This tells us that the standard divisor cannot be the modified divisor, and it gives us an idea of which divisors to try. Because using the standard divisor leads to too few seats allotted, we need to try a divisor that will give larger modified quotas. The population of each state does not change with the choice of divisor, so the way to get larger modified quotas is to divide by a smaller number. Therefore, we should try a number less than the standard divisor of 2,198,437.316 as our first guess for modified divisor. Let's try $D = 2,000,000$. Then the modified quotas for each state are as follows:

$$\text{Delaware's modified quota} = \frac{897,934}{2,000,000} \approx 0.449$$

$$\text{New Jersey's modified quota} = \frac{8,791,894}{2,000,000} \approx 4.396$$

$$\text{New York's modified quota} \approx 9.689$$
$$\text{Pennsylvania's modified quota} \approx 6.351$$

Rounding down the modified quotas (with the exception of Delaware that has to get at least one seat), we obtain 1, 4, 9, and 6. These add up to 20, which is too large. Now we know that the divisor must be greater than 2,000,000 and less than 2,198,437.316. If we try

$D = 2,150,000$, the modified quotas (rounded to three decimal places) are 0.418, 4.089, 9.013, and 5.908, respectively. Rounding down all but Delaware's, we obtain 1, 4, 9, and 5. Because their sum is 19, we have found the modified divisor $D = 2,150,000$ that allows us to allot 1 seat to Delaware, 4 to New Jersey, 9 to New York, and 5 to Pennsylvania.

The following table summarizes our work:

State	Standard Quota ($D =$ 2,198,437.316)	Standard Quota Rounded Down	Modified Quota ($D =$ 2,000,000)	Modified Quota Rounded Down	Modified Quota ($D =$ 2,150,000)	Number of Seats by Jefferson's Method
Delaware	0.4084	1	0.449	1	0.418	1
New Jersey	3.9992	3	4.396	4	4.089	4
New York	8.8145	8	9.689	9	9.013	9
Pennsylvania	5.7779	5	6.351	6	5.908	5
Total		**17**		**20**		**19**

As discussed in Example 14.5, the modified divisor that results when applying Jefferson's method is a number less than the standard divisor; this smaller modified divisor produces greater modified quotas.

Adam's Method

Similar to Jefferson's method, Adam's method uses the same idea of finding a modified divisor that would lead to the total number of seats. The only difference between these two methods is that in Adam's method the modified quotas are *rounded up*. This means that the modified divisor will be greater than the standard divisor and will produce smaller modified quotas.

Webster's Method

Webster's method is another divisor method, also called the method of major fractions. In this method, modified quotas are rounded off to the nearest integer in the standard way. (That is, if its fractional part is 0.5 or larger, the modified quota is rounded up; otherwise, it is rounded down.) The divisor D is chosen so that the modified quotas, when rounded to the nearest integer, add up to the total number of seats.

Huntington–Hill Method

The Huntington–Hill method, also called the method of equal proportions, is the method currently used to apportion seats in Congress. In this method, the modified quotas are rounded to an integer value using the **geometric mean**. The geometric mean of two numbers a and b is the square root of their product:

$$\text{geometric mean of } a \text{ and } b = \sqrt{a \cdot b}$$

If q is a modified quota, and q is between the integers n and $n + 1$, then the Huntington–Hill method uses the geometric mean of n and $n + 1$ as the cutoff point to decide whether q will be rounded up to $n + 1$ or rounded down to n. That is, if $q \leq \sqrt{n \cdot (n + 1)}$, then q is rounded down to n, and if $q > \sqrt{n \cdot (n + 1)}$, then q is rounded up to $n + 1$.

Example 14.6

Suppose the Huntington–Hill method is used to apportion the 19 seats of the council discussed in Example 14.5. Will the modified divisor $D = 2{,}150{,}000$ result in a correct apportionment of the 19 seats? If the answer is no, would a correct divisor be greater than or less than D?

Solution

In Example 14.5, we computed the modified quotas corresponding to the divisor $D = 2{,}150{,}000$, which are as follows: Delaware, 0.418; New Jersey, 4.089; New York, 9.013; and Pennsylvania, 5.908. Because a state cannot have 0 seats, Delaware will get 1 seat (independently of the geometric mean). To see how many seats New Jersey will obtain, we compute the geometric mean of 4 and 5 since 4.089 is between 4 and 5. This geometric mean is $\sqrt{4 \cdot 5} \approx 4.472$. Because New Jersey's modified quota is 4.089, and $4.089 < 4.472$, the number of seats for New Jersey is rounded down to 4. Similarly, we compute the corresponding geometric means for the remaining states, obtaining $\sqrt{9 \cdot 10} \approx 9.487$ for New York and $\sqrt{5 \cdot 6} \approx 5.477$ for Pennsylvania; we compare each of the geometric means with the modified quota of that state. Because the modified quota for New York is 9.013, and $9.013 < 9.487$, New York gets 9 seats. Because Pennsylvania's modified quota is 5.908, and $5.908 > 5.477$, Pennsylvania will get 6 seats. The total number of seats, using this method and the data given, is $1 + 4 + 9 + 6 = 20$. So, the modified divisor $D = 2{,}150{,}000$ does not give an apportionment totaling 19 seats. To get smaller modified quotas, we need to use a divisor that is greater than the standard divisor. We explore this method further in Activity 14.2.

Summary

In this topic, we discussed six methods of apportionment. Two of the methods, Hamilton's and Lowndes', are quota methods that use the standard divisor to obtain the standard quotas. The other four methods discussed, Jefferson's method, Adam's method, Webster's method, and the Huntington–Hill method, use modified divisors.

Explorations

1. Consider Lowndes' method for apportioning representatives.

    a. Describe this method.
    b. Does this method favor larger or smaller states?
    c. Explain in detail your answer to part (b) of this Exploration and how it works mathematically.

2. For each of the following divisor methods, indicate whether the divisor is greater than or less than the standard divisor and explain why.

    a. Adam's method
    b. Jefferson's method
    c. Webster's method
    d. Huntington–Hill method

3. Three colleges in the same region, colleges A, B, and C, decide to form a 10-member committee of students representing the three colleges. It is decided that representation will be proportional to the number of students in each college in such a way that each college will have at least one representative. College A has 1,455 students; College B has 1,683 students; and College C has 2,706 students.

    a. Find the standard divisor and each college's standard quota.
    b. Use Hamilton's method to apportion the number of committee members to each college.
    c. Use Lowndes' method to apportion the number of committee members to each college.
    d. Which of the two methods would the students of College A prefer? Explain your answer.
    e. If we use the Huntington–Hill method with a modified divisor of 750, how many members would each college have?
    f. Is 750 a correct choice of modified divisor to apportion the 10 committee positions using the Huntington–Hill method?
    g. Use Jefferson's method with a modified divisor of 600 to assign the number of members from each of the three colleges. Is 600 a correct choice of modified divisor? If not, would the correct number to use as a modified divisor be greater than or less than 600?

4. Suppose the three states of Maine, Massachusetts, and New Hampshire want to form a tri-state council with 11 members. The organizers want to establish representation based on

population. According to the 2010 census, the population of Maine is 1,328,361; the population of Massachusetts is 6,547,629; and the population of New Hampshire is 1,316,470.

a. Find the standard divisor and each state's standard quota.

b. If we apportion representatives using Hamilton's method, how many representatives would each state have?

c. If we apportion representatives using Lowndes' method, how many representatives would each state have?

d. If we use Adam's method with a modified divisor of 830,000, how many representatives would each state have?

e. If we use Adam's method with a modified divisor of 950,000, how many representatives would each state have?

f. Which of the two modified divisors from parts (d) and (e) of this Exploration would be the one to use with Adam's method?

g. How many representatives would each state have if we use Webster's method with the standard divisor? Can we apportion the 11 representatives in this manner?

h. Can we apportion representatives if we use Webster's method with a modified divisor of 880,000? Why or why not?

5. Suppose the three states of Maine, Massachusetts, and New Hampshire want to have 30 members in their tri-state council.

a. Find the standard divisor and each state's standard quota.

b. If we apportion representatives using Hamilton's method, how many representatives would each state have?

c. If we apportion representatives using Lowndes' method, how many representatives would each state have?

d. Which of those two methods would you prefer if you wanted Maine to have as many representatives as possible?

6. Suppose the tri-state council described in Exploration 4 will have 11 representatives, but the population of each state is double what it was in the 2010 census. How will each state's standard quota change? Explain your reasoning.

7. Find the history of apportionment methods used to apportion seats in the U.S. House of Representatives. Find which methods have been used, when they were used, and why these methods have been challenged.

8. The countries of Argentina, Chile, Paraguay, and Uruguay form the geographic region known as the Southern Cone (Cono Sur) because of their location on the southernmost area of South America. Suppose those four countries decide to form an economic council of 12 members, with representation proportional to their populations. Estimates for each country's population (as of July 2010) are given as follows:

Country	Population
Argentina	41,343,201
Chile	16,746,491
Paraguay	6,375,830
Uruguay	3,510,386

a. Find the standard divisor and each country's standard quota.

b. If we use Hamilton's method to apportion representatives, how many representatives would each country have?

c. If we use Lowndes' method, how many representatives would each country have?

d. If we use Adam's method with a modified divisor of 6,000,000, how many representatives would each country have? Is 6,000,000 a good choice in this case? Explain why or why not.

e. If we use the Huntington–Hill method with the modified divisor of 6,000,000, how many representatives would each country have? Is 6,000,000 a good choice in this case? Explain why or why not.

f. If we use Webster's method with a modified divisor of 5,500,000, how many representatives would each country have? Is 5,500,000 a good choice in this case? Explain why or why not.

g. If we use the Huntington–Hill method with a modified divisor of 5,500,000, how many representatives would each country have? Is 5,500,000 a good choice in this case? Explain why or why not.

9. Suppose the countries of Argentina, Chile, Paraguay, and Uruguay decide to have 25 members in their council. Use the population information given in Exploration 8 above.

a. Find the standard divisor and each state's standard quota.

b. If we use Hamilton's method to apportion representatives, how many representatives would each country have?

c. If we use Lowndes' method to apportion representatives, how many representatives would each country have?

d. Use Jefferson's method with a modified divisor of 2,000,000 to assign the number of members from each country. Is 2,000,000 a correct choice of modified divisor? If not, would the correct modified divisor be greater than or less than 2,000,000?

e. Use the Huntington–Hill method with modified divisor 2,500,000 to assign the number of members from each country. Is 2,500,000 a correct choice of modified divisor? If not, would the correct modified divisor be greater than or less than 2,500,000?

14-1

Methods of Apportionment: Quota Methods

In this activity, you will work with the two quota methods of apportionment: Hamilton's method and Lowndes' method. You will use each of these methods to determine how many representatives in the U.S. House of Representatives would correspond to each state if the method were currently used. All apportionment methods assign to each state a number of representatives based on the state's population.

1. The Excel file "EA14.1 State Population.xls" contains the population of each state, as given by the 2010 census figures. (*Source:* Census Bureau, www.census.gov.)

   a. Open the file "EA14.1 State Population.xls," and then find the total U.S. population. (Instructions to remind you how to do this follow.)

      Total Population : _____

Instructions to Add Consecutive Elements of a Column

To find the total population, you want to add all the state's populations, which in the file "EA14.1 State Population.xls," are the numbers in cells B2 through B51. To do this, enter **=SUM(B2:B51)** in cell B53. Name this cell using the **Name box** above column A on the spreadsheet. (You could use the name **totalpopulation**, or any other appropriate name.) Also add the label **Total Population =** in cell A53.

b. In cell A54, enter the label **Total Seats =** and in cell B54, enter the number **435**; this is the total number of seats in the House of Representatives. Name this cell. You might want to call it **totalseats**, or another appropriate name. Recall that the standard divisor is given by

$$\text{standard divisor} = \frac{\text{total population}}{\text{total number of seats}}$$

Ask Excel to find the standard divisor in cell B55 and name this cell. Add the label **Standard Divisor =** in cell A55.

Write the standard divisor here: _____

c. Recall that a state's standard quota is defined by the equation

$$\text{standard quota} = \frac{\text{state's population}}{\text{standard divisor}}$$

In column C, enter each state's standard quota. First, enter the column title **Standard Quota** in cell C1. Then ask Excel to do the computations for you by entering the appropriate formula in cell C2. Drag-and-fill to enter the rest of the values.

d. Can each state's standard quota be the apportioned number of representatives for that state? Why or why not?

In both quota methods, apportionment is done by rounding up or down each state's standard quota.

In Hamilton's method, each state is initially given the number of representatives equal to the integer value of the state's standard quota, unless this number is 0. For this case, the initial and final apportioned number of seats for that state is 1 (so that every state is represented).

After the initial apportionment is done using the integer value (or 1 when the integer value is 0), the remaining seats are assigned to the states with a standard quota larger than 1. The states with the largest **fractional part** of the standard quota are each given an extra seat.

2. Suppose that three states, State A, State B, and State C, have standard quotas as given here:

	Standard Quota	Initial Number of Representatives	Number of Seats by Hamilton's Method
State A	3.56		
State B	2.1		
State C	0.38		

a. Record in the table the initial number of representatives apportioned to those states when using Hamilton's method.

b. Suppose the total number of representatives that correspond to the three states is 7. Record in the table the number of representatives (seats) that correspond to each of the three states using Hamilton's method.

3. You will now use Excel to determine the number of representatives apportioned to each of the 50 states under Hamilton's method.

a. In column D of your spreadsheet, ask Excel to enter the initial number of seats for each state; that is, enter the integer part of the state's standard quota, unless this is 0. If the integer part of the standard quota is 0, then you'll enter **1**. To do this, enter the integer value of the first state's standard quota in cell D2 by typing **= INT(C2)**; then drag to auto-fill. If column D contains any zeros, go through and replace each 0 with **1**. Label column D with an appropriate label.

b. Use an appropriate formula to enter in column E the difference of each entry in column C minus the corresponding entry in column D. Label column E **Fractional Part**. List here the states for which the entries in column E are negative. Explain why these are negative.

c. Add the numbers in column D (use Excel) to decide how many seats have already been apportioned. Write that number here: _____.

Determine how many seats remain after the initial apportionment. Write the number of remaining seats here: _____.

This is the number of seats you will apportion using the fractional part of the standard quota.

d. To apportion the remaining seats using the fractional part of the standard quota, sort the data in columns A, B, C, D, and E by "Fractional Part" in descending order.

In column F, enter the final number of seats apportioned by Hamilton's method to each state. (*Note:* You do not need to enter these numbers one by one. Enter =**D2 + 1** in cell F2 and drag until you have apportioned all the remaining seats; then have Excel copy the numbers in column D for the remaining states.) Label column F **Seats Using Hamilton's Method**. Then sort the data (remember to highlight all columns) by "Seats Using Hamilton's Method," to help answer parts (i) and (ii) of this question.

i. List the four states that get the most number of seats and say how many seats each of them receives.

ii. List the states that receive just one seat.

Lowndes' method also uses the standard quota to apportion representatives. However, Lowndes' method uses the relative fractional part of the standard quota instead of the fractional part. The relative fractional part is used to determine which states' standard quotas will be rounded up to fill out the remaining seats.

The relative fractional part is the following quotient:

$$\text{relative fractional part} = \frac{\text{fractional part}}{\text{integer part}}$$

4. Assuming that State A and State D have the following standard quotas, record their relative fractional parts in the table.

	Standard Quota	Relative Fractional Part
State A	3.56	
State D	7.56	

5. To apportion the number of seats for each of the 50 states by Lowndes' method, you will first compute the relative fractional part of each state's standard quota.

   a. Insert two columns to the left of column F in your spreadsheet. In the new column F, recompute the integer part of the standard quota for each state. In cell G2, compute the relative fractional part of the standard quota for the state in row 2. Drag-and-fill to compute the relative fractional part for each state, and label column G **Relative Fractional Part**.

   b. Notice that you see #DIV/0! in several places in column G. Explain why this error message occurred and what it means.

   c. Label column I **Seats Using Lowndes' Method** and sort the data by "Relative Fractional Part," in descending order. Give one seat to each state showing the error message and explain why that's needed.

   d. Then apportion the rest of the seats, using the relative fractional part.
      i. List the four states that get the most number of seats and say how many seats each of them receives.

      ii. What is the smallest number of seats assigned? Which states get the smallest number of seats?

6. Name the states that would prefer Hamilton's method and those that would prefer Lowndes' method. Explain your answer.

7. Which states, those with a large population or those with a small population, are favored by Lowndes' method? Explain.

Summary

In this activity, you used Excel to investigate two quota methods of apportionment and compared the results of using them to apportion the seats of the House of Representatives. You learned the Excel command that gives the integer part of a real number and used it to find fractional parts.

14-2

Apportionment: Divisor Methods

In this activity, you will work with two divisor methods of apportionment, Adams' method and the Huntington–Hill method. With the help of Excel, you will determine how many representatives in the House of Representatives would correspond to each state if these methods were used.

Each method uses a modified divisor, that is, a divisor other than the standard divisor, to determine each state's modified quota. If the divisor D is used, then a state's modified quota is

$$\text{modified quota} = \frac{\text{state's population}}{D}$$

Then depending on the method used, all modified quotas are rounded up or down. In Adams' method, they are all rounded up; in Jefferson's method, they are rounded down. In Webster's method, each modified quota is rounded up or down to its nearest integer. The method currently used in the United States to apportion representatives is the Huntington–Hill method. This method uses the geometric mean to round off quotas.

Adams' Method

1. In this method, modified quotas are rounded up; thus, the modified divisor needs to be greater than the standard divisor. Why? Explain what would happen if the modified divisor were less than the standard divisor.

2. Now we will find how many representatives in the U.S. House each state would have when Adams' method is used.

   a. Open the "EA14.1 State Population.xls" file. Compute the total population in cell B53 and record the total population here: _____. To label the "total" row, enter **total =** in cell A53.

   b. Compute the standard divisor in cell B55 using this relationship:

   $$\text{standard divisor} = \frac{\text{total population}}{\text{total number of seats}}$$

   Record it here: _____. Enter the label **Standard divisor =** in cell A55.

   c. In column C, enter each state's standard quota using the relationship

   $$\text{standard quota} = \frac{\text{state's population}}{\text{standard divisor}}$$

   Use cell C53 to find the sum of all states' standard quotas and write the sum here: _____. Label column C by entering **Standard Quota** in cell C1.

   d. In cell A56, enter **Divisor =** and in cell D56 enter a number greater than the standard divisor. (Pick any number greater than the standard divisor—you'll change it later to find the "best" one.)

   Record the number you chose here: _____.

   e. Label column D by entering **Modified Quota** in D1. In cell D2, enter an appropriate formula and then drag it down to compute the modified quota for each state by using the modified divisor you entered in cell D56. (Make sure you enter a formula that will keep the location "D56" fixed when you drag the formula. Recall that you can do this by either naming cell D56 or by using $ signs.)

   f. In each cell of column E, enter the integer number obtained by rounding up the modified quota. (To accomplish this, enter **=INT(D2)+1** in cell E2 and drag to auto-fill the column.) Label the column **Number of Seats Using Adams' Method**.

   g. In cell E53, enter the sum of the numbers in cells E2 through cell E51. Record the sum here: _____, and explain what this number represents.

h. If the previous sum is exactly 435, you have succeeded in using Adams' method to apportion House seats. Explain why this is so.

i. If the sum is not 435, then you need to try other numbers for the modified divisor until 435 is reached. Note that when you change the number in cell D56 (the modified divisor), the numbers in column E, including the sum in E53, will also change automatically.

Using this table, record the divisors you've tried and the total number of House seats resulting from each divisor. Indicate which one is the modified divisor that actually works.

Divisor					
Total No. of Seats					

j. List the four states that would receive the largest number of seats if you used Adams' method and the number of seats each would get.

k. List the four states that would receive the smallest number of seats if you used Adams' method and the number of seats each would get.

Huntington–Hill Method

In this method, each modified quota is rounded up or down according to the **geometric mean**. The geometric mean of two numbers x and y is the square root of the product of the two numbers. That is,

$$\text{Geometric mean of } x \text{ and } y = \sqrt{xy}$$

3. Find the geometric mean of 30 and 17, and compare it with the average of the same two numbers.

4. In the Huntington–Hill method, the geometric mean of the integer part of the modified quota and the next integer is calculated. If the modified quota is greater than the geometric mean, then the modified quota is rounded up. Otherwise, it is rounded down. For example, if the modified quota is 12.742, the geometric mean of 12 and 13 is calculated: $\sqrt{12 \cdot 13} = 12.490$. Because the modified quota 12.742 is greater than 12.490 (the geometric mean of 12 and 13), the modified quota is rounded up to 13.

   Suppose the modified quota is 7.49. Would the modified quota be rounded up or down in this method? Show your computations and explain your conclusion.

5. Next, you will find the number of representatives each state receives when using the Huntington–Hill method.

   a. In cell F56, enter **660,000** (this will be the initial trial divisor). In column F of the spreadsheet, enter the new modified quotas using the number in F56 as the divisor. Label the column **New Modified Quotas** and record what you entered in F2 here:

   _____.

   b. Label column G by entering **Geometric Mean** in cell G1. Use the following Excel instructions to compute, for each state, the geometric mean of the integer part of the modified quota and the next integer.

   ### Instructions to Compute the Geometric Mean

   In cell G2, enter = **SQRT(INT(F2)*(INT(F2)+1))**. (This is the square root of the product obtained by multiplying the integer part of the number in cell F2 times the next integer.) Then drag the formula down to compute the appropriate geometric mean for each state.

c. Record the number you obtained in G2: _____. Using a
calculator, verify that it is the geometric mean used by the Huntington–Hill method to
round off the modified quota in F2:

Modified Quota in F2: _____   Integer Part: _____

$$\sqrt{\underline{\hspace{3cm}} \cdot \underline{\hspace{3cm}}} = \underline{\hspace{3cm}}$$

d. Label column H **Number of Seats Using Huntington–Hill Method** and tell Excel to
round each modified quota up or down, according to the geometric mean, using the
following instructions.

Instructions to Round Numbers Using an "IF Statement"

In cell H2, enter **=IF(F2 < G2, INT(F2),INT(F2)+1)**. (This formula tells Excel to see if the
number in F2 is less than the number in G2; if it is, it instructs Excel to enter the integer part
of F2; otherwise, to enter the integer part plus 1, which is the next largest integer.) Then drag
the formula down to obtain the number of seats for all 50 states.

e. Record the number obtained in H2 here: _____. Explain
how this is consistent with your computations in Question 5(c).

f. In cell H53, enter the sum of the numbers now in H2 through H51. If this number is
435, you have apportioned the House seats using the Huntington–Hill method. If the
number is not 435, change the modified divisor used (which is in cell F56) until the sum
of apportioned seats is 435.

Record the divisors you've tried and the total number of House seats resulting from
each divisor. Indicate which one is the modified divisor that actually works.

Divisor					
Total No. of Seats					

g. List the four states that have the largest number of seats and the number of seats each has.

h. List all states that have just one seat.

6. Name the states that would prefer Adams' method and those that would prefer the Huntington–Hill method. Explain your answer.

Summary

In this activity, you used Excel to investigate two divisor methods of apportionment: Adam's method and the Huntington–Hill method, which is the method currently used to apportion seats in the House of Representatives. Using each of these methods, you found the number of representatives that corresponds to each state. You learned how to use Excel to compute the geometric mean and to round numbers according to the geometric mean.

More on Problem Solving

Tanning salons raise skin cancer risk, study indicates

Paul Recer,
Associated Press

Washington—Regularly baking to a golden tan under sun lamps can increase the risk of malignant melanoma, a sometimes fatal skin cancer, and the younger a woman starts the greater the risk, a study says.

P	Q	not P	not Q	P and Q	not (P and Q)	(not P) or (not Q)
T	T	F	F	T	F	F
T	F	F	T	F	T	T
F	T	T	F	F	T	T
F	F	T	T	F	T	T

State	Population	Standard Q...
Delaware	897,934	0.4084
New Jersey	8,791,894	3.9992
New York	19,378,102	8.8145
Pennsylvania	12,702,379	5.7779
Total	**41,770,309**	

Country	Per Capita Total Spending on Health (in U.S. dollars at average exchange rates)	Per Capita Daily Calorie Supply	Infant Mortality Rate per 1,000 Births
Australia	3,986	3,674	5
Cambodia	36	2,046	62
China	108	2,951	23
France	4,672	3,654	4
Germany	4,209	3,496	4
India	40	2,954	55
Italy	3,136	3,617	4

After completing this topic, you will be able to:

* Identify general problem-solving techniques.

* Use a variety of techniques to solve problems.

* Approach complex problems with appropriate strategies to make them more manageable.

Problem solving involves a variety of types of reasoning. In Topics 11–14, we discussed types of reasoning and used some problem-solving techniques beyond those discussed in Topic 10. In this topic, we review Pólya's problem-solving process and expand the list of problem-solving techniques begun in Topic 10. We then apply the process and techniques to realistic and complex problems.

Problem solving often involves both inductive and deductive reasoning. For example, we use inductive reasoning when we look at special cases or similar cases and deductive reasoning when we apply a formula for a specific calculation.

The form of inductive reasoning we use when we solve a problem by looking at special cases and then drawing a conclusion about the general situation is **generalization**, and when we look at similar cases, we use **analogy** to find the solution to a similar problem. Sometimes we can use deductive reasoning in the form of indirect reasoning when solving a problem. For example, a particular student may have calculated her average in a particular class and determined that if she obtained a C on the final exam, she would get a B as a final grade in the class. But when final grades are posted, she notes that she didn't obtain a B. She can conclude that she didn't get a C on the final exam. (See Example 15.2 for another illustration of indirect reasoning.)

Another technique used in problem solving is that of **estimate and check**. This is a form of prediction and applies to questions where there is a numerical answer that can be checked easily. It is a good practice to use this problem-solving technique, in addition to others, when looking for a numerical solution to a problem. In such a case, we could estimate the answer before we solve the problem. (For example, should the answer be in the thousands or the millions?) After solving the problem using appropriate techniques, we should use the estimate to check if our solution is reasonable.

Finally, after we have a solution to a problem, we "look back" and ask ourselves if the process we used to solve the problem was valid. We can also ask questions that help us reflect on and learn from our solution and the process we used to obtain that solution. For example, we might ask the following:

- Could we have solved the problem using a different approach?
- Can we apply the approach we used to a different problem?

These problem-solving techniques, together with those listed in Topic 10, allow us to give this list based on:

Pólya's Four-Step Problem-Solving Process

A. Understand the problem.

    1. Ask questions to clarify the problem.

    2. Decide what information is relevant.

    3. Represent the information in a different form.

        a. Make a table or chart.

        b. Draw a picture, graph, or diagram.

        c. Write an equation or formula.

B. Devise a plan.

    4. Examine a simple case or try several special cases.

    5. Break a problem into smaller problems or identify a subgoal or subproblem.

C. Carry out the plan.

    6. Work backward.

    7. Look for a pattern.

    8. Generalize.

    9. Look at a related problem (sometimes simpler) or look for analogies.

    10. Use indirect reasoning.

D. Look back.

    11. Estimate and check.

    12. Validate the process used and ask questions to help us reflect on and learn from the process.

We start with an example that shows a common instance when we use problem-solving technique 11. *Estimate and check.*

Example 15.1

A student goes to the bookstore and buys a textbook that costs $89, a notebook for $4.75, and a T-shirt for $20. Sales tax is 8%, but there is no sales tax on clothing. When the cashier rings the sale, the total adds up to $149. Use problem-solving technique 11. *Estimate and check*, to answer the following questions:

a. Using estimation, does the total seem correct, or has an error occurred?

b. Check the correct sale total.

Solution

a. Because the textbook is approximately $90, and the notebook is almost $5, our taxable items are about $95. The tax should then be approximately $8 (the tax on a purchase of $100), and the total sale will not exceed $100 + $8 + $20 = $128. A mistake has definitely been made if the total rung up is $149.

b. The exact amount of the sale should be $4.75 + $89 + $20 + 0.08($4.75 + $89) = $121.25. We can be fairly confident of this result because it is relatively close to our rough estimate in part (a) of this example.

In the following example we see the importance of "looking back."

Example 15.2

Your friend tells you that at his college, there are 12 times as many students as there are professors. He claims that he wrote an equation to represent the relationship between number of students S and number of professors P. The equation he produced is $12S = P$. Give your friend advice about how he can use the strategies to "look back" at his solution.

Solution

First, your friend might want to use technique 11. *Estimate and check*, to see if the equation makes sense for a particular number of students. For example, if the school has 4 professors, it would have 48 students. Letting $P = 4$ and $S = 48$ in your friend's equation, we have "$4 = 12 \cdot 48$," which isn't true. You suggest to your friend that he consider the equation $S = 12P$. In reflecting on the process, you recommend that your friend substitute specific values for S and P to check his equation. You also point out that S and P represent variables and stand for quantities; they are not labels.

The next three examples illustrate the use of other problem-solving techniques from the list.

Example 15.3   ■ ■ ■ ■ ■ ■ ■ ■ ■

We would like to cover the floor of a 9 ft × 10 ft room with carpet. A neighbor offers us, at a reasonable price, a 7 ft × 12 ft piece of carpeting that he bought for himself but later decided not to install.

a. Use the four steps of Pólya's problem-solving process to answer these questions: Can we use the neighbor's carpet to completely cover our room? (Assume we are willing to cut the rug.)

b. Identify the problem-solving techniques used to answer this question, and explain how you used them.

Solution

a. We first ask some questions to make sure we understand the problem. How could the neighbor's carpet be rearranged to cover the 9 ft × 10 ft room? What is the area of the room? What is the area of the carpet? The room and carpet measurements are relevant information that we need. When devising a plan for the problem, we break the problem into two subproblems. We use the formula for area of a rectangle to find that the area of the room is 9 ft × 10 ft = 90 sq ft. The area of the neighbor's carpet is 7 ft × 12 ft = 84 sq ft. To carry out the plan, we compare the size of the available carpet with the area of the room, and see that the carpet is less than we need, so we cannot accept the neighbor's offer. We look back over our calculations and make sure we used the correct formula and did the calculations correctly.

b. We used problem-solving technique 3(c). *Use an equation or formula*, to find the area of the room and the area of the neighbor's carpet. We used problem-solving technique 10. *Use indirect reasoning*, to conclude that we could not use the neighbor's carpet. (The reasoning is this: If we are to cover the room, then we need at least 90 sq ft of carpeting. The neighbor's piece of carpeting is less than 90 sq ft, so it will not cover the room.)

■ ■ ■ ■ ■ ■ ■ ■ ■ ■

Example 15.4   ■ ■ ■ ■ ■ ■ ■ ■

Sam is getting a promotion and is negotiating a new contract with his employer. His yearly salary is now $35,000 and has not changed in the last two years so his purchasing power has decreased because of inflation. Sam wants his new salary to be at least the equivalent of his salary two years ago, updated for inflation, plus $6,000. He lives in a fast-growing town where the price of housing has increased 5% in the last two years. Sam's electricity bill increased 2% in the last two years, and his grocery expenses are about 3% more than they were two years ago.

a. Use the four steps of Pólya's problem-solving process to answer the question: What salary should Sam suggest in his negotiations?

b. Identify the problem-solving technique you used to answer this question, and explain how you used it.

Solution

a. We ask questions to help understand the problem. How can Sam estimate the effect of inflation on all of his expenses? The information about Sam's current salary and the estimates of increases in housing, electricity, and grocery expenses are relevant information that we need. To help devise a plan, we break the problem into two subproblems. First, we need to find one estimate of overall inflation, knowing the effect of inflation on three areas only. Then we figure out what his salary request should be. To carry out the plan and estimate the effect of inflation on all of Sam's expenses, we might make an estimate based on this information and then generalize. The average of the three rates of increase given is $\frac{5 + 2 + 3}{3} \approx 3.33$. Because $35{,}000 + 0.0333 \cdot 35{,}000 + 6{,}000 = 42{,}165.50$, Sam should try to negotiate a salary of no less than $42,165.50. We look back over our calculations and make sure we calculated correctly. We also reflect on the processes we used to see what we can learn from them for future problems.

b. We used problem-solving technique 8. *Generalize*, when we estimated the effect of inflation overall, based on knowing the rate of inflation on three items.

Example 15.5

Suppose you win $20,000,000 in the lottery. You will receive your prize in 20 yearly installments of $1,000,000 each and will pay a 35% tax per year, so your net income will be $650,000 each year. Because of inflation, the value of the $650,000 will be, in today's dollars, less each year.

a. Assuming inflation will be 2% per year during the next 20 years, find the value in today's dollars of $650,000 after 20 years. Identify the problem-solving techniques you used to answer this question, and explain how you used them.

b. Use problem-solving technique 9. *Look at a related problem (sometimes simpler) or look for analogies,* to estimate the value, in today's dollars, of the total amount received after 20 years.

Solution

a. If v_1 is the value in today's dollars of the $650,000 installment after one year, and inflation is 2%, then $v_1 + 0.02v_1 = 650,000$. So, $v_1 \cdot (1 + 0.02) = 650,000$ or $v_1 \cdot (1.02) = 650,000$. Solving for v_1, we see that the value in today's dollars of the $650,000 installment after one year is $v_1 = \dfrac{650,000}{1.02}$. The following year the value in today's dollars will be

$$v_2 = \frac{v_1}{1.02} = \frac{\left(\frac{650,000}{1.02}\right)}{1.02} = \frac{650,000}{(1.02)^2}$$

At the end of the third year, the value in today's dollars of $650,000 will be

$$\frac{\left(\frac{650,000}{(1.02)^2}\right)}{1.02} = \frac{650,000}{(1.02)^3}$$

We summarize these values in a table:

Years from today	Value in today's dollars
0	650,000
1	$\dfrac{650,000}{1.02}$
2	$\dfrac{650,000}{(1.02)^2}$
3	$\dfrac{650,000}{(1.02)^3}$

We now see a pattern: To find the value in today's dollars of $650,000 after 20 years, we calculate $\dfrac{650,000}{(1.02)^{20}} \approx 437,431.37$ and find that the value of the installment, in today's dollars, will be approximately $437,431.37. We used problem-solving technique 4. *Examine a simple case or try several special cases,* when we looked for the value after one, two, and three years. These special cases helped us see a pattern, so we also used problem-solving technique 7. *Look for a pattern.*

b. To find the value, in today's dollars, of the total amount received after 20 years, we would need to find the value, in today's dollars, of $650,000 for each of the next 20 years, and then add those values. We will instead solve a simpler similar problem: Assume that the installment received each year is always equivalent to $540,000 (this is roughly the average between the current value and the value after 20 years). To solve this simpler problem, we need to find only the total of 20 payments of $540,000 each, which is $20 \cdot (\$540,000) = \$10,800,000$.

Most "real-life" problems require the use of several problem-solving techniques to solve them. The following example shows a situation in which we can use several of the techniques listed at the beginning of this topic.

Example 15.6

Suppose you are a homeowner and need to decide which of three options for purchasing house heating oil would be best for you. The company offers two plans: (1) the "Buy-now Plan" at $3.49 per gallon, in which you pay for oil up front for the amount you estimate you will use; (2) the "Economy Plan" at $3.59 maximum per gallon, in which you pay monthly payments spread out over 12 months for the estimated amount of oil you will need (same as last year's usage). According to the company, this plan is best for the consumer. If the market price drops, your price will also drop, and if the market price rises, your price will not go above $3.59. You may choose not to select one of the two plans so another option is (3) no special plan; you pay for the amount of oil used as it is delivered and pay the price at the time of delivery. Explain which problem-solving techniques, from the list given in this topic, would help you decide which plan to use and give the order in which you would use those techniques.

Solution

First, we need to decide what information is relevant; some of the relevant information is given here and other information we would need to collect. Relevant information includes the description of the company's different plans, the records of amount of oil delivered to the household, and when it was delivered throughout the last year. (Because this will be the estimated amount of oil used in the first two plans, we need to know the total, but we may also need to know more about the pattern of oil expenditure in this household.) We also need information on oil prices throughout the last two years (to make an educated guess of what prices we should expect).

Next we divide the problem into two different subproblems: (1) an analysis of the pattern of oil consumption throughout the year (or last two years, if possible); (2) an analysis of the pattern of oil prices for the last one or two years. To perform each of these two pattern analyses, we use tables and graphs. We also use generalization when, based on the pattern for the last one or two years, we assume (predict) that the pattern will continue.

We look for analogies by looking for similarities between last year's and the previous year's patterns. Finally, we examine three different cases (one for each of the company's plans).

In summary, by proceeding in the way described, we would be using the following techniques from the list: 2. *Decide what information is relevant*; 3. *Represent the information in a different form*; 4. *Examine a simple case or try several special cases*; 8. *Generalize*; 5. *Break a problem into smaller problems or identify a subgoal or subproblems*; 9. *Look at a related problem or look for analogies.*

Example 15.7

In a recent study, researchers randomly assigned adults with year-round allergies to have either one dose of carbon dioxide, delivered via a nosepiece, or a placebo treatment. The real treatment group was divided into four subgroups, receiving either a higher or lower dose of carbon dioxide for either 10 or 30 seconds. Thirty minutes later, the study participants reported on their nasal congestion.

a. Describe how the researchers can look back at the work they did to investigate the question "Can nasal doses of carbon dioxide relieve nasal congestion in adults with year-round allergies?"

b. Identify the problem-solving techniques used in this study.

Solution

a. First, researchers should check the process they used. They need to make sure the participants in the study were randomly assigned to the groups. They should also check to make sure that the carbon dioxide was indeed delivered in the correct doses, and that the participants didn't know whether they received carbon dioxide or the placebo. When asking if the problem could be solved using a different approach, they might consider whether it would be possible to measure participants' nasal congestion using an objective measure, rather than subjective self-reports from the participants.

b. In setting up the study, the researchers had to decide what information was relevant to collect from the study participants: technique 2. *Decide what information is relevant.* They also examined a simpler case (technique 4. *Examine a simple case or try several special cases*) when they decided how many participants to involve in the study. If the researchers want to see approval for carbon dioxide as a treatment for nasal allergies, they would likely need to conduct additional studies to replicate the results.

Summary

In this topic, we reviewed the problem-solving techniques discussed in Topic 10 and discussed these new ones: 8. *Generalize*; 9. *Look at a related problem (sometimes simpler) or look for analogies*; 10. *Use indirect reasoning*; 11. *Estimate and check*; 12. *Validate the process used and ask questions to help us reflect on and learn from the process.* We practiced using Pólya's problem-solving process, solved some problems, and identified the problem-solving techniques used.

Explorations

1. For each of the problem-solving techniques 8 through 12 in the list given in this topic, find an example of an activity you did in this course that used the technique.

2. In Topic 5, Exploration 10, you were given the following information about three rental car companies:

   An airport has three rental car companies that rent a particular type of car. Company I offers a $23.50 daily rate with unlimited mileage (that is, there is no additional charge, regardless of how many miles are driven). Company II charges $19.99 per day and $0.15 for each mile driven; Company III charges $15.50 per day and $0.22 cents for each mile driven. Suppose you need to rent a car for 7 days.

   Describe how Pólya's problem-solving process could be used to decide which car company you should use to rent a car for a week.

3. On September 21, 2011, the current U.S. public debt (which includes savings bonds, treasury notes, and government securities) was $14,705,188,992. The government is paying interest on this debt at various interest rates. Suppose the average interest rate on the different types of securities is 3% per year, compounded monthly, and assume that the government does not sell any securities for the rest of the year. Consider the question: How much should the government pay each month from May to December so the debt is less than $7 trillion? Identify the problem-solving techniques you would use to answer the question. (*Source*: Bureau of the Public Debt, www.publicdebt.treas.gov.)

4. Tim is planning to open a savings account to deposit $7,000 he collected from his summer job. He finds two banks that offer interest of 2.4% per year. At the first bank, interest is compounded quarterly, and at the second bank, interest is compounded daily. Tim will not make any withdrawals during the next two years. By how much will the amounts in the two accounts differ at the end of two years? Identify the problem-solving techniques you used to answer the question.

5. You are asked by the student government at your school to devise a way to rate clubs and organizations on your campus to help inform students and potential students about the organizations. Describe how you will use Pólya's problem-solving process to help you solve this problem.

6. You are charged with helping your college organization develop a year-long project that will serve the community in which your college is located and make a significant contribution to the community. Determine how you will understand the problem, devise a plan, carry out the plan, and look back. Describe the problem-solving techniques you will use in the problem-solving process.

7. You need to design a study to test the effectiveness of caffeine, napping, and a combination of caffeine and napping on alertness and performance. Determine how you will understand

the problem, devise a plan, carry out the plan, and look back. Describe the problem-solving techniques you will use in the problem-solving process.

8. Suppose you open a savings account at an interest rate of 1.5% per year, compounded monthly. You make an initial deposit of $1,000 and, starting one month after you open the account, you deposit $100 each month and do not make any withdrawals.

   a. Identify which problem-solving techniques you would use to answer the question: How long would it take for the account to reach $3,000?

   b. Explain how you would use the techniques you identified in part (a).

   c. Use the techniques identified in part (a) to answer the question.

9. Suppose you won $20,000,000 in the lottery. You may choose to receive the prize in a lump sum now and pay 50% of it in taxes, or in 20 yearly installments of $1,000,000, paying 35% tax per year. If you choose to receive the prize in a lump sum, you will invest it in an account and expect a return of 5% per year. You will make yearly withdrawals to cover all your expenses (living expenses and donations to charity and family and friends). If you decide to receive the prize in installments, you think that after covering all your expenses, you will have a reasonable amount to invest and expect a return of 4.5% per year. Explain which problem-solving techniques, from the list given in this topic, would help you decide which of the two options will give you the largest amount of savings at the end of 20 years, and list the order in which you would use those techniques.

Making a Purchase Decision

In this activity, you will use a variety of problem-solving techniques along with the decision-making process you learned in Topic 11 to make a purchase decision of your choice.

You are planning (either for real or hypothetically) a major purchase in the near future. (You may want to look into purchasing a car, a stereo, a computer, a television set, a grill, or another item of your choice.) You need to do the analysis to ensure the purchase you are contemplating will be carried out in optimal fashion.

Follow the guidelines described here, which divide the problem into subproblems, and address the issues raised in a coherent narrative. You can use the spaces provided to note your ideas, but you should write a single essay that incorporates thoughtful responses to all the questions.

1. Describe the product on which you have decided to focus and explain how/why you made that particular choice.

2. Collect information about products in your category of interest. You may need to consult a variety of sources. Include at least two websites and give the URL of each site; you might also use consumer periodicals featuring your product, newspapers, almanacs, and possibly personal visits to "test use" a product. Make certain your sources are up-to-date and relevant.

3. List all the criteria (at least six) that might affect your decision and assign relative priority to those criteria. Justify why you chose the criteria you did. Determine if the criteria are readily measurable and decide how you will measure or rank your choices relative to them. You should use at least five choices of the item you are planning to purchase to compare against each other according to your selected criteria.

4. Decide how you will organize your information and make a table or organize it using an Excel spreadsheet.

5. Use the two processes (cutoff screening and weighted sum methods) discussed in Topic 11 and make a decision about which product to purchase. Explain these processes fully.

6. Reflect on your entire process. What problem-solving techniques did you use? How does this approach compare with the way you have made important purchase decisions in the past or with how you have rated things?

Summary

In this activity, you used problem-solving techniques to help you make a major purchase decision.

Averages and Five-Number Summary

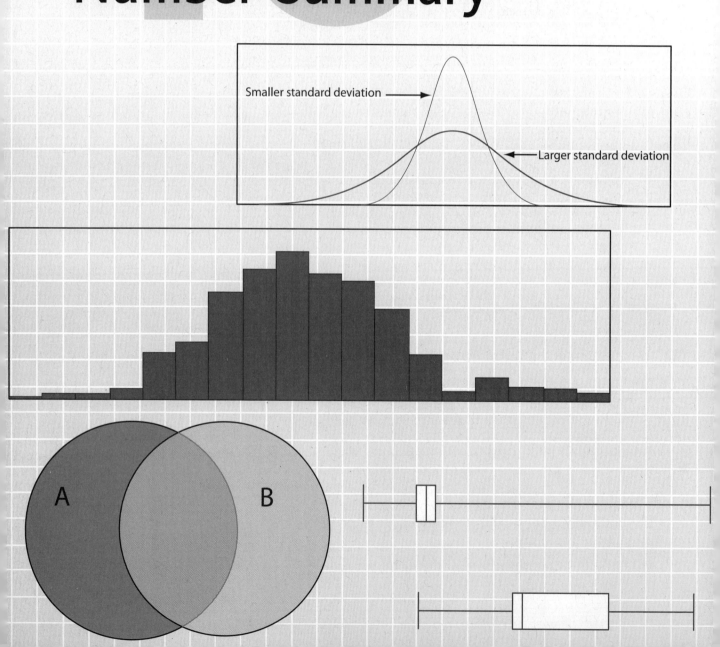

After completing this topic, you will be able to:

- Compute the mean (or average), median, and mode of a data set.

- Use measures of center to analyze and compare data sets.

- Find and interpret numerical measures of spread including the range and quartiles.

- Use the five-number summary and a boxplot to analyze data.

Suppose you scored 82 points on the first assignment and 75 points on the second assignment of the same course. If both were graded on a scale of 100 points, on which of the two assignments did you do better? Because your grade for the first assignment is higher, you would probably say you did better on the first one. But did you really do better on the first assignment? Suppose that the average grade on the first assignment was 85 points and the average grade on the second one was 67. Would this change your answer? Suppose you learned that half of the students in the class scored above 80 on the first assignment, while on the second assignment only one student scored above 69. Wouldn't you think that the second assignment was much harder than the first and that you did fairly well on it? In this topic, we discuss measures of center and spread that can be used to compare different sets of data.

Pictures of a data distribution give us a quick view of the distribution, but we often need to use a single number to summarize a collection of data or to compare two or more collections of data. For example, in Topic 1, Exploration 12, we looked at the average critical reading SAT test scores for high school seniors for each state in the United States. We were interested in using a *single* critical reading SAT number (the average) to associate with each of the states.

Two commonly used measures of center are mean and median. The **mean** is the arithmetic average, commonly called the **average**, and is found by adding all the data values and then dividing that sum by the *number* of data values. If we have n data values, denoted $x_1, x_2, \ldots, x_n$, then the mean is: $\dfrac{x_1 + x_2 + \cdots + x_n}{n}$.

The **median** is the middle observation in an ordered list of data. The median is found by first putting the data values in numerical order. If there is an odd number n of data values, the median is the middle data value, the data value in *position* $\dfrac{n+1}{2}$ in the ordered list. If there is an even number n of data values, the median is the mean of the two middle values, that is, the arithmetic average of the data values in *positions* $\dfrac{n}{2}$ and $\dfrac{n}{2} + 1$. For example, if a data set contains six test scores: 65, 72, 75, 78, 87, 95, then $n = 6$ is even and the median is the average of data values in positions $\dfrac{n}{2} = \dfrac{6}{2} = 3$ and $\dfrac{n}{2} + 1 = \dfrac{6}{2} + 1 = 4$. These values are 75 and 78, so the median is $\dfrac{75 + 78}{2} = 76.5$.

The **mode** is the most frequently occurring data value; that is, it is the data value with the highest frequency. If every data value has the same frequency, then a data set has no mode. If there are two values of the variable that occur the same number of times in the data set and more frequently than the other data values, then a distribution is **bimodal**; it has two modes. Similarly, a data set can have three or more modes. The mode is appropriate for categorical data as well as for numerical data, while the mean and median can only be calculated for numerical data.

More important than knowing how to compute these measures of center is understanding how to use and interpret them by knowing their properties. In the following example, we compute measures of center and analyze how they are affected when one data value is changed.

Example 16.1

The following table gives the number of hazardous waste sites on the U.S. National Priority List that are in each of 15 centrally located states, as of August 2011. (The National Priorities List is a list of hazardous waste sites that are eligible for extensive long-term clean-up procedures.)

State	Number of Sites
Colorado	14
Illinois	52
Indiana	42
Iowa	24
Kansas	17
Minnesota	46

Missouri	36
Montana	17
Nebraska	14
North Dakota	0
Ohio	49
South Dakota	2
Utah	16
Wisconsin	43
Wyoming	2

Source: U.S. Environmental Protection Agency, www.epa.gov.

a. Find the mean, median, and mode of the number of sites, and explain what these measures of center show about this data set.

b. Suppose the number of sites in Illinois was incorrectly recorded as 352 (instead of 52). How do the mean, median, and mode of the data set change?

c. Find how many of the observed data values (called **observations**) fall above the mean and median in each of parts (a) and (b) of this example and comment on these results.

Solution

a. To compute the mean, we add all the data values and divide by the number of data values, which is 15. The mean is

$$\frac{\text{sum of no. of sites}}{15} = \frac{14+52+42+24+17+46+36+17+14+0+49+2+16+43+2}{15}$$

$$= \frac{374}{15} \approx 24.9$$

To compute the median, we need to order the data. The table below shows the data values in increasing order. Since there is an odd number of data values (15), the median is the middle value, that is, the value in position $\frac{n+1}{2} = \frac{16}{2} = 8$. Counting eight data values from the smallest shows that the median is 17.

State	Number of Sites
North Dakota	0
South Dakota	2
Wyoming	2
Colorado	14

Nebraska	14
Utah	16
Kansas	17
Montana	17
Iowa	24
Missouri	36
Indiana	42
Wisconsin	43
Minnesota	46
Ohio	49
Illinois	52

Because the data values 2, 14, and 17 each occur two times, and no other data value occurs two or more times, this data set has three modes: 2, 14, and 17. The mean of approximately 24.9 is more than 40% larger than the median of 17.

b. If the largest data value were 352 instead of 52, the median would still be 17 and the modes would still be 2, 14, and 17. However, the mean $= \dfrac{\text{sum of no. of sites}}{15} = \dfrac{674}{15} \approx 44.9$. Here, the median and mode are unchanged, but the mean is greatly affected by the one (incorrect) value.

c. In part (a) of this example, seven data values lie below the position of the median and seven lie above it. Nine of the data values lie below the mean and six lie above it. In part (b) of this example, it is still true that seven data values lie on either side of the position of the median, but only three data values (46, 49, and 352) lie above the mean of 44.9; 12 lie below the mean.

■ ■ ■ ■ ■ ■ ■ ■ ■ ■ ■

In the hypothetical situation created in Example 16.1(b), 352 would be an **outlier.** Although there are precise ways to quantify outliers, we'll use that term generally to mean a data value that is outside the general pattern of the data. It is important to identify outliers and why they occur, if possible. The mean is more affected by outliers than are the other measures of center, as Example 16.1 shows.

If the distribution of a set of data is **symmetric** (or nearly symmetric), that is, a histogram or a stemplot of the data looks the same to the left and to the right of the "center," then the mean and median will be close together. If the distribution of a data set is **skewed,** that is, if on a histogram of the data, one tail end is longer than the other, then the mean will be farther out on the long tail than the median. Note that in Example 16.1(b), with the value

352 in the data set, the data values are skewed toward larger values and the mean is considerably greater than the median.

Here are two histograms, one of a hypothetical set of freshmen math and critical reading SAT scores that is approximately symmetric and the other of another hypothetical set of freshmen math and critical reading SAT scores that is skewed to the right, or positively skewed.

Measures of center give us a single number (or category in the case of categorical data and the mode) to summarize a data set. But we often need more information about a data set than a single value can convey. In the following examples, we look at the relationship between the mean, median, and mode and a stemplot of a data set.

Example 16.2

Create a stemplot of the hazardous waste-site data given in Example 16.1. Explain where the mean, median, and mode fall on this plot.

Solution

Because the data values go from 0 to 52, the digit in the 10s place will be the stem and the digit in the units place will be the leaf. The stemplot appears next:

Number of Waste Sites (stem = tens digit; leaf = units digit)

```
0 | 0 2 2
1 | 4 4 6 7 7
2 | 4
3 | 6
4 | 2 3 6 9
5 | 2
```

If we picture a smooth curve drawn along the right edge of the leaves, we can see that the curve has two peaks, one in the 10s and one in the 40s. The median falls on the 10s stem and the mean falls on the 20s stem. From the stemplot, we can see that there are three modes, 2, 14, and 17. Because the data are in numerical order, it is straightforward to find the median of such an ordered data set, which is 17.

Example 16.3

The following table shows the 2010 annual salary, rounded to the nearest dollar for each of the U.S. state governors:

State	Salary ($)	State	Salary ($)	State	Salary ($)
Alabama	112,895	Louisiana	130,000	Ohio	144,269
Alaska	125,000	Maine	70,000	Oklahoma	147,000
Arizona	95,000	Maryland	150,000	Oregon	93,600
Arkansas	87,352	Massachusetts	140,535	Pennsylvania	174,914
California	173,987	Michigan	177,000	Rhode Island	117,817
Colorado	90,000	Minnesota	120,303	South Carolina	106,078
Connecticut	150,000	Mississippi	122,160	South Dakota	115,348
Delaware	171,000	Missouri	133,821	Tennessee	170,340
Florida	130,273	Montana	100,121	Texas	150,000
Georgia	139,339	Nebraska	105,000	Utah	109,900
Hawaii	117,312	Nevada	141,000	Vermont	142,542
Idaho	115,348	New Hampshire	113,834	Virginia	175,000
Illinois	177,500	New Jersey	175,000	Washington	166,891
Indiana	95,000	New Mexico	110,000	West Virginia	95,000
Iowa	130,000	New York	179,000	Wisconsin	137,092
Kansas	110,707	North Carolina	139,590	Wyoming	105,000
Kentucky	145,885	North Dakota	105,036		

Source: Council of State Governments.

a. Find the mean, median, and mode of these data.

b. Create a stemplot of the data.

c. Could you obtain the mean, median, and mode from the stemplot created? Explain why or why not.

Solution

a. The mean is the sum of all the salaries divided by 50, which is approximately $130,596. In order to find the median, we need to sort the data in order. The next table shows the data sorted by increasing salary:

State	Salary ($)	State	Salary ($)	State	Salary ($)
Maine	70,000	Idaho	115,348	Ohio	144,269
Arkansas	87,352	South Dakota	115,348	Kentucky	145,885
Colorado	90,000	Hawaii	117,312	Oklahoma	147,000
Oregon	93,600	Rhode Island	117,817	Connecticut	150,000
Arizona	95,000	Minnesota	120,303	Maryland	150,000
Indiana	95,000	Mississippi	122,160	Texas	150,000
West Virginia	95,000	Alaska	125,000	Washington	166,891
Montana	100,121	Iowa	130,000	Tennessee	170,340
Nebraska	105,000	Louisiana	130,000	Delaware	171,000
Wyoming	105,000	Florida	130,273	California	173,987
North Dakota	105,036	Missouri	133,821	Pennsylvania	174,914
South Carolina	106,078	Wisconsin	137,092	New Jersey	175,000
Utah	109,900	Georgia	139,339	Virginia	175,000
New Mexico	110,000	North Carolina	139,590	Michigan	177,000
Kansas	110,707	Massachusetts	140,535	Illinois	177,500
Alabama	112,895	Nevada	141,000	New York	179,000
New Hampshire	113,834	Vermont	142,542		

From the table, we can see that the salaries range from $70,000 to $179,000. Because there are 50 data values, the median is the arithmetic average of the data values in positions $\frac{50}{2} = 25$ and $\frac{50}{2} + 1 = 26$. Thus, the median is $\frac{\$130,000 + \$130,000}{2} = \$130,000$. Each of the data values $95,000 and $150,000 occurs three times, and no other data value occurs that often or more often, so the two modes are $95,000 and $150,000.

b. We have several choices of how to split the data into a stem portion and a leaf portion. It looks as though the stemplot will have enough stems, but not too many if we create the stemplot with the digits in the ten thousands place as the stem and then use the digit in the thousands place as the leaf. To do this, we choose to truncate each salary and look at salary in thousands. By truncating the data, we simplify the stemplot and still retain approximate values of the data. The resulting stemplot is shown here. (Note that we no longer have the exact values of the data in the stemplot.)

U.S. State Governors' Salaries (stem = ten thousands digits, leaf = thousands digit)

```
 7 | 0
 8 | 7
 9 | 0  3  5  5  5
10 | 0  5  5  5  6  9
11 | 0  0  2  3  5  5  7  7
12 | 0  2  5
13 | 0  0  0  3  7  9  9
14 | 0  1  2  4  5  7
15 | 0  0  0
16 | 6
17 | 0  1  3  4  5  5  7  7  9
```

c. We could not get the exact mean, median, or mode using the data in this form because we do not have the exact data values but only approximations.

A histogram of the data on governors' salaries (see Example 16.3), which appears next, shows approximately the same shape that the stemplot shows. Both pictures of the data show a peak in the largest group of data, the $170,000 to 179,000 group. The data do not appear to be skewed either to the right or to the left nor are they symmetric (because of the large peak on the right).

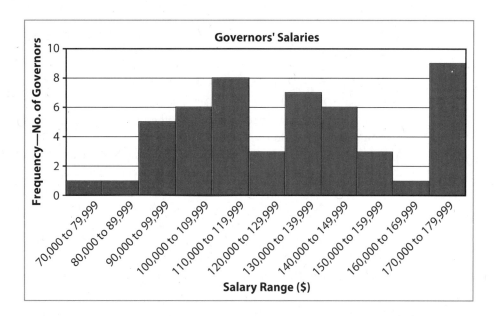

We can picture the median of a data set as the value on the horizontal axis for which half of the total area of the histogram lies to the left of the value and half lies to the right of the value. The mean can be visualized by thinking about a playground ride. Picture the histogram balanced on a seesaw. The mean is the position of the center of the seesaw along the horizontal axis of the histogram where the seesaw is perfectly balanced.

Because a measure of center alone does not give enough information, we often need an additional numerical measure that gives us an idea of how spread out the data are. We can get a visual measure of the spread of a set of data by looking at a graph of the data, but there are three commonly used **numerical measures of the spread** of a data set: the range, the interquartile range, and the standard deviation. We will look at the first two of these in this topic and the third in Topic 17. The **range** of a data set is the difference between the maximum (the largest) data value and the minimum (the smallest) data value. The **interquartile range** is the length of the middle half of the data and is found by first identifying the first and third quartiles of the data.

To find the first quartile and the third quartile, we order the data values in increasing order, find the median, and divide the data into a lower half and an upper half. The **first quartile, Q_1,** is the median of the lower half of the data values. The **third quartile, Q_3,** is the median of the upper half of the data values. So, finding the first and third quartiles involves finding two additional medians.

Note that if the number of data values is odd, the value in the position of the median is not in either the lower or the upper half. For example, if a data set consists of nine exam scores: 67, 68, 72, 75, 76, 79, 83, 90, 96, then the median is the score in position $\frac{9+1}{2} = 5$. The median is 76. The first quartile Q_1 is the median of the four values in the lower half of the data set: 67, 68, 72, 75. So, $Q_1 = \frac{68+72}{2} = 70$. Similarly, the third quartile is the median of 79, 83, 90, 96, so $Q_3 = \frac{83+90}{2} = 86.5$. The interquartile range, denoted **IQR**, is the difference $Q_3 - Q_1$. For our example, IQR $= 86.5 - 70 = 16.5$.

We list the first and third quartiles, together with the median and the minimum and maximum data values, to give the **five-number summary** of a set of data. The five-number summary gives us a quick view of both the center and the spread of a data set and divides the data set into four parts; approximately one-fourth of the data values are between the minimum and Q_1, approximately one-fourth of the data values are between Q_1 and the median, approximately one-fourth of the data values are between the median and Q_3, and approximately one-fourth are between Q_3 and the maximum. So, the quartiles are appropriately named because they divide the data, roughly, into quarters. We present the five-number summary of a data set as the ordered list of the following five values: minimum data value, Q_1, median, Q_3, maximum data value.

Example 16.4

a. Find the five-number summary, the range, and the interquartile range for the hazardous waste-site data from Example 16.1, and explain what the five-number summary shows about the data set.

b. Repeat part (a) of this example for the governors' salaries data set from Example 16.3.

Solution

a. The data, given in numerical order, are

$$0, 2, 2, 14, 14, 16, 17, \underline{\mathbf{17}}, 24, 36, 42, 43, 46, 49, 52$$

The hazardous waste-site data set has 15 observations. Thus, as we noted before, the data value in position 8 in the ordered list is the median, shown underlined in bold. To get Q_1, we need to find the median of the data values in positions 1 through 7, shown in brackets:

$$[0, 2, 2, \underline{\mathbf{14}}, 14, 16, 17,] \, \underline{\mathbf{17}}, 24, 36, 42, 43, 46, 49, 52$$

The median of these seven data values is the data value in position 4; so $Q_1 = 14$, shown underlined in bold. Similarly, Q_3 is the data value in position 12:

$$0, 2, 2, 14, 14, 16, 17, \underline{\textbf{17}}, [24, 36, 42, \underline{\textbf{43}}, 46, 49, 52]$$

Thus, $Q_3 = 43$. The five-number summary for this data set is 0, 14, 17, 43, 52. The range is $52 - 0 = 52$ and the interquartile range is $43 - 14 = 29$. The five-number summary shows that there are 14 units between the minimum and Q_1 and almost 20 units between Q_3 and the maximum. The distance from Q_1 to the median is 3, which is considerably less than the distance of 26 from the median to Q_3. These distances imply that the data are skewed to the right.

b. The governors' salaries data set has 50 observations. The median is the average of the data values in positions 25 and 26 in the ordered list; thus, the median is $130,000. The first quartile is the median of the data values below the position of the median, so it is the median of the data values in positions 1 through 25. Thus, Q_1 is the data value in position 13, and $Q_1 = \$109,900$. Similarly, Q_3 is the median of the data values above the position of the median, so it is the median of data values in positions 26 through 50. Therefore, Q_3 is the data value in position 38, and $Q_3 = \$150,000$. The five-number summary for this data set is $70,000, $109,900, $130,000, $150,000, $179,000. The range is $179,000 - \$70,000 = \$109,000$, while the interquartile range is $150,000 - \$109,900 = \$40,100$. The five-number summary shows that a gap of $39,900 exists between the minimum and the first quartile and there is a smaller gap ($29,000) between the third quartile and the maximum. There is a similar gap between the median and the third quartile and between the first quartile and the median (approximately 20,000). The middle 50% of the data is roughly symmetric, and the largest 25% of the data is more concentrated, while the data in the smallest 25% lie in a larger interval.

We can sketch a graph of the five-number summary of a data set, called a **boxplot**, by creating a horizontal (or vertical) number line that spans an interval just a bit larger than the interval from the minimum to the maximum of the data set. We put a short mark perpendicular to the number line and slightly above it (or to the right of it, if we use a vertical number line), where each of the numbers in the five-number summary is located. We then draw a box from the first quartile to the third quartile and draw lines to the minimum and maximum as shown in the boxplot of the hazardous waste-site data pictured here.

Number of Hazardous Waste Sites

We can use the same scale for two boxplots to compare two data sets.

Example 16.5

Brownies and ice cream bars are two popular desserts. Sketch side-by-side boxplots of the following two data sets that give the calorie content of different brands of brownies and ice cream bars. What do the boxplots show?

Brownie, Prepared According to Package Directions	Calories in One Piece
Pillsbury caramel fudge chunk	170
Betty Crocker caramel swirl	120
Betty Crocker German chocolate	160
Nestle double chocolate chip	150
Pepperidge Farm hot fudge	400
Duncan Hines milk chocolate	160
Betty Crocker microwave frosted	180
Robin Hood/Gold Medal pouch mix	100
Duncan Hines peanut butter	150
Pillsbury triple, chunky	170

Ice Cream Bar, One Piece	Calories per Bar
Good Humor Fat Frog	154
Good Humor Halo Bar	230
Heath	170
Haagen-Dazs, caramel almond	230
Good Humor chip candy crunch	255
Nestle premium milk chocolate w/almonds	230
Haagen-Dazs vanilla w/dark chocolate coating	390
Nestle crunch vanilla w/white chocolate coating	350
Klondike Krispy	290
Oh, Henry vanilla with chocolate coating	320

Source: The Corinne T. Netzer Encyclopedia of Food Values.

Solution

We order the data in each of the sets to find the five-number summary:

Brownie	100	120	150	150	160	160	170	170	180	400

Ice Cream	154	170	230	230	230	255	290	320	350	390

For the calories-of-brownies data, the five-number summary is 100, 150, 160, 170, 400. For the calories-of-ice cream bars data, the five-number summary is 154, 230, 242.5, 320, 390.

The following two boxplots show that the brownie calories have a larger range but a much smaller interquartile range. The middle half of the brownie-calorie data lies below the first quartile of the ice cream bar-calorie data set. The minimum value of the ice cream data lies at approximately the same point as the first quartile of the brownie data. The maximum value of the brownie data lies far above the third quartile of the ice cream data set.

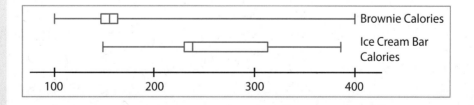

Summary

In this topic, we used data sets on number of waste sites by state, state governors' salaries, and calorie content in popular brands of brownies and ice cream to explore measures of center and spread. We discussed the concepts of mean, median, mode, quartiles, range, and interquartile range. We also calculated the five-number summary and graphed the boxplot for data sets.

Explorations

1. Create a set of exam scores (for a 100-point exam) for a hypothetical class of ten students in which

   a. The mean, median, and mode are all the same.

   b. The mean is at least 10 points greater than the median.

   c. The median is at least 10 points greater than the mean.

2. A particular ATM machine dispenses $20 bills as requested by customers in amounts up to a maximum of $400. The following table shows the frequency of transactions in which customers withdrew specific amounts during a particular month. (Note that this particular machine has a $60 "quick cash" option.)

Amount Withdrawn ($)	Number of Customers
20	32
40	44
60	93
80	43
100	56
120	31
140	33
160	25
180	16
200	45
220	23
240	18
260	21
280	11

300	27
320	14
340	13
360	10
380	7
400	18

a. How many transactions were there in which customers withdrew cash at this ATM machine during the month?

b. Find the mode for the amount of cash withdrawn from this particular ATM machine during the month.

c. Find the median amount of cash withdrawn from this ATM machine during the month.

d. Find the mean amount of cash withdrawn from this ATM machine during the month.

e. Create an appropriate graph for this data set and explain what it shows.

3. The following table gives estimates of the percent change in population for each U.S. state over the years 2000 to 2010. (These data were used in Example 1.8.)

State	Percent Change	State	Percent Change
Alabama	7.5	Montana	9.7
Alaska	13.3	Nebraska	6.7
Arizona	24.6	Nevada	35.1
Arkansas	9.1	New Hampshire	6.5
California	10.0	New Jersey	4.5
Colorado	16.9	New Mexico	13.2
Connecticut	4.9	New York	2.1
Delaware	14.6	North Carolina	18.5
Florida	17.6	North Dakota	4.7
Georgia	18.3	Ohio	1.6
Hawaii	12.3	Oklahoma	8.7
Idaho	21.1	Oregon	12.0
Illinois	3.3	Pennsylvania	3.4
Indiana	6.6	Rhode Island	0.4
Iowa	4.1	South Carolina	15.3

Kansas	6.1	South Dakota	7.9
Kentucky	7.4	Tennessee	11.5
Louisiana	1.4	Texas	20.6
Maine	4.2	Utah	23.8
Maryland	9.0	Vermont	2.8
Massachusetts	3.1	Virginia	13.0
Michigan	−0.6	Washington	14.1
Minnesota	7.8	West Virginia	2.5
Mississippi	4.3	Wisconsin	6.0
Missouri	7.0	Wyoming	14.1

a. Find the five-number summary for this data set and identify which states fall in the lowest quartile and which states fall in the highest quartile.

b. Explain what the five-number summary shows about this data set.

c. Find the average of the percent change for all the states. Is this the same as the overall percent change for the total United States? Explain why or why not.

d. Suppose you include the District of Columbia, which had a percent population change over the time period 2000 to 2010 of 5.2. How does the five-number summary change? Find the new five-number summary and compare it to the five-number summary for the 50 states.

4. The following tables give median annual family income in 2009 for families of different sizes in 15 eastern states:

Family Size (no. of people)	CT	DE	ME	MD	MA	NH	NJ	NY
2	70,800	61,424	50,767	73,291	67,142	62,509	69,539	56,845
3	82,305	67,412	58,097	85,746	82,385	81,134	84,192	67,292
4	101,647	83,928	67,361	101,693	100,462	88,538	99,474	82,587
5	100,989	73,851	68,365	98,508	103,475	83,081	102,931	80,441
6	103,804	70,269	74,776	95,648	99,351	94,246	95,995	77,582
7	105,588	78,579	55,433	89,529	85,558	74,257	89,623	79,704

Family Size (no. of people)	NC	OH	PA	RI	VT	VA	WV
2	49,813	50,491	52,839	57,567	57,013	62,586	41,919
3	54,573	59,275	66,030	71,019	64,767	72,078	50,521
4	66,487	71,453	78,626	87,163	77,127	85,586	59,307
5	59,925	66,204	76,895	83,848	70,777	81,234	58,004
6	52,913	63,512	71,035	59,839	72,307	81,611	52,450
7	49,349	54,910	66,099	103,575	65,422	96,185	52,307

Source: U.S. Census Bureau, www.census.gov.

a. Describe what trends this table shows and what might explain these trends.

b. Explain why the median income was given (instead of a different "representative" value for earning power).

c. What additional information might help you understand the differences?

d. What kind of graph or graphs could you use to present the information in the table? Create one or more graphs to present these data.

5. A pamphlet published by the Commonwealth of Pennsylvania, "Use Water Wisely," contains the following information: "Be aware of how much water you use! Awareness is the first step in conservation. The following table indicates how much water the average person uses each day." Explain how this "average" might have been obtained.

Use	Gallons per Day
Toilet	19
Bathing and hygiene	15
Laundry	8
Kitchen	7
Housekeeping	1
Total	**50**

6. A recent newspaper article reported on the consumption of alcohol by college students. The report indicated that the average amount of alcohol consumed per college student has decreased over the last 10 years. The report went on to discuss the rise among college students of "binge drinking," which is defined as consuming large amounts of alcohol at one sitting.

a. Explain how both of these observations can be true.

b. What data would you want to collect, in addition to the average amount of alcohol consumed per college student, to understand how to address the problem?

7. The sale of special, simpler Internet names has become a big business as unique Internet names become more scarce. On August 12, 2011, the website www.GreatDomains.com, a retailer of domain names and websites, reported on recently sold website names. The median selling price for the 13 sites in this group was $10,700, but the mean price was $17,189. Explain how such a median and mean are possible.

8. Next you will find a table containing the number of hazardous waste sites for a group of 12 New England and mid-Atlantic states (as of August 2011):

State	Number of Sites
Connecticut	18
Delaware	15
Maine	14
Maryland	18
Massachusetts	31
New Hampshire	21
New Jersey	144
New York	116
Pennsylvania	95
Rhode Island	13
Vermont	13
Virginia	30

Source: Environmental Protection Agency, www.epa.gov.

a. Find the five-number summary for this data set and sketch a boxplot.

b. Explain what your five-number summary and boxplot show about this data set.

9. The following excerpt is from an AP news article dated April 30, 2001, entitled "Study: 'Safe' Levels of Lead Still Harm IQ," (http://archives.cnn.com). It describes the results of a study of the relationship between blood lead concentration in children and IQ scores. In particular, the second paragraph describes differences in mean IQ test scores for two groups of children.

a. Describe how the children were grouped into the two groups.

b. Is the description of the mean given in the article correct? Explain your answer.

BALTIMORE — Children exposed to lead at levels now considered safe scored substantially lower on intelligence tests, according to researchers who suggest one in every 30 children in the United States suffers harmful effects from the metal.

Children with a lead concentration of less than 10 micrograms per deciliter of blood scored an average of 11.1 points lower than the mean on the Stanford-Binet IQ test, the researchers found. The mean is the intermediate value between the lowest and highest scores.

"There is no safe level of blood lead," said Dr. Bruce Lanphear, lead author of the lead study presented Monday at the Pediatric Academic Societies annual meeting.

Children are most commonly exposed to lead by inhaling lead-paint dust or eating paint flakes. Lead-based paint was widely used in homes throughout the 1950s and 1960s until it was banned in 1978.

At high levels, lead can cause kidney damage, seizures, coma and death.

10. A study to determine how fast cars travel on Main Street, which has a speed limit of 25 mph, tracked cars traveling at the following speeds: 24, 20, 32, 25, 52, 35, 28, 26, 29, 30.

  a. What measure should be used to identify a "typical" speed on this street?

  b. Explain why you made the choice you did in part (a) of this Exploration.

11. The five-number summary of the 2003 governors' salary data for the 50 states is $70,000, $95,000, $108,565, $127,303, $179,000.

  a. Use these data and the five-number summary of the 2010 governors' salary data found in Example 16.3 to sketch comparative boxplots of the 2003 and 2010 governors' salary data sets. Use a single scale for both plots.

  b. Explain what story these boxplots tell about the changes in governors' salary over the period from 2003 to 2010.

12. The following table gives the 30-year average monthly rainfall, in millimeters, over Bristol, United Kingdom, for three 30-year periods:

	Jan	Feb	Mar	Apr	May	Jun	Jul	Aug	Sept	Oct	Nov	Dec
1961–1990	78.4	56.1	63.1	49.6	59.9	61.3	52.3	67.5	73.1	71.9	74.7	87.0
1971–2000	88.1	61.3	67	54.3	58	61.4	47.6	74	79.1	84.8	81.3	96
1981–2010	91.1	61.7	67.5	56.3	64.8	57.4	65.5	72.6	70.3	98.6	92.3	96

Source: Bristol UK Weather Station, www.afour.demon.co.uk/all_pptn.htm.

  a. Explain why considering average rainfall over a long period of time is useful.

  b. Identify any trends shown by these 30-year averages for different months of the year.

c. Why is it important to identify these trends?

d. Use these data and a reasonable method to estimate the 1991–2020 30-year rainfall in Bristol for each of the months October, November, and December. Explain the method you used for your estimation.

13. The following table gives the mean daily energy intake, in kilocalories (kcals, commonly called calories), for selected four-year periods over the years 1971 to 2008 for U.S. males and for U.S. females ages 20 to 39 years of age:

Sex and Age	1971–1974	1976–1980	1988–1994	1999–2002	2005–2008
Male 20 to 39 years	2,784	2,753	2,964	2,854	2,946
Female 20 to 39 years	1,652	1,643	1,956	2,031	1,973

Source: U.S. Centers for Disease Control, www.cdc.gov.

a. Explain why researchers reported the mean energy intake as the measure of center, rather than another measure of center.

b. Identify any trends in the mean energy intake over the given four-year periods and explain what these trends show.

14. Explain how each of the following averages might be found. Be sure to include a description of how you would collect any needed data. Also identify any difficulties you might encounter in either collecting the data or computing the average.

a. Average mileage for a particular model of car

b. Average temperature in a particular city

c. Mean cholesterol level for U.S. adults ages 20 and older

d. Mean energy intake in kilocalories for U.S. adults

Visualizing Football Scores: Measures of Center and Spread

In this activity, you will examine some data from the National Football League, and investigate several numerical measures of center and spread and a graphical method particularly useful for comparing data sets.

1. The following table gives the points scored by the Philadelphia Eagles in their sixteen 2010–2011 regular season games:

Date	Points Scored— Eagles
12-Sep	20
19-Sep	35
26-Sep	28
3-Oct	12
10-Oct	27
17-Oct	31
24-Oct	19
7-Nov	26

15-Nov	59
21-Nov	27
28-Nov	26
2-Dec	34
12-Dec	30
19-Dec	38
28-Dec	14
2-Jan	13

Source: National Football League, www.nfl.com.

a. Without doing any computations, look at the "points scored" values and estimate the "center" of the data set. Why did you choose this value?

b. Enter the dates and points scored and the column titles, as they appear in the table, into columns A and B of an Excel worksheet.

c. Sort the data by "points scored" (in ascending order) and find the median number of points scored; write it here: _____.

d. Use the following Excel instructions to find the mean and median points scored by the Eagles in 2010−2011.

Instructions to Find the Mean and Median of a List of Data

1. In cell A19, type the word **mean**, and in cell A20 type the word **median**.
2. In cell B19, instruct Excel to calculate the average (that is, mean) points scored by entering the command =**average**(**B2:B17**); in cell B20, calculate the median points scored by entering the command =**median**(**B2:B17**).

   e. Record the average and median points scored per game by the Eagles in 2010−2011 here:

   average = _____

   median = _____

   Describe how these values compare with your estimate of the "center" of this data set in Question 1(a).

   f. Looking at the center of a distribution doesn't give the whole picture; you might also want to measure the spread of the data. One way to do this is to give the range of the data; that is (the maximum score) − (the minimum score) = (max) − (min). Give the range of the Eagles football data here: _____. (This shows the spread of the data, but may be misleading if one or both of these values are outliers.)

   g. To get a better sense of the spread, you'll now look at the **quartiles**. The first quartile, denoted Q_1, is the median of the lower half of the data values.[*] The third quartile, denoted Q_3, is the median of the upper half of the data values. The following list of numbers is the **five-number summary** of a data set: minimum, Q_1, median, Q_3, maximum. Use Excel to find Q_1 and Q_3 for the data set of Eagles'

---

[*]Note that some texts and software give slightly different descriptions of how to find the first and third quartiles. Excel contains a built-in function **QUARTILE** that could also be used to find the quartiles. In this activity, you are asked to find quartiles using the description of quartiles as medians of subsets of the data, because doing so helps enhance understanding of what quartiles represent.

points scored by using the median command and the appropriate halves of the data for each. Record the five-number summary of this data set:

_____.

h. Sketch a **boxplot** for this data set in the space provided. First, set up a horizontal axis. Next, mark a scale on your axis, then locate each of the five numbers in the five-number summary on the axis, and complete the boxplot.

2. The following table gives the points scored in the 2010−2011 regular season games by the Washington Redskins and the N.Y. Giants teams:

Date	Points Scored—Giants	Date	Points Scored—Redskins
12-Sep	31	12-Sep	13
19-Sep	14	19-Sep	27
26-Sep	10	26-Sep	16
3-Oct	17	3-Oct	17
10-Oct	34	10-Oct	16
17-Oct	28	17-Oct	24
25-Oct	41	24-Oct	17
7-Nov	41	31-Oct	25
14-Nov	20	15-Nov	28
21-Nov	17	21-Nov	19
28-Nov	24	28-Nov	13
5-Dec	31	5-Dec	7
13-Dec	21	12-Dec	16
19-Dec	31	19-Dec	30
26-Dec	17	26-Dec	20
2-Jan	17	2-Jan	14

a. Look at the "points scored" by each team and estimate the "center" of the data for each data set.

Your estimate of "center" for the Giants: _____.

Your estimate of "center" for the Redskins: _____.

b. Retrieve the file "EA16.1 Points Scored 2010 2011.xls" (which contains the data shown in the previous tables) from the text website or WileyPLUS website, or enter the data into Excel for the New York Giants and Washington Redskins NFL teams.

c. Find the five-number summary for points scored by each of these teams and record them here.

   Giants: _____.

   Redskins: _____.

d. Go to www.nfl.com/teams and click the "Stats" link below the team you want, to get the most recently completed season points scored for one additional NFL team. Find the five-number summary for your team and sketch comparative boxplots showing the five-number summaries for points scored by the Eagles and the three additional football teams on the same scale.

e. Write a paragraph describing what these four comparative boxplots show about the scores of the four NFL teams. Who do you think would win a game between the Eagles and Giants? Explain your reasoning. Who do you think would win the other five games that could be played between pairs of these four teams?

Summary

In this activity, you used Excel to calculate mean and median. You found the five-number summary for several data sets of football scores. You also drew boxplots and used them to compare the performances of four different football teams.

17.

Standard Deviation, z-Score, and Normal Distributions

Smaller standard deviation

Larger standard deviation

A

B

After completing this topic, you will be able to:

- Compute and interpret the standard deviation of a set of data.

- Use the standard deviation to compare the distributions of several data sets.

- Compute and interpret the z-score of a data value, and use z-scores to compare data from different data sets.

- Visualize data that are normally distributed and use the empirical rule to help understand a normal distribution.

I n Topic 16 we looked at two measures of the spread of a data set: the range and the interquartile range. The range as a measure of spread uses only the maximum and minimum values in the data set. The interquartile range (the third quartile minus the first quartile) uses more information about the data set; it measures the width of the middle 50% of the data and is resistant to the effects of outliers. But the variability of a data set involves the entire collection of data because each data value contributes to the variability of the whole. A third measure of the spread or variability in a data set, which depends on each data value in the set, is the standard deviation.

Suppose we take three samples of seven cars traveling along a residential stretch of Main Street that has a posted speed limit of 25 miles per hour and record the speed of each of the cars. Here are three hypothetical sets of seven sample speeds. (Note that in each sample we have mean = median = 25 and range = 40.)

> Sample 1: 5, 25, 25, 25, 25, 25, 45
> Sample 2: 5, 15, 20, 25, 30, 35, 45
> Sample 3: 5, 5, 5, 25, 45, 45, 45

Although the mean and median are the same in all three samples, there is a real difference in the variability of the data in each of these samples. For example, in Sample 1, five of the cars were traveling at the posted speed limit and only one was traveling over the speed limit; in Sample 3, three cars were traveling 20 miles over the speed limit. We would like to be able to compare the variability of the three samples using a measure that incorporates each data value in the sample. In the context of the setting of this example, how much each individual car's speed deviates from the mean of 25 is important information. (You will be asked to compute the standard deviation of each of these samples in Exploration 1.) The **standard deviation** is a measure of the "typical" deviation from the mean for the values in the data set.

The standard deviation of a data set can be obtained by performing a few computations as described in the following paragraph. We'll illustrate this process using the calories-of-brownies data set introduced in Example 16.5 and reproduced here.

Brownie, Prepared According to Package Directions	Calories in One Piece
Pillsbury caramel fudge chunk	170
Betty Crocker caramel swirl	120
Betty Crocker German chocolate	160
Nestle double chocolate chip	150
Pepperidge Farm hot fudge	400
Duncan Hines milk chocolate	160
Betty Crocker microwave frosted	180
Robin Hood/Gold Medal pouch mix	100
Duncan Hines peanut butter	150
Pillsbury triple, chunky	170

Here are the steps needed to compute the standard deviation of a data set:

1. First compute the mean. (The mean of the brownies data set is 176.)

2. Next calculate the **deviation** of each data value from the mean; that is, subtract the mean from each data value. (These values are shown in the second column of the next table. Note that some of these deviations are positive and some are negative because some of the data values are larger than the mean and some are smaller. In fact, if we add all these deviations, their sum should be zero or very close to zero, allowing for possible round-off errors.)

3. Now square these deviations from the mean, as shown in the third column of the table. (All the values in this column are positive because they are squares.)

4. Next, sum the squares of the deviations; this gives us the total sum of the squared deviations.

5. Divide the sum obtained in Step 4 by 1 less than the number of data values.

6. Finally, take the nonnegative square root of the quotient found in Step 5 to obtain the standard deviation.

Calories per Serving—Brownies	(Calories − Mean)	(Calories − Mean)^2
170	−6	36
120	−56	3,136
160	−16	256
150	−26	676
400	224	50,176
160	−16	256
180	4	16
100	−76	5,776
150	−26	676
170	−6	36
Sum of Columns	**0**	**61,040**

The sum of the squares of the deviations is 61,040; we divide this sum by 9 (1 less than the number of data values) to get approximately 6782.22. The standard deviation of this data set is the nonnegative square root of 6782.22, which is approximately 82.35. Note that to get an average of the squared deviations, it might seem reasonable to divide by the number of data values, because we divide the sum of the data values by the number of data values to get the mean. We divide by $(n - 1)$ rather than n to find the standard deviation of a sample. This is because, on average, the result we get using $(n - 1)$ is closer to the true population sample standard deviation than it would be if we divided by n.

When computing the standard deviation of a sample, we first look at how far above or below the mean each data value falls by computing each data value's deviation from the mean. We then square these deviations and obtain an "average" squared deviation. By taking the nonnegative square root of that "average" squared deviation, we obtain a measure of a "typical" deviation from the mean for the data set. (Recall that the radical sign $\sqrt{x}$ indicates the nonnegative number that when squared gives x. It is defined for $x \geq 0$.)

Example 17.1

Find the standard deviation of the calories-of-ice-cream-bars data set given in Example 16.5 (and repeated here) and compare with the standard deviation of the brownies data set obtained previously. (Keep two decimal places in the calculations.)

Ice Cream Bar, One Piece	Calories per Bar
Good Humor Fat Frog	154
Good Humor Halo Bar	230
Heath	170
Haagen-Dazs caramel almond	230
Good Humor chip candy crunch	255
Nestle premium milk chocolate w/almond	230
Haagen-Dazs vanilla w/dark chocolate coating	390
Nestle Crunch vanilla w/white chocolate coating	350
Klondike Krispy	290
Oh, Henry vanilla w/choc coating	320

Solution

We again use a table to organize the work required to calculate the standard deviation. The first column of the table contains the original data. We set up the other columns as we did for the brownies data.

Looking at the data set before we do the computations, we can predict the standard deviation of the ice cream bar data to be less than 82.35 (the standard deviation of the brownie-calorie data set) because the ice cream bar-calorie data values appear to deviate less from the "center" or mean of their data set. The mean of this data set is 261.9, which is larger than the mean of the brownie-calorie data set.

Calories per Serving—Ice Cream Bar	(Calories − Mean)	(Calories − Mean)²
154	−107.9	11,642.41
230	−31.9	1,017.61
170	−91.9	8,445.61
230	−31.9	1,017.61
255	−6.9	47.61
230	−31.9	1,017.61
390	128.1	16,409.61
350	88.1	7,761.61
290	28.1	789.61
320	58.1	3,375.61
Sum of Columns	**0**	**51,524.9**

The sum of the squares is 51,524.9, and 51,524.9 divided by 9 is approximately 5724.99. The standard deviation is the square root of 5724.99; $\sqrt{5724.99} \approx 75.66$, which is smaller than the standard deviation of the brownie-calorie data set.

■ ■ ■ ■ ■ ■ ■ ■ ■ ■

The standard deviation, like the mean, is greatly influenced by outliers, as the next example shows.

Example 17.2

Suppose the calorie content of the Pepperidge Farms hot fudge brownie is only 200 calories instead of 400; recompute the mean and the standard deviation of the calories-of-brownies data set and compare with the previous results.

Solution

We will again set up a table to see how the mean and standard deviation are affected by not having the one large data value in the data set. The mean of the new data set is 156 (20 less than the mean of the original data set), so we see that the mean is significantly affected by this change.

Calories per Serving–Brownies	(Calories – Mean)	(Calories – Mean)^2
170	14	196
120	−36	1,296
160	4	16
150	−6	36
200	44	1,936
160	4	16
180	24	576
100	−56	3,136
150	−6	36
170	14	196
Sum of Columns	**0**	**7,440**

The standard deviation is the square root of $\frac{7440}{9}$, which is $\sqrt{826.67} \approx 28.75$. The standard deviation was reduced drastically (from 82.35 to 28.75) by a change in one extreme data value.

A greater variability in a sample of data is indicated by a larger value of the standard deviation of the sample, while a smaller value of the standard deviation indicates less variability. The standard deviation of a particular data set is estimated to be large or small by comparing it to the standard deviation of another set of data.

The standard deviation is fairly tedious to compute using a table, as we did for the previous example. Organizing the computation of the standard deviation with a table is a good way to get a sense of what the standard deviation is; however, it is impractical to use except for small examples. Most calculators and computers have a built-in function that can be used to compute the standard deviation.

The median, five-number summary, range, and interquartile range are useful to describe a data set that contains outliers or a data set that is skewed. The mean and standard deviation are more frequently used for roughly symmetric data sets that have no outliers. We will look at one particular type of distribution that fits the description of "symmetric without outliers"; this type of data set occurs frequently with measurements and test scores and other naturally occurring phenomena and is called a **normal distribution**.

Consider the following frequency histogram of critical reading Scholastic Aptitude Test scores for a recent freshman class at a small college. The data appear to be fairly symmetric, with the mean and median occurring at approximately the center peak. This peak is at approximately 550.

If we were to sketch a smooth curve, tracing along the tops of the frequency bars of the histogram of the data set, it would roughly resemble a bell-shaped curve, as shown in the next graph. This type of curve is called a **normal curve**. Data that are normally distributed have a single peak and follow roughly a bell shape; a normal curve is sometimes called a

bell curve because of the bell shape. The curve sketched here is the "idealized" version of the curve of a data set that exactly follows a normal distribution and allows us to visualize quickly the shape of a data set known to be approximately normal.

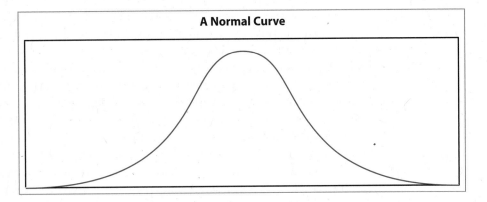

A data set that is normal or approximately normal is characterized by its mean and standard deviation. On the sketch of any normal curve, the mean is the point along the horizontal axis at which the peak occurs. In general, the larger the standard deviation, the more spread out the curve is and the less high the peak is. The following graph shows two normal curves with the same mean. The curve with the higher peak has a smaller standard deviation; the data values with a resulting histogram that gives rise to the taller and narrower normal curve are less spread out along the horizontal axis than those values leading to the shorter curve.

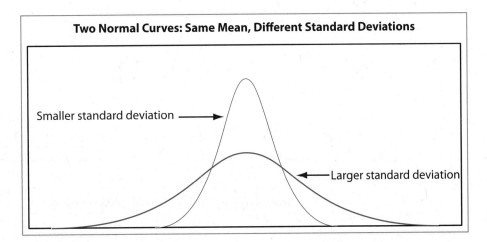

If we trace a normal curve, beginning at the left end of the graph and proceeding to the right, the curve is concave upward initially. The point where the curve changes from concave upward to concave downward lies at the mean minus the standard deviation along the horizontal axis. If we continue tracing along the curve, the curve is concave downward (around the peak) until we reach the mean plus the standard deviation. The curve changes to concave upward at that point. We also see that the curve is symmetric with respect to the mean and that at the left and right ends, it gets closer and closer to the horizontal axis. Although the curve continues indefinitely in both directions, the two values where the curve gets very close to the horizontal axis are: mean − 3 · (standard deviation); mean + 3 · (standard deviation).

We can sketch a normal curve by marking several key points along the horizontal measurement axis. We'll describe how with an example. Suppose we know that grades in a large freshman course are approximately normally distributed, with a mean of 80 and a standard deviation of 5. We want to sketch this distribution of grades. First, we mark the mean of 80 on the horizontal axis, where the highest peak of the curve occurs. Then we identify the numbers that represent grades that are one standard deviation above the mean and one standard deviation below the mean because these points are where the normal curve changes concavity. One standard deviation above the mean is $80 + 5 = 85$; one standard deviation below the mean is $80 - 5 = 75$. These are two key points to locate on the horizontal axis. Other points to mark on the horizontal axis are values two standard deviations above and below the mean (which help us set up the scale), and values three standard deviations above and below the mean (which show us where the curve gets very close to the horizontal axis). For this example, these values are $80 + 2 \cdot 5 = 90$ and $80 - 2 \cdot 5 = 70$, and $80 + 3 \cdot 5 = 95$ and $80 - 3 \cdot 5 = 65$. After we mark these values on the horizontal axis, we sketch in the curve, showing that the graph is symmetric about the mean of 80; it changes concavity at 75 and 85, and it gets very close to the horizontal axis for values greater than 95 and for values less than 65. Here is a graph of this distribution.

The basic characteristics of a normal curve (bell-shaped curve) are:

- The peak occurs above the value on the horizontal axis that corresponds to the mean of the data set.

- The curve is symmetric with respect to the mean.

- The curve changes concavity above the two points on the horizontal axis located one standard deviation from the mean.

- The curve practically touches the horizontal axis at the two points on the horizontal axis that are located three standard deviations from the mean.

- All of the observations (that is, 100% of them) lie above the horizontal axis and under the curve.

Example 17.3

A particular test is used to determine IQ scores (for Intelligence Quotient), and among a certain group of middle school students, IQ scores are known to have an approximately normal distribution with a mean of 100 and a standard deviation of 15.

a. Find the two values that represent IQ scores that lie one standard deviation above and one standard deviation below the mean. Also find values that represent scores that fall two and three standard deviations above the mean and those that fall two and three standard deviations below the mean.

b. Mark the mean and the six values you found in part (a) of this example on a horizontal axis.

c. Sketch the graph of the normal curve that represents IQ scores for the group of middle school students.

Solution

a. One standard deviation above the mean is the IQ value of $100 + 15 = 115$. One standard deviation below the mean is the value $100 - 15 = 85$. The value $100 + 2 \times 15 = 130$ is two standard deviations above the mean, and $100 + 3 \times 15 = 145$ is three standard deviations above the mean. The value $100 - 2 \times 15 = 70$ is two standard deviations below the mean, while $100 - 3 \times 15 = 55$ is three standard deviations below the mean.

b. We mark the horizontal axis, which is the axis representing the variable IQ scores, with the mean (100) in the center as shown here:

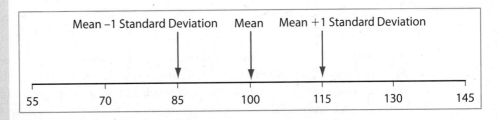

Mean −1 Standard Deviation   Mean   Mean +1 Standard Deviation

55   70   85   100   115   130   145

c. Tracing from left to right, the curve is very close to the horizontal axis for an IQ score of 55. The curve increases and changes from concave upward to concave downward above the IQ value of 85 (the mean minus the standard deviation). The curve reaches its peak above 100, and changes from concave downward to concave upward above the IQ value of 115 (the mean plus the standard deviation). The curve decreases and gets very close to the measurement axis for IQ values greater than 145.

Distribution of IQ Scores

Concavity Changes — Concavity Changes

55   70   85   100   115   130   145

IQ Score

There is a useful **empirical rule** that applies to any data set that follows an approximately normal distribution. Here, the phrase *empirical rule* refers to a rule that is derived from practical experience. The rule says that if the observations in a data set can be approximated by a normal curve, then:

■ **Approximately 68%** of the data values are within one standard deviation of the mean.

■ **Approximately 95%** of the data values are within two standard deviations of the mean.

■ **Approximately 99.7%** of the data values are within three standard deviations of the mean.

For the normal curve shown in Example 17.3 and representing IQ values of a group of middle school students, with a mean of 100 and a standard deviation of 15, this rule says that approximately 68% of the data values fall between 85 and 115; approximately 95% of the IQ values fall between 70 and 130; and approximately 99.7% (that is, almost all) of the data values fall between 55 and 145. This rule is also sometimes called the "68−95−99.7" rule.

In a histogram, the area of a frequency bar corresponds to the proportion of observations that fall into that category; we can then interpret the area under the normal curve between two measurement values as the proportion of observations that fall between those values. For example, on the graph with a mean of 100 and a standard deviation of 15, 95% of the area under the normal curve (and hence 95% of the observations) falls between 70 and 130.

Example 17.4

Suppose the critical reading SAT scores for freshmen at a certain college follow an approximately normal distribution with a mean of 550 and standard deviation of 79.

a. Use the empirical rule to find an interval of values within which approximately 68% of the freshmen critical reading SAT scores fall.

b. Use the empirical rule to find an interval of values within which approximately 95% of the freshmen critical reading SAT scores fall.

c. Use the empirical rule to find an interval of values within which approximately 99.7% of the freshmen critical reading SAT scores fall.

d. If there are 482 students in the freshman class, approximately how many will fall into each of the intervals computed in parts (a–c) of this example?

Solution

a. The empirical rule says that we need to find the interval that covers all values within one standard deviation of the mean, either one standard deviation greater than the mean or one standard deviation less than the mean. Thus, we need to calculate: mean $- 1 \cdot$ (standard deviation); mean $+ 1 \cdot$ (standard deviation). This is $550 - 1 \cdot 79 = 471$ and $550 + 1 \cdot 79 = 629$. So approximately 68% of the freshman critical reading SAT scores will fall between 471 and 629.

b. For the "95% rule," we need to calculate the interval that covers all values within two standard deviations of the mean; that is, we calculate: mean $- 2 \cdot$ (standard deviation) and mean $+ 2 \cdot$ (standard deviation). This is $550 - 2 \cdot 79 = 392$ and $550 + 2 \cdot 79 = 708$. We can say that approximately 95% of the freshmen have critical reading SAT scores between 392 and 708.

c. To calculate the interval into which approximately 99.7% of freshmen critical reading SAT scores fall, we calculate $550 - 3 \cdot 79 = 313$ and $550 + 3 \cdot 79 = 787$. So 99.7% of the freshman at this college have critical reading SAT scores between 313 and 787.

d. We calculate 68% of 482 to get $(0.68) \cdot 482 = 327.76$, or approximately 328 of the freshman have a critical reading SAT score between 471 and 629. Ninety-five percent of 482 is 457.9, so approximately 458 freshmen have a critical reading SAT score between 392 and 708. This also means that approximately 24 freshmen have a critical reading SAT score that falls outside of this range, either greater than 708 or less than 392. The "99.7% rule" says that 480.55 or approximately 481 freshmen have a critical reading SAT score between 313 and 787. Thus, one student would be expected to have a score outside of this range.

Sometimes we want to compare values from different data sets. One way to do this is to **standardize** the values; that is, evaluate how many standard deviations each observation lies away from the mean of its data set. We can then compare the standardized values and assess which observation is more extreme, taking into account the mean and standard deviation of the data set from which the observation was obtained. The standardized value of an observation is called a **z-score**. The z-score tells us how many standard deviations the observed data value lies away from the mean. To evaluate the z-score for an observation from a data set with a known mean and standard deviation, we subtract the mean from the observation and then divide this difference by the standard deviation. Thus, we compute

$$z = \frac{\text{observation} - \text{mean}}{\text{standard deviation}}$$

For example, suppose that critical reading SAT scores of a particular college's freshman class are normally distributed with a mean of 550 and a standard deviation of 50. A student from this class who scored 620 on the critical reading SAT test has a z-score of $z = \frac{(620 - 550)}{50} = \frac{70}{50} = 1.4$. This means that the student's critical reading SAT score lies 1.4 standard deviations above the mean of the data. It tells us where this student's score falls relative to the mean and standard deviation of the scores of all students in the class. A student from this class who scored 520 on the critical reading SAT test would have a z-score of $z = \frac{(520 - 550)}{50} = -\frac{30}{50} = -0.60$.

Note that the standard deviation of any data set is zero (if all the data values are the same) or a positive number. So if an observation is greater than the mean, its z-score is positive because, in this case, (observation − mean) is positive. If an observation is less than the mean, its z-value is negative because, in this case, (observation − mean) is negative.

Example 17.5 ■ ■ ■ ■ ■ ■ ■ ■ ■

The distribution of IQ scores for a group of middle school students, as described in Example 17.3, has a mean of 100 and a standard deviation of 15. Assume that the distribution of IQ scores for college students is also approximately normally distributed and has a mean of 115 with a standard deviation of 18. A middle school student's IQ score is reported to be 134, while her older brother, who is a college student, has an IQ score of 144. How do these scores compare?

Solution

It's clear that the older brother's score is higher. But because the two scores come from different distributions, we can compare them by using the standardized scores. The middle school student's standardized score is $z = \dfrac{(134 - 100)}{15} \approx 2.27$, while the older brother's score is $z = \dfrac{(144 - 115)}{18} \approx 1.61$. The middle school student's IQ score lies 2.27 standard deviations above the mean of her distribution, while her older brother's IQ score lies 1.61 standard deviations above the mean of his distribution. With respect to their own age groups, the middle school student has the higher IQ score.

Summary

In this topic, we looked at how to define and interpret a measure of the spread of a data set, the standard deviation. We saw that the standard deviation can be changed significantly by altering one extreme data value. We investigated characteristics of a normal distribution, including the empirical rule. Finally, we looked at z-scores as a way to compare values that come from different data sets.

Explorations

1. Consider again the three samples of the speeds of seven cars given at the beginning of this topic: Sample 1: 5, 25, 25, 25, 25, 25, 45; Sample 2: 5, 15, 20, 25, 30, 35, 45; and Sample 3: 5, 5, 5, 25, 45, 45, 45.

   a. Look at the data in each sample and without doing any calculations, rank the samples in order from smallest to largest standard deviation. Give reasons for your ranking.

   b. For each of the samples, create a table like those in Examples 17.1 and 17.2; for each sample, compute the deviation of each data value from the mean, the squared deviations, and finally the standard deviation of the data set.

   c. Explain what the three standard deviations tell you about the data sets.

2. The following tables provide information about the number of women-owned businesses and their revenues for 2002 for a group of northwestern and Pacific states and for a group of northeastern states of the United States.

Northwestern/Pacific State	Number of Businesses	Revenue (million $)
Alaska	16,308	2,348
California	870,496	137,692
Hawaii	29,943	4,594
Idaho	28,824	3,216
Montana	24,519	2,139
Oregon	88,317	10,608
Washington	137,394	17,368
Wyoming	12,945	1,130

Northeastern State	Number of Businesses	Revenue (million $)
Connecticut	82,118	12,216
Maine	32,512	3,282
Massachusetts	161,918	23,134
New Hampshire	31,024	4,665
New Jersey	185,197	35,573
New York	505,077	70,838
Pennsylvania	227,117	38,998
Rhode Island	23,195	3,641
Vermont	18,989	1,454

Source: U.S. Census Bureau, www.census.gov.

a. Use a calculator or computer to compute the mean and standard deviation of the year 2002 revenue for women-owned businesses in the northwestern and Pacific states.

b. Use a calculator or computer to compute the mean and standard deviation of the year 2002 revenue for women-owned businesses in the northeastern states.

c. Explain what the values you calculated in parts (a) and (b) of this Exploration tell you about the data sets.

d. How would the means and standard deviations calculated in parts (a) and (b) change if the largest data value in each data set were removed?

e. Find the mean and standard deviation of the number of women-owned businesses in the northwestern and Pacific states.

f. Find the mean and the standard deviation of the number of women-owned businesses in the northeastern states and compare with your results in part (e) of this Exploration.

g. How do the four standard deviations that you computed compare, and what does this tell us about the data?

3. Consider the following data, collected from a sample of five male college students, giving the estimated number of haircuts they had in one year: 12, 7, 12, 30, 24.

   a. Find the standard deviation of the haircut data.

   b. Here is similar data, collected from a sample of seven female college students, giving the estimated number of haircuts they had in one year: 2, 2, 12, 5, 45, 9, 4. Do you think the standard deviation of this data set will be larger, smaller, or the same as the standard deviation you calculated in part (a)? Give reasons for your answer.

   c. Calculate the standard deviation of the female haircut data and compare with the standard deviation found in part (a).

4. Consider four sets of test scores:

   Set A: 50, 50, 80, 100, 100, 100
   Set B: 78, 79, 80, 80, 81, 82
   Set C: 20, 80, 80, 100, 100, 100
   Set D: 60, 80, 80, 80, 80, 100

   a. Without actually computing the standard deviation, rank the four sets of scores in order of smallest standard deviation to largest. Give reasons for your ranking.

   b. Find the standard deviation of each set of test scores and compare with your answer in part (a) of this Exploration.

5. The following tables provide information about a collection of states and their population and population density, from the 2010 census. The first table gives the information for a group of eight southern states. The second table gives similar information for a group of eight states in the northeast.

State	Population	People per Square Mile
Arkansas	2,915,918	56
Georgia	9,687,653	168.4
Kentucky	4,339,367	109.9
Louisiana	4,533,372	104.9
Mississippi	2,967,297	63.2
North Carolina	9,535,483	196.1
South Carolina	4,625,364	153.9
Tennessee	6,346,105	153.9

State	Population	People per Square Mile
Connecticut	3,574,097	738.1
Delaware	897,934	460.8
Maine	1,328,361	43.1
Maryland	5,773,552	594.8
Massachusetts	6,547,629	839.4
New Jersey	8,791,894	1195.5
New York	19,378,102	411.2
Rhode Island	1,052,567	1018.1

Source: U.S. Census Bureau, www.census.gov.

a. Use a calculator or computer to compute the mean and standard deviation of the population of the southern states in the first table.

b. Use a calculator or computer to compute the mean and standard deviation of the population of the northeastern states in the second table.

c. Explain what the values you calculated in parts (a) and (b) of this Exploration tell you about the data sets.

d. Find how the mean and standard deviation change if the largest population data value in each set were removed. Explain what happens in each data set and why that happened.

e. Find the mean and standard deviation of the population density (people per square mile) of the southern states in the first table.

f. Find the mean and standard deviation of the population density (people per square mile) of the northeastern states in the second table and compare your results to those you obtained for part (e) of this Exploration.

g. Find how the mean and standard deviation of the population density change if the smallest population density data value in each set were removed. Explain what happens in each case and why that happened.

6. The following tables give the points scored by the scoring leaders in each NBA season during the decades 1950–1960, 1970–1980, and 2000–2010.

Season	Player	Team	Points Scored
1950–51	George Mikan	Minneapolis Lakers	1932
1951–52	Paul Arizin	Philadelphia Warriors	1674

1952–53	Neil Johnston	Philadelphia Warriors	1564
1953–54	Neil Johnston	Philadelphia Warriors	1759
1954–55	Neil Johnston	Philadelphia Warriors	1631
1955–56	Bob Pettit	St. Louis Hawks	1849
1956–57	Paul Arizin	Philadelphia Warriors	1817
1957–58	George Yardley	Detroit Pistons	2001
1958–59	Bob Pettit	St. Louis Hawks	2105
1959–60	Wilt Chamberlain	Philadelphia Warriors	2707

Season	Player	Team	Points Scored
1970–71	Kareem Abdul-Jabbar	Milwaukee Bucks	2596
1971–72	Kareem Abdul-Jabbar	Milwaukee Bucks	2822
1972–73	Tiny Archibald	Kansas City-Omaha King	2719
1973–74	Bob McAdoo	Buffalo Braves	2261
1974–75	Bob McAdoo	Buffalo Braves	2831
1975–76	Bob McAdoo	Buffalo Braves	2427
1976–77	Pete Maravich	New Orleans Jazz	2273
1977–78	George Gervin	San Antonio Spurs	2232
1978–79	George Gervin	San Antonio Spurs	2365
1979–80	George Gervin	San Antonio Spurs	2585

Season	Player	Team	Points Scored
2000–01	Allen Iverson	Philadelphia 76ers	2207
2001–02	Allen Iverson	Philadelphia 76ers	1883
2002–03	Tracy McGrady	Orlando Magic	2407
2003–04	Tracy McGrady	Orlando Magic	1878
2004–05	Allen Iverson	Philadelphia 76ers	2302
2005–06	Kobe Bryant	Los Angeles Lakers	2832
2006–07	Kobe Bryant	Los Angeles Lakers	2430
2007–08	James LeBron	Cleveland Cavaliers	2250
2008–09	Dwyane Wade	Miami Heat	2386
2009–10	Kevin Durant	Oklahoma City Thunder	2472

a. Before doing any computations, do you think the standard deviation of the points scored by the leading NBA player during the decade 1950–1960 is larger, smaller, or about the same as the standard deviation of the points scored by the leading player during the decade 2000–2010?

b. Before doing any computations, do you think the standard deviation of the points scored by the leading NBA player during the decade 1970–1980 is larger, smaller, or about the same as the standard deviation of the points scored by the leading player during the decade 2000–2010?

c. Use the given data to calculate the standard deviation of the points scored by the leading players for each of the three decades:

  i. 1950–1960

  ii. 1970–1980

  iii. 2000–2010

d. How do the standard deviations you found in part (c) compare with your answers for parts (a) and (b)?

e. If we calculated the standard deviation of the points scored by the leading players during the decade 2000–2010, using the data in thousands (for example, 2.472 instead of 2472 for Kevin Durant's points scored in the 2009–10 season), would the standard deviation be the same? Why or why not? Confirm your answer using the values 2.207, 1.883, 2.407, and so on, to calculate the standard deviation for the points scored by the leading scorers during the decade 2000–2010.

7. Suppose that IQ scores of students at a particular college are known to be approximately normally distributed with a mean of 112 and a standard deviation of 12.

a. What does the empirical rule tell us about the IQ scores of students at this college? Be specific.

b. Sketch a curve that represents the IQ scores of students at this college. On the curve, mark values that are one, two, and three standard deviations above and below the mean.

c. On the same graph, draw a normal curve with mean 120 and standard deviation 12.

8. Consider the distributions of IQ scores described in Example 17.5.

a. Sketch the normal curve for the distribution of middle school student IQ scores with a mean of 100 and standard deviation of 15. Mark the measurements on the horizontal axis.

b. Sketch the normal curve for the distribution of college student IQ scores with a mean of 115 and standard deviation of 18. Mark the measurements on the horizontal axis.

c. Describe how the two curves you sketched are similar and how they differ.

d. On the normal curve for the middle school students, identify the area below the curve and to the right of the student's score of 134. Then, on the college curve, identify the area below the curve and to the right of her brother's score of 144. Describe how the areas show that, with respect to their age groups, the middle school student has a higher IQ than her brother.

9. Critical reading SAT test scores follow approximately a normal distribution and math SAT test scores also follow approximately a normal distribution.

   a. A particular freshman student at a certain college has a standardized critical reading SAT test score of $z = -1.3$ and a standardized math SAT test score of $z = 2.0$. Explain what these values tell you.

   b. Another freshman has a standardized math SAT test score of 1.5. How does the student described in part (a) of this Exploration compare to this student?

   c. Suppose you also know that the mean math SAT test score for freshmen students at this college is 547 and the standard deviation is 74. What can you say about the math SAT test score of the student with a standardized math SAT test score of 2.0?

   d. Use the empirical rule to find an interval of values within which approximately 68% of the freshmen math SAT scores at this college fall.

   e. Use the empirical rule to find an interval of values within which approximately 95% of the freshmen math SAT scores at this college fall.

10. The following table gives the number of home runs hit by the American League home-run champions for the years 1949 to 1987, given in chronological order beginning with 1949 and proceeding left to right across the rows:

43	37	33	32	43	32	37	52	42	42	42	40	61
48	45	49	32	49	44	44	49	44	33	37	32	32
36	32	39	46	45	41	22	39	39	43	40	40	49

Source: *The World Almanac and Book of Facts 2011*, p. 920.

   a. Find the mean and standard deviation of number of home runs hit by the American League home-run champions during the years given in the table.

   b. Assuming that the data are approximately normally distributed, use the empirical rule to find an interval of values within which approximately 68% of the data falls. Count the number of data values that actually fall within that interval. Is it close to 68%?

   c. Use the empirical rule to find an interval of values within which approximately 95% of the data falls. Count the number of data values that actually fall within that interval. Is it close to 95%?

   d. Use the empirical rule to find an interval of values within which approximately 99.7% of the data falls. Count the number of data values that actually fall within that interval. Is it close to 99.7%?

11. The following table gives the number of home runs hit by American League home-run champions from 1988 to 2010, in chronological order left to right across the rows:

| 42 | 36 | 51 | 44 | 43 | 46 | 40 | 50 | 52 | 56 | 56 | 48 |
| 47 | 52 | 57 | 47 | 43 | 48 | 54 | 54 | 37 | 39 | 54 | |

Source: The World Almanac and Book of Facts 2011, p. 920.

a. Compute the mean and standard deviation of number of home runs for the years 1988 to 2010.

b. If you were to compute the mean number of home runs hit by the American League home-run champions during the years 1949 to 2010, do you expect the mean to increase, decrease, or stay about the same as the mean found in part (a) of this Exploration? Explain your reasoning.

c. Compute the mean and standard deviation of number of home runs for the years 1949 to 2010 (use the data from Exploration 10) and compare with your results for part (a) of this Exploration.

12. The following table gives typical monthly temperatures, in degrees Fahrenheit, in each of three U.S. cities. Find the mean and standard deviation of the typical monthly temperatures for each of the cities. What do these values tell you?

	Jan	Feb	Mar	Apr	May	June	July	Aug	Sept	Oct	Nov	Dec
San Diego, CA	54.9	55.9	57.6	59.9	62.2	65.1	68.7	70.2	68.5	64.8	60.1	56.1
Orlando, FL	59.5	61.2	66.6	71.1	76.8	81.0	82.2	82.4	81.0	75.2	68.0	62.1
Philadelphia, PA	30.6	32.9	42.3	52.3	62.8	71.8	76.6	75.4	68.2	56.3	46.4	35.8

13. One professor teaches a large section, section A, of a particular class, and on the first test of the term, the test scores in section A were approximately normally distributed with a mean of 78 and a standard deviation of 6. Another professor also teaches a large section, section B, of the same class, and on the first test of the term, the test scores in section B were also approximately normally distributed with a mean of 74 and a standard deviation of 10. Two students, one from each section, earned a grade of 92 on the exam. The student from section B claims that he did better because the section B test, with a mean of 74, was obviously more difficult than the section A test with a mean of 78. However, the student from section A claims that because she has a higher z-score, she actually performed better. Calculate the z-scores for each student's test grade and settle their dispute; that is, decide who had the superior performance on this test.

14. Colleges typically use grade point average cutoffs to decide who graduates with honors and who is accepted into certain programs (such as teacher education, for example). Suppose at a particular college, a GPA of 3.0 is the cutoff for such a decision. The students in department A have a grade point average of 3.77 with a standard deviation of 0.43 (for all course grades in classes taken in the department in a particular academic year). The students in department B have a grade point average of 2.65 with a standard deviation of 1.16 for that same academic year. One student has taken most of his courses from department A; another student has taken most of her courses from department B. Both students have a GPA of 3.0.

a. Compare the z-scores of the two students.

b. Interpret what these z-scores mean.

15. The first table below gives per capita personal income for each state in a group of eight southern states. The second table gives similar information for each state in a group of eight states in the northeast. (These same data were used in Activity 2.2.)

State	Per Capita Personal Income 2010
Arkansas	33,150
Georgia	35,490
Kentucky	33,348
Louisiana	38,446
Mississippi	31,186
North Carolina	35,638
South Carolina	33,163
Tennessee	35,307

State	Per Capita Personal Income 2010
Connecticut	56,001
Delaware	39,962
Maine	37,300
Maryland	49,025
Massachusetts	51,552
New Jersey	50,781
New York	48,821
Rhode Island	42,579

a. Find the mean and standard deviation of the per capita personal income of the sample of southern states data set.

b. Find the mean and standard deviation of the per capita personal income for the sample of northeastern states data set.

c. Compare the means and standard deviations of the two data sets. What do these values tell us about the data?

d. Find the z-score of Louisiania's per capita personal income of $38,446. What does this z-score tell us?

e. Find the z-score of Delaware's per capita personal income of $39,962. What does this z-score tell us and how does this z-score compare with the z-score you calculated in part (d) of this Exploration?

Coins, Presidents, and Justices: Normal Distributions and z-Scores

In the first part of this activity, you will generate some data that should have an approximately normal (or bell-shaped) distribution. In the second part, you will use the definition of standard deviation and compare the standard deviations for two different data sets.

1. Work with a partner to generate the following data:

   a. Toss 10 coins and record the number of heads you obtained.

   b. Repeat part (a) 24 more times until you have a list of 25 numbers, each between 0 and 10.

c. Count how many times each number of heads occurred in your data and fill in the following table:

Number of Heads	Frequency
0	
1	
2	
3	
4	
5	
6	
7	
8	
9	
10	

2. Retrieve the file "EA17.1 Coins and Presidents.xls" from the text website or WileyPLUS, and you will find the results of 35 tosses of 10 coins that someone else carried out. When you first retrieve the file, column B contains the number of times 0 heads was obtained in the 35 tosses of 10 coins, the number of times 1 head was obtained in the 35 tosses, and so on, up to the number of times 10 heads was obtained.

a. Add your results to those in the file so you have a total of 60 in column B.

b. Create a scatterplot of these data, using one of the versions of the scatterplot with the dots connected. Be sure to add axis titles to your scatterplot. Describe what your curve looks like, including information about its center and spread.

c. Change your graph to a bar graph (instructions follow).

Instructions to Change a Scatterplot to a Column Graph

1. Click on the plot area and, from the **Design** tab, select **Change Chart Type** from the **Type** group.

2. From the drop-down menu, select **Column** to change your graph to a column (that is, a bar) graph and click **OK**.

3. Click one of the bars, go to the **Format** tab, and choose **Format Selection** from the **Current Selection** group. From **Series Option,** change the **Gap width** to **0,** by moving the side to the left, so adjacent bars will touch. (This will make it look like a histogram, and then you can sketch a curve along the tops of the bars.) Click **Close.**

    d. Print your bar graph, with appropriate titles on the axes, and by hand draw in a bell-shaped curve that "fits" this data. How does your hand-drawn curve compare with the curve you described in part (b) of this question?

3. In the second part of this activity, you will examine one measure of the spread of a data set, the standard deviation. The standard deviation plays an important role in understanding the spread of a distribution, especially a bell-shaped or **normal distribution**. You'll use the data set of ages of U.S. presidents at their inauguration, which can be found on sheet 2 of the file "EA17.1 Coins and Presidents.xls." (*Source: The World Almanac and Book of Facts 2011,* p. 506.)

    a. First, you will calculate the standard deviation of this data set. To do this, first find the mean (average) of the ages and store that value in cell B47.

        Record the mean age here: _____.

    b. In cell C1, enter the label **Deviation from the mean**. In cell C2, enter an appropriate formula to subtract the mean age from the value in cell B2 and that will allow you to drag down to compute each data value minus the mean. (Remember to use **B47** or name the cell to keep the value of the mean fixed when you drag.) Drag down to compute each data value minus the mean.

How many of the values in this column are negative? _____

What property makes these values negative?

  c. Add the values in column C and record their sum here: _____.

Write this number without using scientific notation: _____.

This number should be 0 or very, very close to 0. Is it? _____.

  d. Now you want to compute the square of the deviation from the mean of each data value. In cell D1, enter the label **Squared deviation** and use the instructions that follow to enter the squares of the deviation values in column D.

Instructions to Compute Squares

In cell D2, enter the formula **=C2^2**. Then drag this formula down to cell D45.

Then add all the values in column D, store that number in cell D47, and record the sum here: _____.

  e. Next you need to divide the sum of the squared deviations (the value in cell D47) by 1 less than the number of data points. There are 44 data points, so divide the value in cell D47 by 43; store this number in cell E47.

  f. Finally, compute the standard deviation by taking the square root of the number in cell E47 (see the following instructions for a reminder of how to compute the non-negative square root of a number). Store this value in cell F47 and record it here:

Standard deviation = _____

Instructions to Compute a Square Root

To compute the nonnegative square root of the value in E47, use the command **=SQRT(E47)**.

g. Check the computations by using the following Excel command to compute the standard deviation of the ages in column B. Enter the value obtained in cell F48, and write this value here: _____.

Instructions to Compute the Standard Deviation

To compute the standard deviation of the values in cells B2 through B45, use the command =**STDEV(B2:B45)**.

h. The standard deviation of a set of data is a measure of the spread of the data values. It incorporates the sum of the squared deviations from the mean, of all data values. Suppose Obama had been 87 years old instead of 47 at his inauguration. What would change in the computations you just performed?

i. Change Obama's age on your spreadsheet from 47 to 87. Notice that all of your computations change automatically. Record the new mean and the new standard deviation.

Mean = _____; standard deviation = _____

j. Change Obama's age back to its correct value of 47.

4. Go to sheet 3 of the file "EA17.1 Coins and Presidents.xls," where you will find another data set. This data set contains the names and ages at time of appointment as chief justice for all the chief justices of the U.S. Supreme Court. (*Source: The World Almanac and Book of Facts 2011,* p. 503.)

a. Compute the mean and standard deviation of these ages, and record these values here:

Mean = _____; standard deviation = _____.

b. Describe how the means and standard deviations of the two data sets, "Presidents' Ages" and "Supreme Court Chief Justices' Ages," compare.

c. Pick the maximum data value in the "Presidents' Ages" data set. Call it x, and compute its z-score by computing: $z = \dfrac{x - \text{mean}}{\text{standard deviation}}$, using the mean and the standard deviation of the presidents' ages.

Record this z-score here: _____

d. Find the z-score for the largest data value in the "Supreme Court Chief Justices' Ages" data set, using the mean and standard deviation of the chief justices' ages.

Record this z-score here: _____

e. What do the two z-scores tell you?

Summary

In the first part of this activity, you generated data and created a graph to see that the data have an approximately normal distribution. In the second part of the activity, you worked with the standard deviation to see the impact of a large data value on this measure of spread. You also compared data from two data sets by looking at z-scores.

18

Basics of Probability

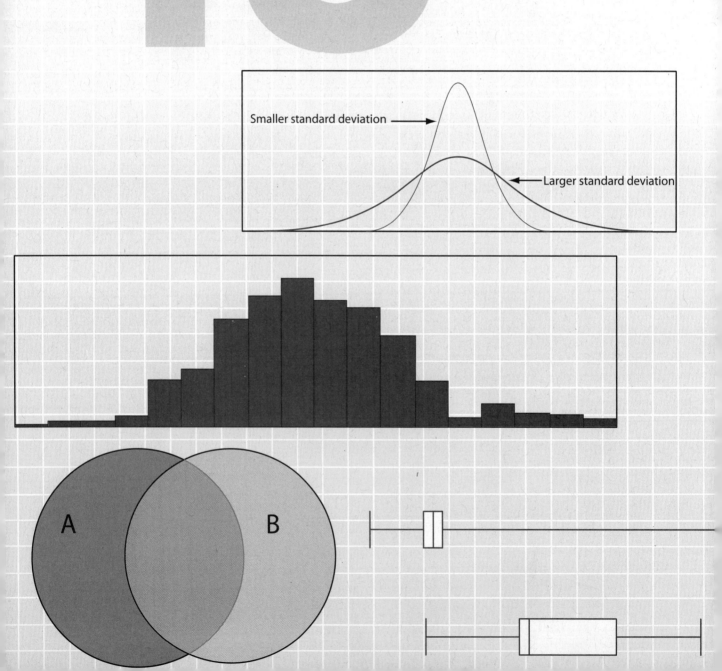

Shortly after the *Challenger* space shuttle accident in 1986, officials at NASA assessed the risk of another such disaster at approximately 1 in 225 for each shuttle mission. Another scientific organization, the National Academy of Sciences, estimated the likelihood of failure for a particular shuttle flight to be 1 in 145. Tragically, the *Challenger* disaster was followed by the *Columbia* shuttle disaster on February 1, 2003. (The space shuttle program, after many successful missions, ended on July 21, 2011. See www.nasa.gov for current information on space programs.) Many practical issues involve an assessment of the likelihood of some event happening or the assessment of risk. Examples include statements such as those given about the shuttle disaster, as well as statements such as, "There is a 50 percent chance that it will rain tomorrow" and "Your chance of being seriously injured in a car accident is reduced if you wear a seatbelt." To understand and evaluate these and other risks and to help make informed decisions that might involve considering such risks, we will investigate the language and methods of assessing the likelihood associated with a chance event.

A **random process** is a situation that can be repeated and for which the set of *possible* outcomes is known. While there is uncertainty about what the outcome will be on any particular repetition, there is a predictable pattern over many repetitions, a regularity that appears only with many repetitions. The launching of a shuttle is a random process because we don't know what the outcome of a particular launch will be. Flipping a coin, tossing a pair of dice, or randomly selecting a set of voters for jury duty are all examples of random processes.

The collection of all possible outcomes of a random process is called the **sample space** for the process. The types of outcomes in a sample space depend on what we are interested in observing for a particular random process. For example, if we choose a set of voters for jury duty, we may want to record how many females are among the set of jurors, or we may want to record how many retirees there are, each juror's political party, or some other characteristic.

Example 18.1

For each of the random processes described here, list all possible outcomes in the sample space.

a. Note people who go through the checkout line at a particular convenience store and record the gender of each.

b. Note people who go through the checkout line at a particular convenience store and record if they buy bread and/or milk or not.

c. Toss a pair of dice, one green and one white, and record the two values on the top faces of the dice.

d. Toss a pair of dice and record the sum of the two values on the top faces.

e. Select graduates of a particular college and record the GPA of each. Assume that at this particular college, the students need a GPA of 2.00 or above to graduate.

Solution

We often denote a collection or set using curly brackets, { }, and if there are a finite number of objects in the collection, we list the objects in the set within the brackets, separated by commas. We will use S to denote the sample space for a random process.

a. There are only two possible outcomes for this process, so there are two elements in the sample space: S = {male, female}.

b. We could represent the sample space for this random process in several ways. If we record which of these two items customers buy, we have S = {bread only, milk only, both bread and milk, neither}. Alternatively, we could use a pair of yes or no replies, where the first reply answers the question "Did the customer buy bread?" and the second reply answers the question "Did the customer buy milk?" In this case, S = {(yes, no), (no, yes), (yes, yes), (no, no)}. Notice that the outcome (yes, no) is different from the outcome (no, yes). For each of the given ways of representing S, the number of elements in S is 4.

c. We will use pairs of numbers to indicate the elements in the sample space, with the first number representing the result on the green die and the second number representing the result on the white die. Thus,

$$S = \{(1, 1), (1, 2), (1, 3), (1, 4), (1, 5), (1, 6), (2, 1), (2, 2), (2, 3), (2, 4), (2, 5), (2, 6),$$
$$(3, 1), (3, 2), (3, 3), (3, 4), (3, 5), (3, 6), (4, 1), (4, 2), (4, 3), (4, 4), (4, 5), (4, 6),$$
$$(5, 1), (5, 2), (5, 3), (5, 4), (5, 5), (5, 6), (6, 1), (6, 2), (6, 3), (6, 4), (6, 5), (6, 6)\}$$

d. In this example, $S = \{2, 3, 4, 5, 6, 7, 8, 9, 10, 11, 12\}$, which represents all possible sums of the two numbers in the pairs given in the solution to part (c).

e. Students need a GPA of 2.00 to graduate, so the sample space contains all numbers between 2.00 and 4.00, given to two decimal places (the accuracy with which GPA's are reported). We could denote S using an ellipsis (the symbol . . .) to indicate that there are missing values not listed, but the pattern continues. Thus, $S = \{2.00, 2.01, 2.02, 2.03, \ldots,$ $3.95, 3.96, 3.97, 3.98, 3.99, 4.00\}$.

Each element in the sample space is an outcome. An **event** is any collection of outcomes from the sample space of a random process. To make it easy to refer to an event, we denote the event by a capital letter, such as A or B or some other appropriate letter chosen to signify that particular event. Because an event is a collection or set, we will again use brackets to enclose the list of outcomes in the event.

Example 18.2

For each event, list the outcomes from the sample space that correspond to the given event.

a. Note people who go through the checkout line at a particular convenience store and record the gender of each. Consider the event that a male is not chosen.

b. Toss a pair of dice, one green and one white, and record the two values on the top faces of the dice. Consider the event where both numbers are greater than 4.

c. Toss a pair of dice and record the sum of the two values on the top faces. Consider the event that the sum is even.

d. We select a college graduate and record his or her GPA. Consider the event that the graduate's GPA is 3.5 or higher.

Solution

We will choose appropriate capital letters to denote each event.

a. If a male is not chosen, then the event consists of one outcome, namely, "female," which we'll denote with F. Thus, the event "male is not chosen" = F = {female}.

b. This event consists of all outcomes where both numbers are either 5 or 6. We'll let this event be the set L (for "large"); L = {(5, 5), (5, 6), (6, 5), (6, 6)}.

c. This event is E = {2, 4, 6, 8, 10, 12}.

d. We call this event H and list all possible GPA scores between 3.5 and 4.0, assuming GPA scores are given to two decimal places. H = {3.50, 3.51, 3.52, ..., 4.00}.

■ ■ ■ ■ ■ ■ ■ ■ ■ ■

Many questions that require us to consider random processes involve choosing one or more items from a collection of items. If we choose more than one item from a collection, we can make our selection in one of two ways. We can replace the first item back into the original set before we choose the second item; this is called choosing **with replacement**. Alternatively, we can pick the second item (and any subsequent items) without returning the already chosen items to the set before picking additional items; this is called choosing **without replacement**. Whether we choose with or without replacement affects the outcomes we include in the sample space.

We also want to think about whether the order in which items are chosen is important to the situation at hand. For example, if we are choosing two people to receive first and second prizes, where first prize is a car and second prize is a book, order is certainly important. If we are choosing two people to form a committee, then which one is chosen first doesn't matter because both will serve on the committee. We explore these ideas in the next example.

Example 18.3

■ ■ ■ ■ ■ ■ ■ ■ ■

There are five students in a class: Aneesa, Benito, Carol, Donovan, and Eve. The professor, wanting to be fair, puts all names in a hat, mixes them up, and picks two to answer questions on the assigned homework.

a. If the professor picks the second name without replacing the first name in the hat before he picks the second, list the elements in the sample space.

b. If the professor picks the two names without replacement, list the outcomes in the event "Benito is chosen." How many of the outcomes in the sample space of part (a) are in the event "Benito is chosen"?

c. If the professor picks the second name after he has replaced the first name in the hat, list the elements in the sample space.

d. How many of the outcomes in the sample space of part (c) of this example are in the event "Benito is chosen"?

e. For each of the two methods of choosing names, list the outcomes in the event "no females are chosen."

Solution

We will represent the students by the first letters of their names.

a. If the second name is picked without replacing the first name, the same person cannot be picked twice. Also, the order in which the names are picked is not particularly important because both students will have similar jobs to do. This means that we don't need to list both pairs, for example, Aneesa-Benito (A-B) and Benito-Aneesa (B-A). So, S = {A-B, A-C, A-D, A-E, B-C, B-D, B-E, C-D, C-E, D-E}.

b. We'll denote this event by N; N = {A-B, B-C, B-D, B-E}. N has four elements (all the elements of S with "B").

c. Now a student can be picked more than once because we are choosing with replacement, but the order in which names are chosen is still unimportant for the situation. Thus, S = {A-A, A-B, A-C, A-D, A-E, B-B, B-C, B-D, B-E, C-C, C-D, C-E, D-D, D-E, E-E}.

d. S has 15 elements; 5 of them correspond to the event "Benito is chosen."

e. Let's denote this event by M and assume that Aneesa, Carol, and Eve are female. If the two names are chosen without replacement, then event M = {B-D}. If we choose with replacement, then M = {B-B, B-D, D-D}.

We can approach assessing the likelihood or **probability of an event** in several ways. The **probability** that an event will occur is the proportion of time the event occurs over the long run, or the **relative frequency** with which the event occurs if we repeat the random process over and over again. Simulation can be a helpful tool to assess probability by creating a model of the random process; we can imitate the process and repeat it a large number of times to assess what proportion of the time a particular event occurs. In this case, the probability of an event A, denoted $\mathcal{P}(A)$, is

$$\mathcal{P}(A) = \frac{\text{number of times A occurs}}{\text{number of times process is repeated}}$$

If the process is repeated a large number of times, the relative frequency (what we will interpret to be the probability of the event) will establish itself around a specific value. This is also referred to as **experimental probability**. For example, we can toss a coin to simulate the playing of a game between two evenly matched teams, A and B. The coin landing heads means team A wins; the coin landing tails means team B wins. We can simulate the playing of many games, identify winning streaks for each team, and keep win-loss records over many games.

The **technique of counting outcomes** for assigning the probability of an event when the outcomes in the sample space S are equally likely allows us to count the number of elements in S and in the event whose probability we want to find. We will refer to this method as the **counting method**. The probability obtained is called **theoretical probability**. Then the probability of an event A, denoted $\mathcal{P}(A)$, is the number of outcomes in the event A divided by the number of outcomes in S:

$$\mathcal{P}(A) = \frac{\text{number of outcomes in A}}{\text{number of outcomes in S}}$$

For example, if we toss two dice, the probability of getting doubles, the same number on each die, is $\mathcal{P}(\text{doubles}) = \frac{6}{36}$. The total number of outcomes in the event "doubles" is 6 (the outcomes in this event are $(1, 1), (2, 2), \ldots, (6, 6)$), whereas there are 36 equally likely outcomes in the sample space. (See Example 18.1.)

Another approach to probability is to use **subjective judgment**. In this situation, probability is interpreted as a measure of the strength of one's belief that a particular outcome will occur. This measure of probability is often personal and thus is subject to personal biases. Some people might be fairly good at subjective probability judgments, but this method of assessing probability is not very useful because different people will make different subjective judgments. It is important to be aware, however, that sometimes given probabilities may be the result of subjective judgments. We will concentrate on the relative frequency interpretation and the counting method for assessing probabilities.

Example 18.4   ▪ ▪ ▪ ▪ ▪ ▪ ▪ ▪ ▪

Describe how one would find or estimate the following probabilities:

a. The probability of winning a particular state's lottery. (There are various ways in which state lotteries are organized, but we'll assume a state lottery in which you pick six distinct numbers, and to win, your numbers must match all six distinct winning numbers. We'll assume each number is between 1 and 70 and the order of the numbers is not significant.)

b. The probability of "snake eyes" on one toss of two dice. (Note that "snake eyes" means both dice show a 1.)

c. The risk of dying in an airplane accident.

d. The chances of getting bitten by a shark.

e. The probability of precipitation on a particular day in a particular city.

Solution

a. To assess the probability of winning the lottery, we need to know how the particular lottery system is organized. We need to figure out how many elements there are in the sample space of sets of six distinct numbers between 1 and 70. We'll call this number N. (N actually is 131,115,985.) Because there is exactly one winning set of numbers, the probability of winning is $\frac{1}{N}$, which is approximately 7.6×10^{-9}. This solution uses the counting method for finding probability.

b. We could estimate the probability of obtaining a 1 on both dice by repeatedly tossing a pair of dice a large number of times and counting the number of times (1, 1) occurred. We then divide that number by the total number of tosses. This solution uses the relative frequency method. Alternatively, we could count the number of elements in the sample space of all possible outcomes of the random process of tossing two dice (which we listed in Example 18.1). There is only one outcome in the event of "snake eyes," so we divide 1 by the number of elements in S, which is 36. Thus, we have $\mathcal{P}(\text{snake eyes}) = \frac{1}{36}$. This alternative solution uses the counting method.

c. We could estimate this probability by finding the total number of airplane passenger fatalities that occurred during a particular time period; we then divide that number by the total number of passengers who traveled by air during that time period. This is the relative frequency method for assessing a probability.

d. As in part (c) of this example, we could divide the number of ocean swimmers/surfers who were bitten by a shark by the total estimated number of swimmers/surfers during the same time period. If we did this, we would be using the relative frequency method for assessing a probability. The estimates sometimes given in news articles for a particular risk such as this might, in fact, be "subjective probabilities" arrived at by an expert who doesn't actually count swimmers and surfers. A 1-in-5 million chance might be a way to indicate something that someone assesses is very, very unlikely to occur.

e. The relative frequency method can be used to estimate the probability of precipitation. Studying other times when weather patterns and conditions were similar and finding what proportion of them resulted in precipitation allow forecasters to model and predict current weather.

We can estimate probabilities from a pie chart, as we see in the next example.

Example 18.5

The following pie chart, created with estimates from the U.S. Census Bureau, represents the total number of families in the United States in 2009, by family income:

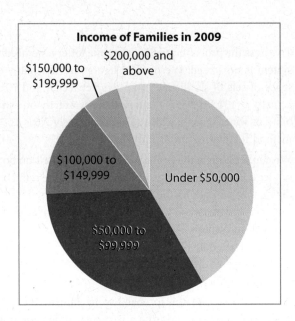

Income of Families in 2009

Suppose we chose at random a family in 2009.

a. Estimate the probability that the chosen family's income was under $50,000.

b. Estimate the probability that the chosen family's income was in the $50,000 to $99,999 interval.

c. Estimate the probability that the chosen family's income was $200,000 or more.

Solution

a. Looking at the pie chart, we see that the portion of the pie that represents the number of families with an income of less than $50,000 is more than one-third but smaller than one-half. We estimate this portion to be 40%, so the probability that the chosen family's income is under $50,000 is approximately 0.4.

b. The portion of the pie that corresponds to families with income in the $50,000 to $99,999 interval is about one-third or 33%, so we estimate the probability that the chosen family has an income in that interval as approximately 0.33.

c. The portion of the graph that corresponds to families with income of $200,000 or higher looks to be about one-fifth of a quarter, so we estimate it to be 5%. So, the probability that the chosen family's income is $200,000 or higher is about 0.05.

PROBABILITY RULES

There are several basic rules that assigned probabilities must satisfy:

1. The probability of an event is always a number between 0 and 1 because it represents a proportion. If an event has probability 0, then that event cannot occur. If an event has probability 1, the event is certain to occur. Thus, for any event A,

$$0 \leq \mathscr{P}(A) \leq 1$$

2. If we consider all possible outcomes associated with a random process, that is, all those listed in the sample space S, the sum of their probabilities must be 1. Thus,

$$\mathscr{P}(S) = 1$$

3. If a random process has a total of n *equally likely* outcomes (and *equally likely* is extremely important!), the probability of each outcome is $\frac{1}{n}$.

4. Either an event A occurs or it does not occur. So, the probability that an event A occurs is 1 minus the probability that A does not occur:

$$\mathscr{P}(A) = 1 - \mathscr{P}(\text{not A})$$

Equivalently, if we want the probability that A does not occur, we could rewrite this rule:

$$\mathscr{P}(\text{not A}) = 1 - \mathscr{P}(A)$$

5. If A and B are any two events that have *no outcomes in common*, then the probability that A or B occurs is the sum of the probability of A plus the probability of B; that is,

$$\mathscr{P}(A \text{ or } B) = \mathscr{P}(A) + \mathscr{P}(B)$$

Events that have no outcomes in common are called **disjoint events** or **mutually exclusive events**. For example, let's consider students at a particular college and ask them to record the average number of hours spent exercising each week and their gender. Suppose A is the event "student spends more than five hours exercising each week," B is the event "student spends fewer than two hours exercising each week," and C is the event "student is female." Then A and B are disjoint events, but A and C are not disjoint (because the chosen student might be a female who exercises more than five hours per week). Similarly, B and C are not disjoint.

Example 18.6

Use the counting method and the basic rules of probability to find the following probabilities:

a. A professor chooses two students' names without replacement from a hat containing the names of **Aneesa**, **Benito**, **Carol**, **Donovan**, and **Eve**. These two students will answer questions on the homework. Find the probability "Benito is not chosen."

b. Two dice are tossed. Find the probability that the "sum of the two dice is less than 4."

c. Two dice are tossed. Find the probability that the "sum of the two dice is less than 4 or greater than 9."

Solution

a. The sample space of this random process contains 10 equally likely outcomes, as seen in Example 18.3. The event "Benito is chosen" contains 4 outcomes. If we let N = Benito is chosen, then $\mathscr{P}(N) = \dfrac{4}{10} = 0.4$. But we want the probability that Benito is not chosen. We compute $\mathscr{P}(\text{not } N) = 1 - \mathscr{P}(N) = 1 - 0.4 = 0.6$.

b. This sample space has 36 equally likely outcomes, as seen in Example 18.1(c). (We need to assume these are fair, balanced dice to ensure that the 36 outcomes are equally likely.) If we denote the event A as "sum is less than 4," then A = $\{(1, 1), (1, 2), (2, 1)\}$. Thus,

$$\mathscr{P}(A) = \frac{\text{number of outcomes in A}}{\text{number of outcomes in S}} = \frac{3}{36} = \frac{1}{12}$$

c. In part (b) of this example, we computed $\mathscr{P}(A)$, where A is the event "sum is less than 4." We will let B represent the event "sum is greater than 9." So, B = $\{(4, 6), (6, 4), (5, 5), (5, 6), (6, 5), (6, 6)\}$ and $\mathscr{P}(B) = \dfrac{6}{36} = \dfrac{1}{6}$. Because A and B have no outcomes in common, that is, a dice toss cannot have a sum both less than 4 and greater than 9, $\mathscr{P}(A \text{ or } B) = \mathscr{P}(A) + \mathscr{P}(B) = \dfrac{3}{36} + \dfrac{6}{36} = \dfrac{9}{36} = \dfrac{1}{4}$.

■ ■ ■ ■ ■ ■ ■ ■ ■ ■

Tables are often used to give information about a certain population or group of individuals. If we consider the individuals to be the outcomes in the sample space, we can use the counting method to answer probability questions about this group of individuals.

Example 18.7   ■ ■ ■ ■ ■ ■ ■ ■

Use the given table to answer the questions. (The information given in the tables was taken from the website of the National Highway Traffic Safety Administration, www.nhtsa.dot.gov.)

a.   The following table gives information on single-vehicle crashes in the United States by size of vehicle during a particular time period:

Passengers	Fewer than 5	5 to 9	10 to 15	More than 15
Crashes	1,815	77	55	10

   i. If an investigator chooses a single-vehicle crash at random to carry out an additional audit of the investigation, what is the probability that the chosen crash involved a vehicle carrying more than 15 passengers?

   ii. If an investigator chooses a single-vehicle crash at random, what is the probability that the chosen crash involved a vehicle carrying 5 or more passengers?

   iii. Interpret what each of the probabilities found in (i) and (ii) means.

b.   The following table gives information on single-vehicle crashes in the United States involving rollovers by size of vehicle during a particular time period:

Passengers	Fewer than 5	5 to 9	10 to 15	More than 15
Rollovers	224	16	16	7

   i. If an investigator chooses at random a single-vehicle crash involving a rollover to carry out an additional audit of the investigation, what is the probability that the crash involved a vehicle that carried fewer than 5 passengers?

   ii. What proportion of single-vehicle rollover crashes involved vehicles carrying 15 or fewer passengers?

Solution

a.   The total number of crashes is $1,815 + 77 + 55 + 10 = 1,957$. Because we assume a crash is selected at random, each is equally likely to be chosen, and the number of equally likely elements in the sample space is 1,957.

   i. There were 10 crashes involving more-than-15-passenger vehicles, so the probability that the chosen crash involved a more-than-15-passenger vehicle is $\frac{10}{1,957} \approx 0.0051$. We could also say that approximately 0.5% of crashes during the time period involved vehicles carrying more than 15 passengers.

   ii. The total number of single-vehicle crashes was 1,957 and 1,815 of them involved vehicles with fewer than 5 passengers. So, the probability that a crash involved a single vehicle carrying fewer than 5 passengers is $\frac{1,815}{1,957} \approx 0.9274$. Therefore, the probability that a single-vehicle crash involved a vehicle carrying 5 or more passengers is

$$1 - \frac{1,815}{1,957} = \frac{142}{1,957} \approx 0.0726.$$

iii. The probability found in part (i) is the proportion of single-vehicle crashes that involved vehicles carrying more than 15 passengers. The probability found in part (ii) is the proportion of single-vehicle crashes that involved 5 or more passengers.

b. The total number of rollover crashes in the table is $224 + 16 + 16 + 7 = 263$.

i. Of these 263 rollover crashes, 224 involved vehicles carrying fewer than 5 passengers. If the investigator chooses one rollover crash at random, the probability that the chosen crash involved a vehicle carrying fewer than 5 passengers is $\frac{224}{263} \approx 0.8517$.

ii. The proportion of rollover crashes involving vehicles with more than 15 passengers is $\frac{7}{263} \approx 0.0266$. Thus, approximately $1 - 0.0266 = 0.9734$ of the rollover crashes involved vehicles carrying 15 or fewer passengers.

Summary

In this topic, we discussed basic concepts of probability: random process, sample space, outcomes and events, probability of an event. We calculated the probability of an event by computing relative frequency and by counting the number of outcomes in the sample space. We also discussed several basic probability rules.

Explorations

1. Give the sample space for each of the following random processes:

a. You survey the customers who go to an automatic teller machine and record the amount of money withdrawn from the ATM in one visit. (You will need to make reasonable assumptions about possible increments of money that can be withdrawn and the total amount that can be withdrawn.)

b. You place pennies, nickels, dimes, and quarters in a jar and pick one coin from the jar.

c. You place pennies, nickels, dimes, and quarters in a jar and pick two coins from the jar. Consider whether it matters if you draw with or without replacement and whether or not the order is important to the situation.

d. You question students at a particular college and record the last digit of their college ID number.

2. You toss two dice and compute the "difference," which you will define to be the larger top number minus the smaller top number if the numbers are different, and zero if the numbers are the same.

a. How many equally likely elements are there in the sample space of all tosses of two dice?

b. List all outcomes in the sample space that correspond to the event Z = the difference is 0.

c. Find $\mathscr{P}(Z) = \mathscr{P}$(the difference is 0).

d. Find the probability that the difference is 6.

e. Find the probability that the difference is greater than 0.

f. Find the probability that the difference is 5.

3. The following table contains information on the 2010 resident population of the United States, by age:

Age	Younger than 18 years old	18 to 24 years old	25 to 44 years old	45 to 64 years old	65 years and older
Number	74,181,467	30,672,088	82,134,554	81,489,445	40,267,984

Source: U.S. Census Bureau, www.census.gov.

a. If a 2010 resident of the United States is chosen at random, find the probability that he or she is 25 to 44 years old.

b. If a 2010 resident is chosen at random, find the probability that he or she is older than 24 years old.

c. In what age category does the median age fall?

4. The pie chart represents the U.S. population that did not have health insurance coverage in 2010, by region. Suppose a person who lived in the United States and did not have health insurance coverage in 2010 is selected at random.

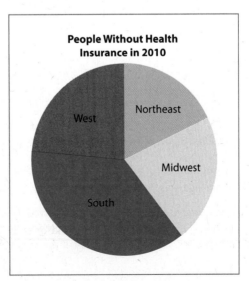

Source: U.S. Census Bureau, www.census.gov.

a. Estimate the probability that the chosen person lived in the South.

b. Estimate the probability that the chosen person lived in the South or in the West.

c. Estimate the probability that the chosen person did not live in the Northeast.

5. The following table gives the number of U.S. residents in 2010, by sex and region:

	Both Sexes	Male	Female
Northeast	55,317,240	26,869,408	28,447,832
Midwest	66,927,001	32,927,560	33,999,441
South	114,555,744	56,134,681	58,421,063
West	71,945,553	35,849,677	36,095,876

Source: U.S. Census Bureau, www.census.gov.

a. If a 2010 U.S. resident is chosen at random, what is the probability that he or she lives in the Midwest?

b. If a 2010 resident is chosen at random, what is the probability that the resident is female?

c. If a 2010 resident is chosen at random, what is the probability that the resident is male and does not live in the Northeast?

6. The following table gives the number of drivers involved in a fatal vehicle accident in 2009, by driver's age. It also gives the number of drivers who were distracted at the time of the accident.

Age	Total Drivers	Distracted Drivers
Under 20	3,967	619
20–29	10,719	1,378
30–39	7,633	832
40–49	7,930	811
50–59	6,559	631
60–69	3,968	367
70+	3,778	408

Source: National Highway Traffic Safety Administration, www.distraction.gov.

a. If we randomly select a 2009 driver involved in a fatal crash, what is the probability that his or her age was between 20 and 29?

b. If we randomly select a 2009 driver involved in a fatal crash, what is the probability that the driver is younger than 20 years old or is 40 years old or older?

c. If we randomly select a 2009 driver involved in a fatal accident, what is the probability that the driver was distracted at the time of the accident?

d. What proportion of 2009 drivers involved in a fatal accident were under 20 years old and distracted at the time of the accident?

e. What proportion of 2009 drivers involved in a fatal accident were between 40 and 49 years old?

f. What proportion of 2009 drivers involved in a fatal accident were not distracted at the time of the accident?

7. The following table gives the number of individuals in the United States that were living in poverty in 2010, by age:

Age	People in Poverty (in thousands)
Under 18 Years	16,401
18 to 64 Years	26,258
65 Years and Older	3,520
Total	**46,179**

Source: U.S. Census Bureau, www.census.gov.

If a person living in poverty in the year 2010 is chosen at random, find each of the following probabilities:

a. What is the probability that the chosen person is 65 years or older?

b. What is the probability that the chosen person is younger than 18 years old?

c. Give the probability that the chosen person is either younger than 18 years old or 65 years or older.

d. Give the probability that the chosen person is 18 to 64 years of age.

8. One card is drawn at random from a well-shuffled standard deck of 52 cards:

a. Find the probability that the card drawn is a heart.

b. Find the probability that the card drawn is an ace.

c. Use the counting method to find the probability that the card drawn is an ace or a heart.

d. Explain why probability rule 5 given previously does not work for this exercise; that is, explain why $P(\text{heart or ace})$ does not equal $P(\text{heart}) + P(\text{ace})$.

e. Explain why $P(\text{ace or 2}) = P(\text{ace}) + P(2)$.

9. Suppose you first draw two cards (drawn without replacement) out of a well-shuffled standard deck of 52 cards. Then without replacing them, you draw a third one.

a. If the first two cards are aces, what is the probability that the third card you draw is another ace?

b. If the first two cards are face cards (a jack, queen, or king), what is the probability that the third one is also a face card?

c. If the first two cards are face cards, what is the probability that the third card is not a face card?

10. Two fair dice are tossed and we record the sum of the two top numbers:

a. Find $\mathscr{P}$(sum is 7).

b. Find $\mathscr{P}$(sum is 11).

c. Find $\mathscr{P}$(sum is odd).

d. Explain why $\mathscr{P}$(sum is odd) $= \mathscr{P}$(sum is 3) $+ \mathscr{P}$(sum is 5) $+ \mathscr{P}$(sum is 7) $+ \mathscr{P}$(sum is 9) $+ \mathscr{P}$(sum is 11).

11. If a state lottery involves matching five numbers between 1 and 99, where the order in which the numbers are chosen is not important, there are 71,523,144 possible choices for the five winning numbers.

a. If you buy one lottery ticket, what is the probability that you will pick the five winning numbers? What if you buy two lottery tickets?

b. Explain in words what 1 minus the probability calculated in part (a) of this exploration represents.

12. Describe how the following probabilities might have been obtained:

a. "A 1-in-25,000 risk of being hit while crossing the street."

b. "A 1-in-a-million chance of being hit by lightning."

13. A jar contains five quarters and eight silver dollars. You pick a coin at random from the jar.

a. Give the sample space for the random process.

b. Find the probability that the coin picked is a silver dollar.

c. You pick a second coin from the jar without replacing the first coin in the jar. Give the sample space for the random process of picking two coins from the jar without replacement.

d. Are the elements listed in your sample space in part (c) of this Exploration equally likely? Why or why not?

14. We toss a coin four times and record the outcome using H for head and T for tail. (For example, if the first outcome is a head, followed by three tails, you would record HTTT.)

a. Give the sample space for the random process.

b. Find the probability that the number of heads is 4.

c. Find the probability that the number of heads is 3.

d. What is the probability of getting only one head?

15. Three friends, Matt, Nick, and Otto, attend an event together and place their three coats on a chair together. When the event is over, in a hurry to leave, they each grab one of the three coats at random and put it on.

a. Give the sample space for the random process.

b. Find the probability that none of the three friends gets his own coat.

c. Find the probability that they all get their own coats.

d. Find the probability that exactly two of them get their own coats.

16. A cooler contains cans of six different canned beverages. You have four favorites among the six beverages. You randomly pick two beverages out of the cooler.

a. Give the outcomes in the sample space.

b. How many outcomes in the sample space are in the event "two of your favorite beverages are chosen"?

c. How many outcomes in the sample space are in the event "exactly one of your favorite beverages is chosen"?

d. How many outcomes in the sample space are in the event "none of your favorite beverages is chosen"?

17. The following table contains information about the percentage of people in the United States, 15 years old and older, who fall into each of the following marital status classes for 2010: never married, married, widowed, divorced, or separated. Assume these categories are disjoint; that is, each person is in one and only one of these categories.

Never Married	Married	Widowed	Divorced	Separated
32.1%	48.8%	6%		2.2%

Source: U.S. Census Bureau, http://factfinder2.census.gov.

a. Fill in the appropriate percent for "divorced."

b. If a person 15 years old or older is chosen from this population at random, what is the probability that he or she is either married or divorced?

18. Refer to Example 18.3.

a. Assume that the professor picks the second name without replacing the first name in the hat before he picks the second. What is the probability that Carol is chosen?

b. Assume that the professor picks the second name after he has replaced the first name in the hat. What is the probability that Carol is chosen?

c. Suppose the class has six students: Aneesa, Benito, Carol, Donovan, Eve, and Frank. If the professor picks the second name without replacing the first name in the hat, what is the probability that Frank is chosen?

d. Suppose the class has six students: Aneesa, Benito, Carol, Donovan, Eve, and Frank. If the professor picks the second name after replacing the first name in the hat, what is the probability that Frank is chosen?

e. In Example 18.3, you assumed that the order in which the students were chosen was not important. Describe a similar scenario where the order in which the students are chosen would be important.

ACTIVITY

18-1

Simulations

In this activity, you will explore some ideas of probability by using Excel to simulate tossing a coin 1,000 times and throwing 100 free throws in basketball.

1. We first gather some data using a real coin.

    a. Toss a coin 10 times and, after each toss, record in the following table the result of the toss and the proportion of heads so far. For example, suppose you obtain the following sequence of heads and tails for the first five tosses: H T T T H. After the first toss, the proportion of heads so far is one out of one: $\frac{1}{1}$ or 1. After the second toss, the proportion of heads so far is one out of two: $\frac{1}{2}$. After the third toss, the proportion of heads is one out of three: $\frac{1}{3}$. After the fourth toss, the proportion of heads is one out of four: $\frac{1}{4}$. After the fifth toss, the proportion of heads is two out of five: $\frac{2}{5}$.

Toss #	1	2	3	4	5	6	7	8	9	10
H or T?										
Prop of H So Far										

b. On the following axes, plot the proportion of heads so far, for each toss from your table. What does the graph show?

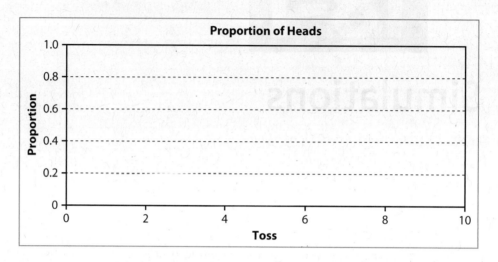

2. Now, you will use Excel to simulate 1,000 independent tosses of a fair coin and plot on a graph the proportion of heads so far after each toss using the instructions that follow part (a).

a. In Excel, the function **RAND()** (that is, RAND followed by two parentheses) produces a decimal number between 0 and 1, in such a way that every decimal number between 0 and 1 is equally likely to be produced. You will use the **RAND()** function to generate integers 0 or 1 with equal probability. The integer 1 will signify "heads" and the integer 0 will signify "tails." To get a 0 or 1 with equal probability, you'll multiply the random number by 2 and then take the integer part of it; that is, you will drop all digits after the decimal point.

Suppose the decimal number produced is 0.13061. What value do you get if you multiply that number by 2 and then take the integer part of it?

_____

Suppose the decimal number produced is 0.78934. What value do you get if you multiply that number by 2 and then take the integer part of it?

_____

The Excel formula to do this is **=INT(2*RAND())**.

Instructions to Simulate Tossing a Coin

1. Open Excel and start with a blank worksheet. Enter the label **Results from 1,000 Tosses** in cell A1.

2. Enter the formula **=INT(2*RAND())** in cell A3. Drag this formula down to cell A1002 to generate a column of 1,000 0s and 1s, representing 1,000 tails and heads.

3. Enter the label **Heads, So Far** in cell B1. In cell B2, enter the value **0**, and in cell B3, enter the formula **=A3+B2**. (This formula will keep a running count of the number of heads so far.) Drag this formula down to cell B1002.

4. Because you want to keep a running count of the proportion of heads, you'll start by recording the number of tosses so far. In cell C1, enter the label **Tosses, So Far**. In cell C3, enter **1**, and in cell C4, enter **2**. Highlight cells C3 and C4; then drag down to cell C1002. You should have a column of integers 1 through 1,000.

5. Now fill the cells D3 to D1002 with the value **0.5**, so the graph you construct will have a horizontal line at the height 0.5. (Do this efficiently using auto-fill.)

6. In cell E1, enter the title **Proportion of Heads, So Far**, and in cell E3, enter **=B3/C3**. Drag this formula down to cell E1002.

7. To display the results of the coin toss simulation on a graph, first highlight the cells in columns D and E, from row 3 to row 1002. Then click the **Insert** tab. Select **Line** from the **Charts** group and select the first subtype choice.

8. With the graph still "selected" go to the **Layout** tab. From the **Labels** group, select **Chart Title** and enter a graph title. Then from the **Labels** group, select **Axis Titles** and enter labels for the axes.

9. Right-click on the legend and select **Delete** to remove it. Also, move the cursor until it is an arrow pointing to one of the horizontal gridlines. Right-click and select **Delete** to remove the gridlines.

   b. Write a paragraph explaining what your graph shows.

   c. Put the cursor in any blank cell near your graph. Press **Ctrl=** to change the random numbers and your graph. Do this several times and describe how the graph changes.

3. Shaquille O'Neal, a successful NBA player who retired in 2011, was not particularly good at free throws. He made about 50% of his free throws over an entire season.

   a. Go to sheet 2 and set up a new worksheet to simulate 100 free throws shot independently by a player who has probability 0.5 of making each shot. In column A, keep a record of the attempt number by generating integers 1, 2, . . . , 100 in cells A2 through A101. (Enter an appropriate title in cell A1.)

   b. Label cell B1 **Hit or Miss** and generate a random sequence of 1s (each 1 represents a hit) and 0s (each 0 represents a miss) in cells B2 through B101.

   c. Compute the overall proportion of hits by calculating the mean of the B column values. Also look at your data and identify the length of the longest streak of hits and the length of the longest streak of misses. Write a paragraph commenting on the proportion of hits and the "streaks."

4. A certain college's men's basketball team is quite accomplished at making free throws. According to the coach, in their most recent season the team made approximately 75% of free-throw attempts, and ranked 10th in the country among similar schools in successful free-throw attempts.

   a. Set up another new worksheet to simulate 100 free throws shot by team members who have probability 0.75 of making each shot.

    b. Again, use column A to keep track of the attempt number in cells A2 through A101.

    c. Label cell B1 **Hit or Miss**. Now you will generate a sequence of equally likely occurrences of the numbers 0, 1, 2, or 3, with 0 representing a miss and any of the other three numbers representing a hit. By entering the formula **=INT(4*RAND())** into cell B2, you can set up the sequence so each of the numbers occurs with equal probability. Then auto-fill to cell B101.

    d. In cell C2, enter the formula **=IF(B2=0,0,1)**. This will give you a "1" in cell C2 if cell B2 recorded a hit, and a "0" if cell B2 recorded a miss. Auto-fill the formula down to cell C101.

    e. Find the overall proportion of hits, and identify the length of the longest streak of hits and the length of the longest streak of misses. Write a paragraph commenting on your proportion of hits and your "streaks."

5. Describe how the "streaks" compare for the 50% and 75% scenarios.

Summary

In this activity, you learned how to use Excel to simulate random processes. You simulated the random process of tossing a coin and graphed the frequency of heads to visualize the probability that the toss comes up heads. You also used Excel to simulate basketball free throws and analyzed the proportion of hits and streaks of hits and misses.

18-2

Finding Probabilities

In this activity, you will use various techniques to explore additional probability problems.

1. All human blood can be typed as O, A, B, or AB, but the distribution of the types varies a bit with race. The following table gives the probability model for the blood type of a randomly chosen African American:

Blood Type	O	A	B	AB
Probability	0.49	0.27	0.20	0.04

Source: BloodBook.com, http://bloodbook.com

a. What is the probability that a randomly selected African American does not have type O blood? Identify the probability rule used to answer this question.

b. Maria has type B blood and she can safely receive blood transfusions from people with blood types O and B. What is the probability that a randomly chosen African American can donate blood to Maria? Identify the probability rule used to answer this question.

2. A lottery has a $6,000,000 grand prize and advertises that any individual ticket has a 1 in 3,000,000 probability of winning the grand prize. It also has a $10 consolation prize, with an advertised probability of winning of 1 in 1,000. This means that if there were 3,000,000 tickets sold, one ticket would win the grand prize, there would be 3,000 consolation prize winners, and the rest of the tickets would win $0. We could summarize this in the following table:

Prize	Number of Tickets
$6,000,000	1
$10	3,000
$0	2,996,999
Total	3,000,000

a. Find the average payout per ticket for the 3,000,000 tickets. (Think about what a list of payouts to the 3,000,000 tickets would look like.) How can you interpret this average payout?

b. If you were charged $5 for this lottery ticket, would that be a fair price? Why or why not?

3. A worker has three different letters to put into three different envelopes. (Each letter belongs in a particular envelope.) He is distracted while placing the letters inside the envelopes and randomly puts each letter into an envelope.

a. Explain how you could simulate this random process (using something other than actual letters and envelopes).

b. Find a way to represent the elements in the sample space so that all outcomes in the sample space are equally likely. How do you know they are equally likely?

c. Find the probability that:

    i. All letters are in the correct envelopes.

    ii. None of the letters is in its correct envelope.

    iii. Two of the letters are in the correct envelopes.

4. "Let's Make a Deal." Suppose you're on a game show and you're given the choice of three doors. Behind one door is a car, and goats are behind the other two doors. You pick a door, and the host, who knows what is behind each door, opens another door to show a goat. Although he doesn't show you what is behind the door you picked, he says, "Do you want to switch doors?" Should you switch? That is, do you increase your probability of winning if you switch or is it better to stay with your original door, or doesn't it matter?

a. Figure out how to set up a simulation of this game using cards, dice, coins, or whatever else you have available. Describe your setup.

b. Play the game 10 times, using the strategy of always staying with your original door. Find your experimental probability of winning if you stay with your original pick.

c. Now play the game 10 more times, using the strategy of switching your choice of door. Find your experimental probability of winning if you switch.

d. What strategy should you choose and why does it work?

5. The following table reflects the results of a survey sent to 1,000 randomly chosen individuals under 65 years of age. It gives the number of people who, in the first half of 2011, reported they did not have health insurance for at least part of the past year, by age group.

Age Group	Uninsured for at Least Part of the Past Year	Insured for the Whole Past Year	Total
0–17	30	246	276
18–25	43	71	114
26–64	146	464	610
Total	**219**	**781**	**1,000**

Source: 2011 National Health Interview Survey.

Use the table to find, for a randomly selected individual from this population, the probability that the individual:

a. was in the age interval 18–25.

b. was in the age interval 18–25 and was uninsured for at least part of the past year.

c. was in the age interval 0–17 or in the age interval 18–25.

d. was insured for the whole past year and was at least 18 years old.

e. was uninsured for at least part of the past year and was younger than 18 years.

6. On September 22, 2011, *The New York Times* reported on the pieces of the Upper Atmosphere Research Satellite that were expected to survive and hit the surface of Earth as the orbiting NASA satellite decayed. NASA "calculated a 1-in-3,200 chance of anyone on Earth being hurt by its satellite's death plunge." In addition, NASA scientists reported that "the odds of demise-by-satellite for any particular person among the world's seven billion people are on the order of 1-in-a trillion."

a. How could probabilities like those given in the article be determined?

b. Are the two probability statements compatible? That is, can they both be true?

Summary

In this activity, you practiced reading tables and using the probability rules; you also simulated a chance situation and found an experimental probability.

Conditional Probability and Tables

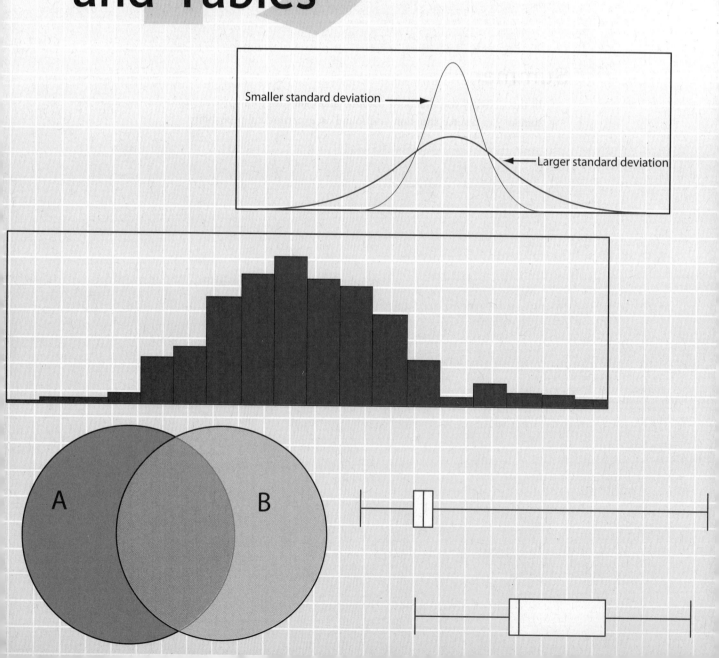

After completing this topic, you will be able to:

- Organize information in two-way tables.

- Analyze data given in two-way tables.

- Look for relationships using conditional probability.

- Understand and calculate conditional probabilities.

- Determine when two events are independent.

A two-way table is a valuable tool for organizing information in which each individual represented in the table is characterized in two different ways. For example, if the individuals are people, they might be characterized by gender and by age, or they might be described by highest educational level and by family income level. Individuals could also be cars, characterized by, for example, color and cost. We represent one variable using the rows and the other variable using the columns, and each cell contains the number of individuals who fall into that column and row classification. Often, looking at data organized in a two-way table allows us to explore the association between the variables and to see relationships that we may have otherwise overlooked.

In Example 18.7, we obtained information about single-vehicle crashes through two separate tables: One table gave the number of rollover accidents, while the other presented information on accidents that did not involve a rollover. We can organize that data in a single two-way table to help analyze it, as the first example shows.

Example 19.1

The following two-way table contains the crash data given in Example 18.7. In this table, each single-vehicle crash is classified in two different ways: by the number of passengers in the vehicle and by whether it involved a rollover or not.

Passengers	Fewer Than 5	5 to 9	10 to 15	More Than 15
Rollover	224	16	16	7
No Rollover	1,591	61	39	3
Total	**1,815**	**77**	**55**	**10**

a. Among the crashes that involved a vehicle carrying more than 15 passengers, what proportion were rollovers?

b. Among the crashes that involved a vehicle carrying 10 to 15 passengers, what proportion were rollovers?

c. Among the crashes that involved a vehicle carrying 5 to 9 passengers, what proportion were rollovers?

d. Among the crashes that involved a vehicle carrying fewer than 5 passengers, what proportion were rollovers?

e. What conclusions might be drawn from the results in parts (a–d) of this example?

Solution

a. To focus on crashes that involved a vehicle carrying more that 15 passengers, we look solely at the last column of the table. There were a total of 10 such crashes, and 7 of them involved rollovers. So, $\frac{7}{10} = 0.70$ or 70% of all crashes involved rollovers.

b. Among the 55 single-vehicle crashes involving vehicles carrying 10 to 15 passengers, 16 of them were rollovers, as seen in the third column of data in the table. This means that $\frac{16}{55} \approx 0.29$ or approximately 29% of the crashes involving vehicles carrying 10 to 15 passengers were rollovers.

c. For those 77 crashes involving vehicles carrying 5 to 9 passengers, 16 involved rollovers. Thus, $\frac{16}{77} \approx 0.21$ or approximately 21% of such crashes involved rollovers.

d. Among vehicles carrying fewer than 5 passengers, $\frac{224}{1,815} \approx 0.12$ or approximately 12% involved rollovers.

e. It appears that vehicles that carry more than 15 passengers have a much higher risk of rollover when fully loaded than do other vehicles. The vehicles least likely to be involved in rollovers are those carrying fewer than 5 passengers.

The data in the first example were organized in a two-way table. Sometimes it is helpful to organize information given in a description in a two-way table to better understand the information. In the example that follows, we consider a study of the occurrence of autism spectrum disorder in siblings.

Example 19.2

A recent study investigated the likelihood of parents having a child with some form of autism, if they already have one or more children with some form of autism. The study involved researchers in 12 locations; they followed 664 infants who had at least one older sibling with some form of autism. The children were evaluated at age 3, and 132 of the 664 were found to have some form of autism; 29 of the children with autism were females. Assume approximately 55.6% of the children in the study were males.

a. Organize the given information in a two-way table, classifying the babies in the study by gender and by whether they developed some form of autism or not.

b. Use the information in the table to compare the rate at which girls in this study developed autism with the rate at which boys in this study developed autism.

c. Among the babies in the study who developed autism, what proportion were boys? How is this value different from the rate of autism among the boys in the study?

Solution

a. We set up a two-way table and fill in the total number of babies in the study, as well as the total number who developed some form of autism:

	Male	Female	Total
Some Form of Autism		29	132
No Autism			
Total			**664**

Next, we use the information about the proportion of males in the study to record the total in the male column: 55.6% of 664 is $0.556 \cdot 664 \approx 369$. We round our calculations to the nearest whole number to estimate the number of boys in the study. Then there are $664 - 369 = 295$ girls in the study and $132 - 29 = 103$ boys who developed autism.

	Male	Female	Total
Some Form of Autism	103	29	132
No Autism			532
Total	369	295	664

We can use the row and column sums to fill in the remainder of the table:

	Male	Female	Total
Some Form of Autism	103	29	132
No Autism	266	266	532
Total	369	295	664

b. The proportion of girls who developed some form of autism is $\frac{29}{295} \approx 0.098$. Approximately 9.8% of the girls in the study developed some form of autism. The proportion of boys who developed some form of autism is $\frac{103}{369} \approx 0.279$ Approximately 27.9% of the boys in the study developed some form of autism. The rate among boys is almost 3 times the rate among girls.

c. Of the 132 children who developed autism, 103 of them were boys, so $\frac{103}{132} \approx 0.780$ or 78% of the children who developed autism were boys. This is much larger than the 27.9% of boys in the study who developed some form of autism.

In Example 19.1, we looked at proportions that involve assumed conditions. We might also think of these as conditional probabilities if we phrased our questions in terms of probabilities. For example, we could ask the following: If we choose a single-vehicle crash at random for an audit and know that it involved a vehicle carrying 15 or more passengers, what is the probability that it was a rollover?

In general, a **conditional probability** asks us to find the probability of some event A, given that an event B has occurred. Conditional probability questions look at distributions of one of the variables in a two-way table for given values of the other variable. If the two variables of interest are gender and income, these questions might contain the following types of phrases: *Assuming that* the person chosen was female, find the probability that the person earned over $100,000 last year; or find the probability that the person chosen earned over $100,000 *given that* the person was female; or *among* females, what proportion earned over $100,000?

When dealing with conditional proportions and conditional probabilities, it is important to know whether we want to look at the proportion of As that occur among the Bs or the proportion of Bs that occur among the As. Equivalently, we might ask for the probability that A occurs given that B has occurred or for the probability of B given A. For example, among crashes involving vehicles carrying more than 15 passengers, $\frac{7}{10}$ or 70% of them were rollovers, which is a large proportion, but only a small proportion of rollovers involved vehicles carrying 15 or more passengers $\left(\frac{7}{263} \text{ or approximately } 2.7\%\right)$.

Example 19.3

The following table gives information about Winter Olympics gold medal winners (1948–2010) for seven events. (Note that there was no four-man bobsled winner in 1960.) The row and column totals are included in the table.

	Men's Downhill Alpine Skiing	Men's 50K Cross-Country Skiing	Men's 500 m Speed Skating	Men's Singles Figure Skating	Women's Singles Figure Skating	Ice Hockey	Four-man Bobsled	Total
United States	2	0	5	7	7	2	2	25
Other Countries	15	17	12	10	10	15	14	93
Total	17	17	17	17	17	17	16	118

Source: The World Almanac and Book of Facts 2011, pp. 868–874.

a. What proportion of gold medals for these seven events were earned by U.S. athletes?

b. What proportion of gold medals for these seven events were for figure skating?

c. Among U.S. gold medals for these seven events, what proportion of them were won for figure skating?

d. Among gold-medal winners from other countries for these seven events, what proportion of them were won for figure skating?

e. What proportion of ice hockey medals went to countries other than the United States?

f. Among Winter Olympics gold medals in these seven events that went to other countries, what proportion were awarded for ice hockey?

g. Compare the answers to parts (e) and (f) of this example.

Solution

a. There are a total of 118 gold medals represented in the table. Of these 118, 25 were from the United States. Thus, $\frac{25}{118}$ or approximately 21.2% of them were awarded to U.S. participants.

b. We first observe that the set of men's figure-skating medalists and the set of women's figure-skating medalists are disjoint. So, we compute the proportion of men's figure-skating medalists and add to it the proportion of women's figure-skating medalists. We have $\frac{17}{118} + \frac{17}{118} = \frac{34}{118}$ or approximately 28.8% of the gold medals were for figure skating.

c. Among the 25 U.S. gold medals, 14 of them were for figure skating (7 for men and 7 for women). Thus, $\frac{14}{25}$ or approximately 56% of the U.S. gold medals for these events were for figure skating.

d. There were 93 gold medals awarded to other countries, of which 20 were for figure skating. So, $\frac{20}{93}$ or approximately 21.5% of gold medals awarded to other countries were for figure skating.

e. Of the 17 ice hockey medals, 15 went to other countries. That's $\frac{15}{17}$ or approximately 88.2% of the ice hockey medals went to countries other than the United States.

f. There were 93 medals that went to other countries; 15 were for ice hockey, so $\frac{15}{93}$ or approximately 16.1% of the medals awarded to other countries were for ice hockey.

g. Both of the fractions used to answer the questions in parts (e) and (f) of this example use the 15 ice hockey medals given to other countries as the numerator. In part (e), we want to know what proportion of all ice hockey medals those 15 medals represent. In part (f), we want to consider what proportion of all medals awarded to other countries those 15 medals represent.

■ ■ ■ ■ ■ ■ ■ ■ ■ ■ ■

If A and B are two specific events, then "A or B" is the event that either A occurs or B occurs, or possibly both occur. "A and B" is the event that both individual events, A and B, occur. It can be helpful to visualize relationships between events with a picture. A **Venn diagram** shows the collection of all possible events as the interior of a rectangle. Other events are typically represented using circles within the rectangle. In the diagram on the left that follows, the shaded region shows the event A or B; it includes outcomes in A and outcomes in B. Note that it also includes outcomes that are in A and B, the slightly darker shaded overlap region. The diagram on the right shows the event A and B as the dark blue overlap region.

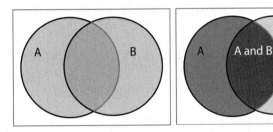

In Topic 18, we considered a probability rule for $\mathscr{P}(A \text{ or } B)$ when events A and B are disjoint events; that is, they do not have any outcomes in common. In that case, the rule is: $\mathscr{P}(A \text{ or } B) = \mathscr{P}(A) + \mathscr{P}(B)$. This rule was used in Example 19.3(b) where we computed the proportion of figure-skating medals as the proportion of men's figure-skating medals plus the proportion of women's figure-skating medals. (Note that if A and B are disjoint events, then $\mathscr{P}(A \text{ and } B) = 0$, because A and B have no outcomes in common and so both cannot occur simultaneously.)

In the Venn diagram that follows, the outside rectangle represents all 104 gold medals in the table (the sample space) and each circle inside the rectangle represents an event as labeled. These events are disjoint, so they are represented by non-overlapping circles.

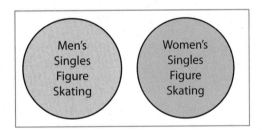

If A and B are not disjoint events, then we cannot compute $\mathscr{P}(A \text{ or } B)$ as the sum of the probability of A plus the probability of B. We must be careful not to "double count" outcomes that lie in both A and B. An additional probability rule that holds, in general, for any two events A and B is

$$\mathscr{P}(A \text{ or } B) = \mathscr{P}(A) + \mathscr{P}(B) - \mathscr{P}(A \text{ and } B)$$

This formula tells us that to compute the probability of the event "A or B," we add the probability of A plus the probability of B and then subtract the probability of "A and B." The area of the shaded region is not the sum of the blue area plus the tan area because that sum includes the area of the overlap region twice, once because it's within A and a second time because it's within B. So, we need to subtract the probability of "A and B" so that we include that overlap region only one time.

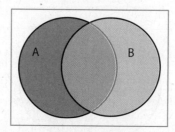

Note that this formula is valid for any two events A and B. If A and B are disjoint, then the last term, $\mathcal{P}(A \text{ and } B)$, is zero and the formula reduces to $\mathcal{P}(A \text{ or } B) = \mathcal{P}(A) + \mathcal{P}(B)$, which we saw earlier.

In the next two examples, we consider how to figure out $\mathcal{P}(A \text{ or } B)$ and $\mathcal{P}(A \text{ and } B)$ from the information in a two-way table.

Example 19.4

The table below gives the numbers of workers age 16 and older who work at home, by gender and by age:

Age	Males	Females	Total
16 and 17 years	18,727	17,517	36,244
18–20 years	46,496	45,416	91,912
21–24 years	59,461	73,204	132,665
25–29 years	98,992	158,990	257,982
30–34 years	152,497	252,991	405,488
35–39 years	213,552	321,703	535,255
40–44 years	252,252	320,584	572,836
45–49 years	255,042	284,668	539,710
50–54 years	245,926	252,094	498,020
55–59 years	208,236	192,225	400,461
60–61 years	67,051	57,880	124,931
62–64 years	85,796	66,821	152,617
65 and 66 years	47,869	34,869	82,738
67–69 years	66,159	43,242	109,401
70–74 years	81,620	47,741	129,361
75–79 years	46,849	26,145	72,994
80–84 years	18,424	10,690	29,114
85 years and over	7,282	5,212	12,494
Total	1,972,231	2,211,992	4,184,223

Source: U.S. Census Bureau, www.census.gov.

a. If a worker age 16 or older who works at home is chosen at random, find the probability that the chosen worker is 21 to 24 years old or is female.

b. If a worker age 16 or older who works at home is chosen at random, find the probability that the chosen worker is male or in his 30s.

Solution

a. Let TW represent the event that a worker is 21 to 24 years old and let F represent the event that a worker is female. We want to find $\mathcal{P}(\text{TW or F})$. We know $\mathcal{P}(\text{TW or F}) = \mathcal{P}(\text{TW}) + \mathcal{P}(\text{F}) - \mathcal{P}(\text{TW and F})$. Using the counting method, $\mathcal{P}(\text{TW}) = \frac{132,665}{4,184,223} \approx 0.032$ and $\mathcal{P}(\text{F}) = \frac{2,211,992}{4,184,223} \approx 0.529$ and $\mathcal{P}(\text{TW and F}) = \frac{73,204}{4,184,223} \approx 0.017$. So, $\mathcal{P}(\text{TW or F}) = \mathcal{P}(\text{TW}) + \mathcal{P}(\text{F}) - \mathcal{P}(\text{TW and F}) \approx 0.032 + 0.529 - 0.017 = 0.544$.

b. Let M represent the event that a worker is male and let TH represent the event that a worker is 30 to 39 years old. We want to find $\mathcal{P}(\text{M or TH})$. To use the probability rule to find $\mathcal{P}(\text{M or TH})$, we need to find $\mathcal{P}(\text{M})$, $\mathcal{P}(\text{TH})$, and $\mathcal{P}(\text{M and TH})$. Using the counting method, $\mathcal{P}(\text{M}) = \frac{1,972,231}{4,184,223} \approx 0.471$ and $\mathcal{P}(\text{TH}) = \mathcal{P}(30 \text{ to } 34 \text{ years}) + \mathcal{P}(35 \text{ to } 39 \text{ years}) = \frac{405,488}{4,184,223} + \frac{535,255}{4,184,223} \approx 0.097 + 0.128 = 0.225$. Here, we add the probability that an individual is between 30 and 34 years plus the probability that he or she is between 35 and 39 years because they are disjoint events. Also, $\mathcal{P}(\text{M and TH}) = \frac{152,497 + 213,552}{4,184,223} \approx 0.087$. So, $\mathcal{P}(\text{M or TH}) = \mathcal{P}(\text{M}) + \mathcal{P}(\text{TH}) - \mathcal{P}(\text{M and TH}) \approx 0.471 + 0.225 - 0.087 = 0.609$.

■ ■ ■ ■ ■ ■ ■ ■ ■ ■

Example 19.5

A measure to authorize extension of nondiscriminatory treatment (normal trade-relations treatment) to the People's Republic of China, and to establish a framework for relations between the United States and the People's Republic of China passed the U.S. Senate on September 19, 2000, with the breakdown of votes as given in the following table:

	Democrat	Republican	Total
Yea	46	37	83
Nay	8	7	15
Not Voting	2	0	2
Total	56	44	100

Source: The Library of Congress, www.congress.gov.

Suppose we pick a senator from the September 2000 U.S. Senate at random.

a. Find the probability that the senator is a Democrat and voted Nay.

b. Find the probability that the senator is a Democrat or voted Nay.

c. Find the probability that the senator voted Nay or was Not Voting.

d. Find the probability that the chosen senator voted Yea, given that he or she was a Democrat.

e. Find the probability that the chosen senator voted Yea, given that he or she was a Republican, and compare this answer with your answer in part (d) of this example.

Solution

a. We want to identify how many "outcomes" (that is, senators) are Democrats *and* voted Nay. These are the outcomes that fall in the "Democrat" column and in the "Nay" row. Eight senators out of the total Senate count of 100 are in both the "Democrat" column and the "Nay" row. Using the counting method, $\mathscr{P}$(Democrat and voted Nay) $= \dfrac{8}{100}$ or 8%.

b. We can't use the rule for $\mathscr{P}$(A or B) for disjoint events A and B to find the probability that the chosen senator is a Democrat *or* voted Nay, because the events "is a Democrat" and "voted Nay" are not disjoint. As we saw in part (a) of this example, there are 8 senators who fall into both categories. So, $\mathscr{P}$(Democrat or voted Nay) $= \mathscr{P}$(Democrat)$+$
$\mathscr{P}$(voted Nay) $- \mathscr{P}$(Democrat and voted Nay) $= \dfrac{56}{100} + \dfrac{15}{100} - \dfrac{8}{100} = \dfrac{63}{100}$ or 63%. We could also use the counting method to count all outcomes that fall into either the Democrat column or the Nay row, taking care not to count any outcomes more than once. There are $46 + 8 + 2 = 56$ Democrats and 7 Nay voters who were not yet counted in the Democrat count, for a total of 63 senators who are Democrats or Nay voters. Thus, there is probability $\dfrac{63}{100}$ of choosing a Democrat or Nay voter.

c. Because events "voted Nay" and "Not Voting" are disjoint (that is, there are no senators counted in both groups), $\mathscr{P}$(voted Nay or Not Voting) $= \mathscr{P}$(voted Nay) $+ \mathscr{P}$(Not Voting)
$= \dfrac{15}{100} + \dfrac{2}{100} = \dfrac{17}{100}$ or 17%.

d. Because we know that the chosen senator is a Democrat, we consider only the column corresponding to "Democrat." Our sample space is reduced to the 56 Democrats. Using the counting method, of these 56, 46 voted Yea, so $\mathscr{P}$(voted Yea given Democrat) $= \dfrac{46}{56} \approx 0.82$.

e. Because we know that the chosen senator is a Republican, we consider only the 44 senators in the Republican column. Of these, 37 voted Yea, so $\mathscr{P}$(Yea voter given Republican) $= \dfrac{37}{44}$ or approximately 84%. So, the proportion of senators voting Yea among Democrats is approximately the same as the proportion of senators voting Yea among Republicans. This is roughly the same as the proportion of senators who voted Yea, which was $\dfrac{83}{100}$ or 83%.

In Example 19.5(d) and (e), we saw that $\mathcal{P}$(voted Yea given Democrat) is approximately equal to $\mathcal{P}$(voted Yea) and $\mathcal{P}$(voted Yea given Republican) is approximately equal to $\mathcal{P}$(voted Yea). In general, when $\mathcal{P}$(A given B) $= \mathcal{P}$(A), for two events A and B, we say that A and B are **independent events**. The probability that A has occurred is the same whether we have information about whether B has occurred or not. In the context of Example 19.5, the probability that a chosen senator voted Yea on the bill is approximately the same whether we know the senator is a Democrat or not.

It is sometimes convenient to use the shorthand notation of a vertical line "|" to denote the word "given" in $\mathcal{P}$(A given B) and write it as $\mathcal{P}$(A|B). We will look further at the notion of independent events in the next two examples.

Example 19.6

On February 15, 2011, the U.S. Senate approved a bill to extend the expiring provisions of the USA Patriot Improvement and Reauthorization of 2005 and Intelligence Reform and Terrorism Prevention Act of 2004 relating to access to business records and roving wiretaps, among other things. The voting tally of the senators is given in the next table:

	Democrat	Republican	Independent	Total
Yea	41	45	0	86
Nay	9	2	1	12
Not Voting	2	0	0	2
Total	**52**	**47**	**1**	**100**

Source: Library of Congress, www.congress.gov.

A senator is chosen at random from those voting on this bill. Let event Y = senator voted Yea; let N = senator voted Nay; let D = senator is a Democrat; and let R = senator is a Republican.

a. Find $\mathcal{P}$(N|D) and compare with $\mathcal{P}$(N). What do these probabilities tell us about the independence of the events N and D?

b. Are the events Y and R independent?

c. What meaning do the answers to parts (a) and (b) have in the context of this example?

Solution

a. $\mathcal{P}$(N|D), which we read as "the probability the chosen senator voted Nay given that he or she was a Democrat," is obtained by considering the 52 Democrats represented in the first column of data. Of these, 9 voted Nay, so $\mathcal{P}$(N|D) $= \dfrac{9}{52}$ or approximately 17.3%. A total of

$\frac{12}{100}$ of the senators voted Nay, so $\mathscr{P}(N) = 0.12$ or 12%. Since $\mathscr{P}(N|D) \neq \mathscr{P}(N)$, the events N and D are not independent events. If we know the chosen senator is a Democrat, it will change our assessment of the likelihood that he or she voted Nay.

b. We will compare $\mathscr{P}(Y|R)$ and $\mathscr{P}(Y)$. $\mathscr{P}(Y|R) = \frac{45}{47}$ or 95.7% but $\mathscr{P}(Y) = 0.86$ or 86%. Because these probabilities are not equal, Y and R are not independent.

c. A larger proportion of the Senate Republicans favored the bill than did Senate Democrats. The likelihood that a senator voted Yea on this bill depends on his or her political party.

There are some scenarios where we know from the setting that events are independent; that is, we know that the outcome of one event does not impact the outcome of the other event. If we conduct an experiment and then repeat it, we often want to set up the conditions so the outcome of the first experiment will not influence the outcome of the repeated experiment; that is, we want to obtain independent results. Other situations in which we can reasonably assume independence include the following: the likelihood of failure of a particular computer system component is independent of any other component's failure; the number of people in the party when making a call to a restaurant for a reservation is independent of the number of people in the party of the previous call.

Example 19.7

Explain how the independence of events enters into the reasoning process in each of the following scenarios:

a. A coin is tossed repeatedly. The first 10 tosses land heads. What is the probability of heads on the 11th toss?

b. We pick a card at random from a standard deck of 52 cards and then pick a second card, also at random, with replacement (after shuffling the cards). Are the events A1, obtaining an ace on the first draw, and A2, obtaining an ace on the second draw, independent?

c. We pick a card at random from a standard deck of 52 cards and then pick a second card, also at random, without replacement. Are the events A1, obtaining an ace on the first draw, and A2, obtaining an ace on the second draw, independent?

d. A family consists of three sons. What is the likelihood that their fourth child will be a girl?

Solution

a. Assuming the coin is a fair coin, each time the coin is tossed, the probability of obtaining heads face up is $\frac{1}{2}$. The coin does not "remember" the result of any prior toss. The frequency interpretation of probability says that as we increase the number of repetitions (in this case, the number of tosses), the proportion of heads gets closer to the theoretical probability of heads. But no matter how many heads have occurred in a row, the probability of heads on the next toss is still $\frac{1}{2}$.

b. If we replace the first card in the deck before we randomly pick the second card, $\mathcal{P}(A2|A1) = \mathcal{P}(A2) = \frac{4}{52}$, and they are independent events. Knowing that an ace was picked on the first draw does not change the likelihood of picking an ace on the second draw.

c. If we pick without replacement, the events A1, obtain an ace on the first draw, and A2, obtain an ace on the second draw, are not independent. In this case, knowing that an ace was obtained on the first draw impacts the probability of obtaining an ace on the second draw because at the time of the second draw only 3 aces and 51 cards are left in the deck.

d. This situation is interpreted to be similar to a coin toss, although the probability of having a boy baby in the United States, is approximately 0.51, while the probability of having a girl baby is 0.49. (This estimate is obtained from the 2012 Statistical Abstract of U.S. Census Bureau.) The probability of having a girl on the fourth try would be 0.49, the same as the probability of having a girl on the first or second or third try.

One particularly important area in which looking at conditional probabilities can help us make decisions is in the area of testing for a condition or disease. We will construct an example within this context. This example will involve setting up a table from given conditional probabilities, which is the reverse of what we did in Examples 19.1 and 19.3 through 19.6, where we were given tables of data and figured out the conditional probabilities from the table.

Example 19.8

Consider a diagnostic test for a rare disease, such as hepatitis. You are being tested, and neither you nor your doctor knows whether or not you have the disease.

a. Assume a total population of 500 people tested and set up a two-way table with the columns representing categories H: "has the disease" and F: "free of the disease," and the rows representing the test results P: "test is positive" and N: "test is negative." To set up the table, assume that approximately 10% of all people have the disease. Also assume that, among people with the disease, the test is known to give positive results in 90% of the cases, and among people free of the disease, the test gives negative results in 98% of the cases. (These last two results are known as the test's **sensitivity** and **specificity**, respectively, and are often available from the results of clinical trials.)

b. Find the probability that if a person tests positive, he or she really has the disease.

c. What proportion of people who test negative have the disease?

d. What implications would these results have?

Solution

a. Because we have a population of 500 people tested, we first fill in the "grand total" of 500 people. We are assuming that 10% of them have the disease, so the total for the H column is 10% of 500 or 50. That leaves 450 for the F column total (the 90% of people who do not have the disease):

	H (has the disease)	F (free of disease)	Row Total
P (test is positive)			
N (test is negative)			
Column Total	**50**	**450**	**500**

Next, we fill in the conditional probabilities given that among people with the disease (the H column), the test is positive for 90% of the cases. For the 50 people with the disease, 90% of 50 or 45 of them test positive; thus, 5 of them test negative. Similarly, among those free of the disease (the F column), 98% of the 450 people, which is $0.98 \times 450 = 441$, test negative. Thus, 9 of the 450 who are free of the disease will test positive. Finally, we fill in the row totals:

True Condition	H (has the disease)	F (free of disease)	Row Total
P (test is positive)	45	9	**54**
N (test is negative)	5	441	**446**
Column Total	**50**	**450**	**500**

b. To find the probability that if a person tests positive, he or she really has the disease, we find a conditional probability given that a person tests positive. In symbols, we want to find $\mathscr{P}(H|P)$. Look at the first row; we find that among those 54 people who tested positive, 45 have the disease. Thus, $\mathscr{P}(H|P) = \frac{45}{54}$, which is approximately 83%. This means that approximately 17% of those who test positive don't have the disease.

c. This question asks us to find $\mathscr{P}(H|N)$, so we look at the second row. Out of the 446 people who tested negative, 5 have the disease. Thus, $\mathscr{P}(H|N) = \frac{5}{446}$ or approximately 1%.

d. Approximately 17 in every 100 people who are identified as having the disease do not really have it. These people might be treated for the disease when they should not, which might cause side-effects and cost money and time. One in 100 people who test negative really do have the disease. They may go untreated. Note that these percentages depend on the proportion of all people who have this particular disease (which we assumed to be 10% in this example) and on the sensitivity and specificity of the particular diagnostic test.

Summary

In this topic, we used a two-way table to represent data in which each individual in the data set is characterized in two different ways. Using two-way tables, we looked at conditional proportions, which are conditional probabilities. We also discussed the concept of independent events. Finally, we explored the use of conditional probability and two-way tables to help make decisions when testing for a condition or disease.

Explorations

1. The following table gives the number, in thousands, of U.S. employed persons 16 or older in professional and related occupations, by occupation category and gender for 2009:

Occupation Category	Men	Women
Computer and Mathematical Occupations	2,618	863
Architecture and Engineering Occupations	2,363	377
Life, Physical, and Social Science Occupations	707	621
Community and Social Services Occupations	868	1,474
Legal Occupations	859	851
Education, Training, and Library Occupations	2,221	6,407
Arts, Design, Entertainment, Sports, and Media Occupations	1,453	1,271
Healthcare, Practitioner, and Technical Occupations	1,968	5,770

Source: *The World Almanac and Book of Facts 2011*, p. 114.

a. Among people in legal occupations, what proportion of them are men?

b. Among people in architecture and engineering occupations, what proportion of them are women?

c. Among men in the occupation categories listed in the table, find the proportion that are in legal occupations.

d. Given that a person is in education, training, and library occupations, find the probability that the person is female.

e. If a person in these occupation categories is chosen at random, find the probability that a woman is chosen and she is in the life, physical, and social science occupations.

f. If a person in the professional and related occupations is chosen at random, find the probability that a man is chosen or the person is in the legal occupations.

2. A report of a study on U.S. emergency room visits by children age 19 and under for sledding-related injuries from 1997–2007 reported on 229,023 patients treated for such injuries. Approximately 59.8% of the cases involved boys. Suppose we also know that 6.9% of the injuries involved children 0 to 4 years of age, 51.2% of whom were boys; 29.5% involved children 5 to 9 years of age, 55% of whom were boys; and 42.5% involved children 10 to 14 years of age, 57% of whom were boys.

a. Use the given information to fill in the following two-way table:

Ages	Boys	Girls	Total
0–4			
5–9			
10–14			
15–19			
Total			

Source: http://healthychildren.org

b. If a child is chosen at random from the children in this study, what is the probability that the child is a girl and 15 to 19 years of age?

c. If a child is chosen at random from the children in this study, what is the probability that she is a girl, given that she is 15 to 19 years old?

d. If a child is chosen at random from the children in this study and is known to be 10 to 14 years old, what is the probability that he is a boy?

e. Of the boys in the study, what proportion are 5 to 9 years old?

f. If a child is chosen at random from the children in this study, what is the probability that the child is over 9 years old or a boy?

3. A newspaper article reported that a woman was misdiagnosed with a rare form of cancer. In fact, she was told by three different doctors that she tested positive for that type of cancer. Weeks later it was determined that she had "false positive" results. Explain what false positive results mean in this context.

4. The next table shows the final results for the vote on the Lawsuit Abuse Reduction Act in the House of Representatives on October 27, 2005. The bill, which originated in the House, is titled

"To amend Rule 11 of the Federal Rules of Civil Procedure to improve attorney accountability, and for other purposes." The bill passed the House and was then sent to the Senate.

	Yea	Nay	Not Voting	Total
Republican	212	5	13	
Democratic	16	178	8	
Independent	0	1	0	
Totals	**228**	**184**	**21**	

Source: Library of Congress, www.congress.gov.

A representative is chosen at random from those considered in the table.
Let Y = representative voted Yea; let N = representative voted Nay;
let D = representative is a Democrat; and let R = representative is a Republican.

a. What is the probability that the chosen representative voted Yea?

b. What is the probability that the chosen representative is a Democrat?

c. Given that the chosen representative voted Yea, what is the probability that he or she is a Republican?

d. Represent the following probability in symbols and find it: Among Nay-voting representatives, what proportion are Democrats?

e. Represent the following probability in symbols and find it: If the chosen representative was Republican, what is the probability that he or she voted Yea?

5. Suppose we pick one card at random from a standard deck of 52 cards. Let K represent the event "the card is a King"; let Q represent the event "the card is a Queen"; let H represent the event "the card is a Heart"; let C represent the event "the card is a Club."

a. Set up a two-way table to represent the 52 cards in the deck. Create three rows that categorize the cards by clubs, hearts, and other suits. Similarly, create three columns that categorize the cards by kings, queens, and other denominations.

b. Find $\mathscr{P}(K)$.

c. Find $\mathscr{P}(H)$.

d. Are the events H and C disjoint? Find $\mathscr{P}(H$ and $C)$.

e. Are the events K and H disjoint? Find $\mathscr{P}(K$ and $H)$.

f. Find $\mathscr{P}(H$ or $C)$.

g. Find $\mathscr{P}(K$ or $H)$.

h. Find $\mathscr{P}(K$ or $Q)$.

6. A new test has been developed for a rare condition that occurs in approximately 1 out of 1,000 people. The test has a sensitivity of 100% and a specificity of 99.9%.

   a. Assume a population of 1,000 people and set up a two-way table, with the columns labeled H (has the condition) and F (free of the condition) and the rows labeled P (test is positive) and N (test is negative).

   b. Fill in the appropriate column totals and use the given sensitivity and specificity to fill in the column amounts. Finally, fill in the row totals.

   c. Suppose an individual tests positive for the rare condition using this new test. What are the chances that the test results are erroneous? What should this individual do?

   d. If this were a test for a condition that could be passed between married individuals, would you argue for or against making the test mandatory for all couples prior to marriage? Why?

7. Explain how independence of events enters into the reasoning process in each scenario:

   a. A college student is chosen at random. Are the events "student has GPA over 3.5" and "student is female" independent?

   b. You are playing *Monopoly* and toss doubles three times in a row and get sent to jail. Are the events "get doubles on first toss" and "get doubles on second toss" independent?

   c. You buy a lottery ticket during the first week of June and you buy a ticket during the first week of July. Are the events "you win the first week of June" and "you win the first week of July" independent?

8. A recent public opinion poll asked people whether they approve or disapprove of cash-prize lotteries. Here are the results:

	Approve	Disapprove	No Opinion
Adults, Age 18 and Over	1,142	366	15
Teens, Ages 13–17	411	90	0

Source: www.libraryindex.com/

   a. Based on this study, what proportion of adults approve of cash-prize lotteries?

   b. How does the proportion of adults who approve of cash-prize lotteries compare with the proportion of teens who approve of cash-prize lotteries, based on this poll?

   c. What proportion of individuals in this poll who approve of cash-prize lotteries are adults?

9. On May 19, 2009, the U.S. Senate approved the Credit Cardholders' Bill of Rights Act. This act is a bill to amend the Truth in Lending Act to establish fair and transparent practices relating to the extension of credit. The voting tally of the senators is given in the next table:

	Democrat	Republican	Independent
Yea	54	35	1
Nay	1	4	0
Not Voting	3	1	0

Source: Library of Congress, www.congress.gov.

A senator from the 2009 U.S. Senate is chosen at random.

a. Find the probability that the chosen senator voted Nay, given that the senator is a Democrat.

b. Find the probability that the chosen senator voted Nay, given that the senator is a Republican.

c. Are the events "senator is a Republican" and "senator voted Yea" independent events? Give a quantitative justification to support your answer.

d. Are the events "senator is a Democrat" and "senator voted Nay" disjoint events? Give a quantitative justification to support your answer.

10. The following table contains information about the number of major (over 1,000 square miles in surface area) natural lakes in the world that lie on each continent, by size:

Size in Square Miles	Africa	Asia	Australia	Europe	North America	South America
1,000 to less than 5,000	2	3	3	2	7	1
5,000 to less than 10,000	0	1	0	1	3	1
Over 10,000	3	3	0	1	5	0

Source: The World Almanac and Book of Facts 2011, p. 693.

a. Find the proportion of major natural lakes that lie in North America.

b. What proportion of the major natural lakes in Europe are over 10,000 square miles?

c. Given that a major natural lake is in the "small" category (1,000 to less than 5,000 square miles), what is the probability that it is in Africa?

d. Are the events "major natural lake is in Asia" and "major natural lake is over 10,000 square miles" independent events? Give a quantitative justification to support your answer.

e. Are the events "major natural lake is in South America" and "major natural lake is over 10,000 square miles" disjoint events? Give a quantitative justification to support your answer.

11. For each of the following, support your explanation with examples:

a. Explain what it means for two events to be independent.

b. Explain what it means for two events to be disjoint.

Diagnostic Testing and Conditional Probability

In this activity, you will investigate how to interpret the results of diagnostic tests and other tests that have some possibility of erroneous results associated with them. You will also explore ways to represent graphically the information given in a two-way table.

1. Suppose you are a varsity athlete at a college that has decided to institute drug testing to screen out those athletes taking drugs. Here are some important questions to consider: How concerned should you be that the test will detect drug use when, in fact, you are not using drugs? What are the chances that the test will report that you are not using drugs if, in fact, you are? How sensitive is the test in picking up real drug usage?

   Here is a review of some terminology associated with such a test.

   A test's **sensitivity** refers to the proportion of drug users that the test detects accurately. These are also referred to as **true positives**.

   A test's **specificity** refers to the proportion of non-drug users (or, in general, cases that do not have the condition) that the test accurately identifies. These are also referred to as **true negatives**.

   Sensitivity and specificity for a particular diagnostic test can be determined from tests run on people known to be using or not using drugs (or known to have a certain condition or not). Of course, the administrators who are testing the student athletes for drug use don't know whether the students are using drugs or not. And if a clinician is testing you for a particular disease, he or she doesn't know whether or not you have the disease. (That's why the test is done!)

Screening tests for drugs, as well as diagnostic tests for diseases, can give false positive or false negative results. A **false positive** occurs when someone is identified by the test as using drugs when, in fact, he or she is not using drugs. A **false negative** occurs when someone is identified by the test as not using drugs when, in fact, he or she is using drugs.

As an athlete, your primary concern would probably be that you might be a false positive; that is, you would be identified as using drugs when you, in fact, do not. The college should be concerned both with false positives (it doesn't want to make false accusations) and false negatives (it wants to identify anyone who is using drugs).

To figure out what proportion of student athletes fall into these categories, it is helpful to set up a two-way table like this:

	Student Uses Drugs	Student Is Drug-Free	Row Total
Test Is Positive			
Test Is Negative			
Column Total			

Now, suppose the total population of student athletes taking the test is N, and values a, b, c, and d have been entered into the cells of the two-way table as follows:

	Student Uses Drugs	Student Is Drug-Free	Row Total
Test Is Positive	a	b	
Test Is Negative	c	d	
Column Total			**N**

a. Explain why the sensitivity is given by the proportion $\dfrac{a}{a+c}$.

b. Explain why the specificity is given by the proportion $\dfrac{d}{b+d}$.

c. Explain why the proportion of false positives is given by $\frac{b}{a+b}$.

d. Explain why the proportion of false negatives is given by $\frac{c}{c+d}$.

e. What fraction gives the proportion of student athletes who use drugs? Explain why this is so.

2. Now you'll enter some values in the table, using the following information:

Assume a population of 1,000 student athletes takes the test.

It is estimated that 9% of all student athletes use drugs.

The test's sensitivity is known to be 95%.

The test's specificity is known to be 96%.

	Student Uses Drugs	Student Is Drug-Free	Row Total
Test Is Positive			
Test Is Negative			
Column Total			

a. First, enter the total number of student athletes being tested into the appropriate cell of the table.

b. Given that 9% of athletes take drugs, fill in the appropriate total number of students who use drugs; then fill in the number who are drug-free.

   c. Use the test's sensitivity to fill in the appropriate cell of the table.

   d. Use the test's specificity to fill in the appropriate cell of the table.

   e. Fill in the rest of the table.

   f. Now use the filled-in table to find the probability that if a student athlete tests positive, he or she is not taking drugs.

   g. Find the probability that if a student athlete tests negative, he or she is using drugs.

   h. What are the implications (for the athletes, for the coaches, for the school's administration) of your answers to parts (f) and (g) of this question?

3. Now suppose that 4% of all athletes take drugs, the test's sensitivity is 98%, and the test's specificity is 99%. Recompute the table for a population of 1,000 athletes using the changed values and find the probabilities asked for in Question 2(f) and (g). What are the implications?

	Student Uses Drugs	Student Is Drug-Free	Row Total
Test Is Positive			
Test Is Negative			
Column Total			

4. Think of additional situations in which the results of tests such as the one described here might be used to make important recommendations and/or decisions. How should the information from such tests be used?

5. You will now look at how to represent conditional probabilities pictorially. Consider the following table of data, which gives the number of physicians in selected specialties in the United States in 2005:

	Female Physicians	Male Physicians
Family Practice	26,305	54,022
Dermatology	3,906	6,535
Ob/Gyn	17,258	24,801
Pediatrics	36,636	33,515
Psychiatry	13,079	27,213

Source: The World Almanac and Book of Facts 2011, p. 152.

Because the two variables (gender and medical specialty) are categorical, you can display these data using a column chart. Retrieve the Excel file "EA19.1 Physicians.xls" from the text website or from WileyPLUS.

   a. Use these instructions to create two-column charts.

Instructions to Construct a Stacked-Column Chart

1. Highlight the rectangle containing the data and labels, and go to the **Insert** tab. From the **Charts** group, choose **Column**, and select the second subtype choice, **Stacked column**.
2. Add an appropriate chart title to your chart.
3. Now you will construct a stacked-column chart with two columns, one for female physicians and the other for male physicians. To do so, repeat the previous steps to create a stacked-column chart as before; then, with the new graph selected, from the **Design** tab, click on **Switch Row/Column** from the **Data** group. Add a title to your chart.

   b. Use your table and charts to answer the following conditional probability questions, and for each question, determine which of your charts would help visualize the probability and describe how it helps visualize the probability. Suppose a doctor from the set of doctors represented by this table is chosen at random.

     i. Among all the doctors in pediatrics, what proportion are female? That is, find the probability that the chosen doctor is female, given that he or she is in pediatrics.

     ii. Among doctors in psychiatry, what proportion are male? That is, find the probability that the chosen doctor is male, given that he or she is in psychiatry.

     iii. Among all male doctors, what proportion are in psychiatry? That is, find the probability that the chosen doctor is in psychiatry, given that he is male.

     iv. Among all female doctors, what proportion are in family practice? That is, find the probability that the chosen doctor is in family practice, given that she is female.

Summary

In this activity, you explored the concepts of sensitivity and specificity of a diagnostic test. You used conditional probability and two-way tables to analyze the accuracy of a diagnostic test. As an aid to visualize conditional probabilities, you used Excel to draw stacked-column charts.

Sampling and Surveys

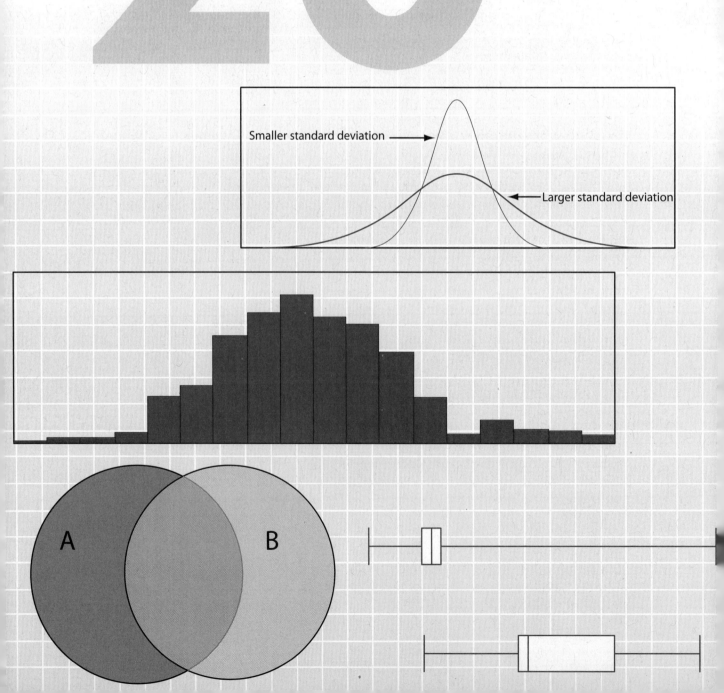

What are dreams? This is the title of a program that was part of the NOVA series that aired on PBS stations on June 6, 2011. The program included interviews with several scientists who are conducting studies to answer questions about dreams and sleep. A related *Newsweek* magazine article in the August 9, 2004, issue addressed "The Mystery of Dreams." The article looked at the history of dream research and cited results from experiments and results from observational studies. In one of the dream-research experiments, one group of subjects was deprived of REM sleep (a phase of sleep characterized by intense brain activity and rapid eye movement), while another group was allowed REM sleep. Differences in dreaming patterns were recorded. In another study, a psychology professor collected 550 dreams from a group of twenty-four 9- to 15-year-olds and analyzed them, looking for age and gender differences in their dreams. In this topic, we will explore issues related to these types of studies.

In an **experiment**, the investigator imposes some treatment or condition on the subjects of the experiment and records how the subjects respond. In an **observational study**, the investigator observes the subjects, without attempting to manipulate or impose any treatment, and records the responses. In both kinds of studies, researchers often will draw conclusions.

In both experiments and observational studies, researchers frequently want to find a relationship between two or more variables. A **response variable** or **outcome variable** is a variable that measures an outcome of a study. An **explanatory variable** is any variable that explains or influences the different responses measured by the response variable. For example, Harvard researchers conducted a ten-year study of more than 70,000 women and found that women getting fewer than five hours of sleep each night are more likely to develop diabetes (*Source: Prevention,* July 2003). Amount of sleep was the explanatory variable and whether or not diabetes developed was the response variable.

For any study, the entire group of individuals about which we want information is called the **population**. The portion of the population that we actually use to collect information is a **sample**, a subset of the population.

Example 20.1

For each of the following scenarios taken from news articles, identify the population and the sample, the explanatory variable(s) and response variables(s), and whether the scenario describes an experiment or an observational study.

a. A study of New Jersey drivers showed that approximately 40% of the 250 drivers surveyed scored highly to moderately stressed when driving. Nearly 58% reported that they had high to moderate levels of anger while driving. Almost half of those surveyed were prone to "take it out" on other drivers. The survey was administered at American Automobile Association offices in New Jersey.

b. *T'ai Chi Chih* (TCC) is an exercise routine that combines physical motion and mental relaxation. A study by Dr. Irwin at the UCLA Neuropsychiatric Institute showed that TCC improves resistance to the VZ virus, which causes shingles. This is a painful disease more common among older people. In the study, 18 people were given TCC classes during a 15-week period, while 18 others were placed on a waiting list. Blood samples of all 36 participants were tested before and after the 15 weeks. Researchers found that after the 15 weeks, resistance to the VZ virus was 50% higher among those who took the classes and there was no change among those wait-listed. (*Source: Prevention*, April 2004.)

c. In a study about walking and cardiovascular health, Harvard researchers found that walking at least 30 minutes a day was linked to an 18% decrease in the risk of coronary artery disease. The study involved 44,452 male health professionals. (*Source:* "Walking: Your Steps to Health," *Harvard Men's Health Watch*, August 2009, www.health.harvard.edu.)

Solution

a. The sample consisted of New Jersey drivers who were willing to participate in the survey at American Automobile Association (AAA) offices around the state. The population would likely consist of New Jersey drivers who visit AAA offices. The explanatory variable was the level of stress and anger people felt while behind the wheel of a car. The response variable was the number of incidents of aggressive acts while driving. This is an observational study because the researchers did not control the explanatory variable but rather "observed" by measuring people's self-perceptions of their anger levels when behind the wheel and incidents of aggressive driving.

b. The sample involved 36 individuals, probably from California, aged 60 and older. We may need more information to determine if men and women and various races were represented in the sample, but the population would likely be Californians aged 60 and older. The explanatory variable was a T'ai Chi class or no class, while the response variable was resistance to the virus as measured by a blood test. This is an experiment because the researchers formed two groups, an experimental group and a control group. They controlled the explanatory variable by having one group take T'ai Chi classes, while a control group did not take the classes.

c. The sample involved 44,452 men. The population would likely consist of all men. The explanatory variable in this study is amount of time spent walking per day. The response variable is occurrence of coronary artery disease. Although the summary did not explicitly say, it appears that this is an observational study. It would be difficult for researchers to perform a controlled experiment involving over 40,000 subjects, but an observational study would be possible.

Whether the study in question is an experiment or an observational study, it is very important that a sample be characteristic of the population that it represents. For example, consider the study on walking and men's heart health involving a sample of over 40,000 men described in Example 20.1(c). We could not expect to draw conclusions about the effect of walking on heart disease risk in women based on this study because the results from the men in the sample may not be applicable to women.

We must also take care when designing an observational study or experiment that we control, as much as possible, for **confounding variables**. A confounding variable is a variable whose influence on the response variable cannot be separated from that of the explanatory variable. For example, a newspaper article titled "Study: Oscar Winners Live Longer" found that, on average, Oscar winners lived to age 79.7, while nonwinners lived to be 75.8 (*Source: The Morning Call*, May 15, 2001). The article reported on a study that included all 762 actors and actresses who had ever been nominated for an Academy Award. Each nominee was paired with an actor of the same gender and age. The researchers investigated whether famous movie stars can benefit from a boost in self-esteem. They tried to control for the potential confounding variables of age and gender by matching each award winner with a "similar" actor. It is easier to control for confounding variables in experiments than in observational studies, but care must be taken with both types of studies.

The observational study involving Academy Award winners indicates one way to try to control for confounding variables in observational studies. When conducting an experiment, randomly assigning subjects to receive the different treatments or to be exposed to different conditions helps to control for confounding. The researcher reduces the chance of introducing any unwanted influences by randomizing. One way to choose subjects at random from a population is to put all the names in a bin, mix up the names, and pick them out one-by-one without looking at any of the names. A more sophisticated method, with a similar result, involves using a table of random digits or a random number generator on a calculator or computer. In the next example, we describe how to use a table of random digits to collect a random sample.

Example 20.2 ■ ■ ■ ■ ■ ■ ■ ■ ■

Consider the following list of U.S. presidents and their ages at inauguration. The list is given in chronological order across the rows. We will consider this group as our population. We will then use this population to illustrate some ideas associated with sampling.

Washington, 57	J. Adams, 61	Jefferson, 57	Madison, 57
Monroe, 58	J. Q. Adams, 57	Jackson, 61	Van Buren, 54
W. H. Harrison, 68	Tyler, 51	Polk, 49	Taylor, 64
Fillmore, 50	Pierce, 48	Buchanan, 65	Lincoln, 52
A. Johnson, 56	Grant, 46	Hayes, 54	Garfield, 49
Arthur, 50	Cleveland, 47	B. Harrison, 55	Cleveland, 55
McKinley, 54	T. Roosevelt, 42	Taft, 51	Wilson, 56
Harding, 55	Coolidge, 51	Hoover, 54	F. D. Roosevelt, 51
Truman, 60	Eisenhower, 62	Kennedy, 43	L. B. Johnson, 55
Nixon, 56	Ford, 61	Carter, 52	Reagan, 69
G. H. Bush, 64	Clinton, 46	G. W. Bush, 54	Obama, 47

a. Choose a sample of six U.S. presidents as follows: Number the presidents from 01 to 44, going across the first row for numbers 01 to 04 and proceeding left-to-right across each subsequent row. Use the following table of random digits; group the digits in the table in pairs (because we want to choose six numbers between 01 and 44). Ignore any pairs of digits outside of the range of 01 to 44; also ignore any pair of digits that has already appeared. Proceed across the first row and to the second row of digits, if necessary, continuing until you have six numbers between 01 and 44. Note that the digits are separated into groups of five to facilitate reading them, but the grouping is not important. Record the six chosen random numbers.

12975	13258	13048	45144	72321	81940	00360	02428	96767	35964	23822
96012	94591	65194	50842	53372	72829	50232	97892	63408	77919	44575
24870	04178	88565	42628	17797	49376	61762	16953	88604	12724	62964
99612	93465	64658	27402	56319	81103	46759	14520	19807	46845	30862

b. Write down the name and age at inauguration of each president corresponding to your six numbers.

c. What is the average age at inauguration of the presidents in your sample and how does it compare to the average of the whole population, which is approximately 54.6?

d. Theodore Roosevelt was the first president elected after 1900. (He was elected in 1904, although he became president in 1901 after McKinley was assassinated.) How many presidents elected after 1900 are in your sample?

e. Repeat parts (a–d) but use the third row in the given table of random digits to identify your sample of six. How do the samples compare?

Solution

a. Grouping the digits in pairs gives the following pairs of numbers: 12 97 51 32 58 13 04 84 51 44 72 32 18 19 40, and so on. We choose 12, and then ignore 97 and 51 because they are outside the range of 01 to 44. The next number chosen is 32; we ignore 58 and then select 13 and 04. Then we ignore 84 and 51, and select 44. We ignore 72, and because we already selected 32, we ignore it also. The sixth number we pick is 18. Thus, the numbers we choose are 12, 32, 13, 04, 44, and 18.

b. The names and ages of the chosen presidents are as follows: Taylor, 64 (he was number 12); F. D. Roosevelt, 51 (number 32); Fillmore, 50 (number 13); Madison, 57 (number 04); Obama, 47 (number 44); and Grant, 46 (number 18).

c. The average age of presidents in the sample is $\dfrac{64 + 51 + 50 + 57 + 47 + 46}{6} = 52.5$. It a bit less than the average age of the population.

d. Just two presidents in the sample were elected after 1900, F. D. Roosevelt and Obama.

e. We again group the digits in pairs and choose six numbers between 01 and 44. These pairs of digits in the third row are as follows: 24 87 00 41 78 88 56 54 26 28 17 79 74 93 76 61 76 21, and so on. The first six two-digit numbers in the range are 24, 41, 26, 28, 17, 21. These correspond to Cleveland, 55; G. H. Bush, 64; T. Roosevelt, 42; Wilson, 56; A. Johnson, 56; Arthur, 50. Three of these presidents were elected after 1900. The average age for the presidents in this sample is $\dfrac{55+64+42+56+56+50}{6} \approx 53.8$. This average is closer to the average of the population than the average we found in part (c).

Nineteen of the 44 U.S. presidents were elected after 1900, which is approximately 43% of them. If, for some reason, we want to be sure we have a proportional representation of presidents elected before and after 1900, we might want to use **stratified random sampling** in which we would pick, in this case, three presidents from the group elected before T. Roosevelt (50% of our sample of six) and three from the remaining group. This would guarantee that the proportion in each of these groups or **strata** is as close as we can get to being approximately the same for both the sample and the population. We illustrate how to collect a stratified random sample in the next example.

Example 20.3

Collect a stratified random sample for the "Presidents" example using "elected before 1900" and "elected 1900 or later" as the strata or classes as follows:

a. Number the presidents elected before T. Roosevelt from 01 to 25 and use the second row (and continue to the third row if necessary) of random digits in the given table to choose three from this group, as we did in Example 20.2.

b. Number the remaining presidents from 01 to 19 and use the third row (and continue to the fourth row if necessary) of random digits to choose three more for a total sample of six. List the names and ages of the six chosen presidents and compute the average age of the six chosen presidents.

c. Why might we want to take a stratified random sample rather than a simple random sample?

Solution

a. We group the numbers from the second row of the table of random digits in pairs, and disregard any that are not in the range 01 to 25. These numbers are 96 01 29 45 91 65 19 45 08 42, and so on. We choose 01, 19, and 08, which gives us Presidents Washington, 57; Hayes, 54; and Van Buren, 54 from the first group of those elected before 1900.

b. Grouping digits from the third row in pairs, we get the following pairs: 24 87 00 41 78 88 56 54 26 28 17 79 74 93 76 61 76 21 69 53 88 60 41 27 24 62 96. So far, we have only one number between 01 and 19 (and it's 17). Using the remaining digit in the third row (which is 4) and the fourth row of the table, we continue our grouping. We get 49 96 12 93 46 56 46 58 27 40 25 63 19 81, and so on. We can stop there because we have our remaining two sample presidents, number 12 and number 19. Our three presidents from this group, "elected 1900 or later," are Clinton, 46 (number 17); Nixon, 56 (number 12); and Obama, 47 (number 19). The average age of this stratified random sample is $\dfrac{57+54+54+46+56+47}{6} = 52.3$.

c. In general, a stratified sample will guarantee that different groups or strata are represented in the sample. If we want to sample attitudes among college students, we might want to be sure to include students from the freshman, sophomore, junior, and senior classes in numbers proportional to their sizes. In addition, we might want to make sure we have a fair (that is, proportional) representation of both men and women.

In the previous examples, we used a table of random numbers to find samples of six out of a total population of 44 presidents. The same table can be used to find samples from larger populations.

Example 20.4

Two studies will be conducted at a college that has a total student population of 935 students and we have been asked to choose a sample of 10 students for each of the two studies.

a. The first study is about parking and the researchers want to collect information from 10 students. Explain how we could use the following table of random numbers (this is the same table given in Example 20.2) to find a random sample of 10 students from the total student population of the college:

12975	13258	13048	45144	72321	81940	00360	02428	96767	35964	23822
96012	94591	65194	50842	53372	72829	50232	97892	63408	77919	44575
24870	04178	88565	42628	17797	49376	61762	16953	88604	12724	62964
99612	93465	64658	27402	56319	81103	46759	14520	19807	46845	30862

b. The second study will look into the amount of writing students are required to do during their first two years in college. We need to select a stratified random sample of 10 students using "first-year student" and "sophomore student" as the strata or classes. Assuming there are 300 first-year students and 220 sophomores, explain how the given table of random digits could be used to select the sample.

Solution

a. We could number the students from 001 to 935 in an orderly fashion (such as in alphabetical order) and choose three-digit numbers from the random digits table. We would ignore the number 000, if it occurred, and numbers greater than 935. Using the first line of the table, with numbers grouped in 3s, we would have the following: 129 751 325 813 048 451 447 232 181 940 003 600 242, and so on. From this list, we take the first 10 numbers that are in the range from 001 to 935. The students in the sample are those with numbers 129, 751, 325, 813, 048, 451, 447, 232, 181, and 003.

b. Among the 520 students in our total population of first- and second-year students, 300 are first-year students, which is approximately 58% of the population. For a sample of 10 to have the same proportion of first-year and sophomore students as the total population, it should have approximately $0.58 \times 10 = 5.8$ first-year students. Rounding to the nearest integer, we decide that we need to choose a sample with 6 first-year students. To obtain the stratified random sample, we first number the first-year students from 001 to 300, and the sophomore students from 301 to 520. (This is a slightly different approach from that used in Example 20.3.) Using the first line of the table, with digits grouped in 3s, we read: 129 751 325 813 048 451 447 232 181 940 003 600 242 896 767 359 642 382, and so on. We need to choose the first six numbers on that list that are between 001 and 300, so we choose 129, 048, 232, 181, 003, and 242. We then choose the first four numbers in the list that are between 301 and 520. These are 325, 451, 447, and 359. So, the students in the sample are the first-year students with numbers 129, 048, 232, 181, and 242, and the sophomore students with numbers 325, 451, 447, and 359.

It is not always possible to obtain results from a truly random sample, because of some practical reasons. If a sampling method produces results that are systematically different from the true results about the population, then the method is **biased**. **Selection bias** occurs when the sampling method either excludes entirely or includes disproportionately some particular section of the population. If there is some variable that is important to the study and on which those included and those excluded differ, the results of the study may be meaningless. In particular, phone-in surveys in which subjects self-select to respond often suffer from selection bias.

Another source of bias, **response bias**, occurs when questions on a survey or the behavior of the interviewer or the situation in which the interview is held influence the responses received. Special interest groups sometimes conduct surveys designed to promote a special cause and will word questions in such a way that the results will be biased in favor of their cause.

Nonresponse bias occurs when some of the subjects selected for the sample choose not to participate and the nonresponders are different from the responders. This might result in the received responses not being representative of the population from which the sample was taken. Many surveys, when done by telephone or mail, will have high nonresponse rates; often only those who care deeply about the results will choose to respond. Personal interview surveys usually result in a lower nonresponse rate.

Example 20.5 ■ ■ ■ ■ ■ ■ ■ ■ ■

For each of the following scenarios, explain the type and source of any potential bias.

a. The magazine *Literary Digest* had been able to predict the results of the U.S. presidential election successfully until 1936. In that year, Republican Alf Landon and incumbent Democrat Franklin Delano Roosevelt were running for president. The *Literary Digest* sent out questionnaires to 10 million people whose names had appeared on lists of car owners, magazine subscribers, telephone directories, and registered voters. Based on 2.3 million responses, the *Literary Digest* predicted that Landon would win with 57% of the vote to Roosevelt's 43%. Roosevelt was reelected with over 60% of the popular vote.

b. The article "Take Your *Kids* to Work?" from the April 2001 issue of *Working Woman* reported on a poll to determine how many Fortune 100 companies allowed parents to take boys to work on Take Our Daughters to Work Day. With a total of 58 companies responding, the results showed that 9% did not participate in the program, 12% had separate programs for boys and girls, and 79% allowed boys and girls to participate together. Some of them had changed the name of the program to Take Our Children to Work.

c. A group of college students were randomly asked one of two questions: One group was asked, "Should this campus allow speeches that might incite violence?" The other group was asked, "Should this campus prohibit speeches that might incite violence?" Even though the two groups were chosen randomly, a larger proportion answered "no" to the first question than answered "yes" to the second question.

d. The article "Most Ex-welfare Recipients Not Better Off, Survey Says" (which appeared in *The Morning Call* on July 26, 2001) reported on a study that surveyed 893 former welfare recipients who were using social service facilities in 10 states. The study found that their lives had not improved much even though the total number of people on welfare was lower.

Solution

a. Selection bias occurred in this survey, because by using telephone, voter, magazine, and car-owner lists, the poor tended to be excluded. The people excluded were more likely to support Roosevelt. Nonresponse bias (only 23% of those polled responded) also influenced the survey results because Roosevelt supporters, who were satisfied with the current situation, tended not to respond.

b. In this scenario, nonresponse bias might be present. Those responding might be more likely to include boys in their programs than the nonresponders, but we cannot know for certain unless we follow up and get information from the 42 companies polled who did not respond. (Note that 100 companies were polled. Because only 58 responded, 42 did not respond.)

c. Because the word *prohibit* in the second question carries negative connotations, people were more reluctant to answer, "Yes, we want to prohibit this" than they were to answer, "No, we don't want to allow this." This is an example of response bias, because the wording of the question may have influenced the responses.

d. Some people criticized the agency for only surveying people at social welfare facilities. They claimed response bias might have been introduced because of the location chosen for the survey. Selection bias might also have occurred because, by going to a social service agency, the respondents might not have been a representative sample of former welfare recipients.

Some reports on studies do not offer complete details about the way the studies were conducted.

Example 20.6

An op-ed article "Our Hidden Government Benefits" (which ran in *The New York Times*, September 20, 2011) referred to a survey by the Cornell Survey Research Institute in which 1,400 Americans were asked if they had ever used a government social program. At the same time, respondents were asked if they had benefited from any of a list of 21 federal programs, such as Social Security, unemployment insurance, a student loan, or a mortgage-tax deduction. The study showed that 57% of the respondents said they had not used a government social program. On the other hand, the study showed that 94% of the respondents had benefited from at least one of the programs on the list.

What else would help you evaluate and draw conclusions from the given information?

Solution

From this description, we know that the study surveyed 1,400 people. It would be useful to know if this is the actual number of respondents or the total number of people asked to respond. It also would be useful to know how the respondents were selected. Did the sample include people of different ages? Were they from diverse educational levels, from diverse economic levels? It would also be useful to know exactly how the questions were phrased to rule out possible biases. In addition, the order in which the questions were asked would help us evaluate whether that might have affected the responses.

When analyzing results from surveys, researchers usually assume that the respondents are telling the truth. However, when asked sensitive questions, those surveyed might be reluctant to answer truthfully, especially in personal interviews or if they think someone might be aware of how they, personally, are responding. An interesting way to elicit candid responses to sensitive questions can be carried out using **Warner's randomized response model**, which is illustrated in the following example.

Example 20.7

We will first ask each college student respondent to toss a penny and a nickel and keep track of the result of each. We will then survey the respondents and ask two questions. Question 1, which we will denote by Q1 is, "Have you ever cheated on a college exam?" Question 2, which we will denote by Q2 is, "Is the outcome on the nickel a head?"

If the penny toss results in a head, the respondent answers Q1. If the penny comes up a tail, the respondent answers Q2. Each respondent tosses the coins so only he or she knows the results of the toss, and then each respondent answers Q1 or Q2 and writes a single answer on his or her paper. Because no one knows which question the respondent is answering, there is no stigma attached to answering yes. We will set up a table to find the proportion who answer yes to Q1, the sensitive question.

a. Assume there are 100 respondents in the survey; we would expect about half of the respondents to toss a head on their penny and half to toss a tail, so half of those surveyed will answer Q1 and half will answer Q2. Fill in the row totals and the row corresponding to "Answered Q2" in the following table:

	Answered Yes	Answered No	Total
Answered Q1			
Answered Q2			
Total			100

b. Suppose that a total of 44 respondents answered yes. Write this number in the table in the appropriate place and fill in the rest of the cells.

c. Find the probability that a respondent chosen at random answered yes given that he or she answered Question 1; that is, find $\mathcal{P}$(Answered yes | Answered Q1). Explain what this probability represents.

Solution

a. We would expect 50 of the 100 respondents to answer Q1 and 50 to answer Q2. Of those 50 respondents who answered Q2 (Is the outcome on the nickel a head?), we expect 25 of them to answer yes and 25 to answer no, assuming fair tosses. We can enter that information in the table as follows:

	Answered Yes	Answered No	Total
Answered Q1			50
Answered Q2	25	25	50
Total			100

b. The "Answered Yes" column total is 44. We can then fill in the rest of the cells by maintaining the row and column totals:

	Answered Yes	Answered No	Total
Answered Q1	19	31	50
Answered Q2	25	25	50
Total	44	56	100

c. This conditional probability is $\frac{19}{50} = 0.38 = 38\%$. This gives us an estimate of the proportion of college students who have ever cheated on an exam, based on the information in the sample of 100 students.

This technique makes it possible for respondents to answer candidly and truthfully sensitive questions that have "yes" or "no" answers. Of course, how good an estimate this proportion is also depends on how representative of the population the sample is.

Summary

In this topic, we examined various issues surrounding studies and experiments. We discussed observational studies and experiments and considered the population and the sample in these studies. We discussed explanatory and response variables and looked at types of bias introduced into studies and experiments. We also investigated how to collect a simple random sample and how to collect a stratified random sample. Finally, we looked at Warner's randomized response model as a way to obtain honest responses to sensitive questions.

Explorations

1. The *HealthDay* article "Eating Berries Might Help Preserve Your Memory" reported on a study published in the *Annals of Neurology* online journal on April 26, 2012. The researchers used data from the U.S. Nurses' Health Study that contained dietary information since 1980. They also measured cognitive function since 1995 in 16,000 females. Women who ate more than one serving of blueberries per week and more than two servings of strawberries per week showed a delay in cognitive aging of as much as 2.5 years.

    How do you think the research was carried out, was it an observational study or an experiment, and what can you conclude from the study description?

2. A news article reported on a study about diet and breast cancer that was conducted in Mexico, with support from the U.S. Center for Disease Control and Prevention, the

Ministry of Health of Mexico, and the American Institute for Cancer Research. The study found that Mexican women who ate a lot of carbohydrates doubled their risk of getting breast cancer.

Dr. Willet, chief of nutrition at the Harvard School of Public Health, who participated in the study, warned that this does not mean that it is safe to eat a diet based on meat and cheese. He said that people should be careful not to increase consumption of unhealthy fat when cutting back the intake of carbohydrates. There are serious health concerns about diets high in animal fat. (*Source: The Morning Call*, August 6, 2004.)

How do you think the research was carried out? Was it an observational study or an experiment, and what can you conclude from the study description?

3. The following article describes a "study":

Hazards: Secondhand Smoke May Affect Hearing

NICHOLAS BAKALAR

Add yet another item to the long list of damaging effects of secondhand smoke: hearing loss in teenagers.

Researchers, writing in the July issue of *The Archives of Otolaryngology—Head and Neck Surgery*, tested more than 2,000 teenagers for cotinine, an indicator of exposure to tobacco smoke. After eliminating smokers from the study, they were left with 799 nonsmokers whose cotinine levels indicated exposure to secondhand smoke, along with 754 who were not exposed to cigarette smoke.

After controlling for many variables they found that the higher the cotinine level in a participant's blood, the greater the likelihood there was some type of hearing loss. More than 17 percent in the highest quartile for cotinine levels had hearing loss at low frequencies.

It is unclear exactly how exposure to secondhand smoke could cause the damage, but tobacco is known to affect blood flow through the smallest blood vessels, the kind the inner ear depends on.

"Most kids, about 85 percent, were unaware of their hearing loss," said Dr. Anil K. Lalwani, the lead author of the study. "You can't rely on self-reports."

Source: The New York Times, July 19, 2011.

    a. Is this an observational study or an experiment? How do you know?

    b. Identify the population and the sample described in the study.

    c. Identify the explanatory variable(s) and the response variable(s).

    d. What does the author mean by "the highest quartile for cotinine levels"?

4. A study by the AAA Foundation found that drivers age 16 or 17 increase the risk of dying in a car crash by 44% when they have one passenger younger than 21. The fatality risk doubles for these drivers when they carry two passengers younger than 21, and quadruples when three or more passengers are younger than 21. The study analyzed data from the National Highway Traffic Safety Administration that included records of car accidents from 2007 to 2010.

   a. Is this an observational study or an experiment? How do you know?

   b. Identify the population in the study.

   c. Identify the explanatory variable(s) and the response variable(s).

5. An article on cell phones and driving in *Prevention,* May 2004, reported that the risk of a traffic accident increases by 38% if the driver is talking on a cell phone while driving. Researchers estimate that using a cell phone while driving contributes to 330,000 injuries and 2,600 deaths per year.

   a. Is this an observational study or an experiment? How do you know?

   b. Identify the population and the sample described in the study.

   c. Identify the explanatory variable(s) and the response variable(s).

6. The following excerpt is taken from the article "The Evidence for Acupuncture" (*Source: Shape,* August 2004):

> In a study of 400 people (about 80 percent women) who suffered from severe headaches, largely migraines, for several days each month, one group received up to 12 acupuncture treatments in three months. Nine months later, the acupuncture group experienced the equivalent of 22 fewer headache days per year.
> They also used 15 percent less pain medication, made 25 percent fewer visits to physicians, missed 15 percent fewer days of work and had a generally improved quality of life.

   a. Is this an observational study or an experiment? How do you know?

   b. Identify the population and the sample described in the study.

   c. Identify the explanatory variable(s) and the response variable(s).

7. The article "Antiobiotics's Role in Heart Attacks to Be Focus of Study" (*The Morning Call,* April 5, 1999) reported on the design of a study to find if antibiotics could save heart patients from future serious heart incidents. The article included the following description of the study:

"Half of the 4,000 heart disease patients to be studied will be given a weekly antibiotic pill, Zithromax, for one year, while the comparison group gets a dummy pill. Neither will know who is taking which pill."

a. Is this an observational study or an experiment? How do you know?

b. Identify the population and the sample described in the study.

c. Identify the explanatory variable(s) and the response variable(s).

d. Identify the source of any potential bias.

e. The "dummy pill" referred to in the article is also called a **placebo**. Why might that be an important consideration in a study such as this one?

8. A report in the March 2004 issue of *Prevention* claims in its title that "Spring Veggies Fight Cancer and Stroke." The study tracked 40,000 people for 18 years and found that stroke risk was reduced by 26% among those who ate green or yellow vegetables almost daily.

a. Identify the explanatory variable(s) and response variable(s).

b. Give other possible explanations for the conclusion given in the quote.

9. An Internet poll posed the question, "What do you use the Internet for most?" The possible answers and the percentage of respondents who gave each answer were as follows: e-mail, 33.15%; surfing the web, 24.9%; education, 5.87%; online chatting, 19.58%; shopping, 0.7%; online games, 6.57%; something else, 9.23% (*Source: Internet Poll Questions*, www.jwen.com/poll/poll.html).

a. Identify the population and sample for this survey.

b. Why might the results of this poll be biased?

10. A poll found that the majority of Americans favored legislation that would require the use of birth control in order for a mother to be eligible to receive welfare benefits. The question posed to the poll participants was as follows: "Today, some pople believe that mothers on welfare continue to have children that they are financially unable to support. There is proposed federal legislation that would require all welfare recipients to use birth control as a condition of their public benefits. Do you strongly support, somewhat support, somewhat oppose, or strongly oppose requiring welfare recipients to use birth control?" (*Source:* http://womensenews.org).

Why might the results of this poll be biased?

11. Use the table of the 50 U.S. states and governors' salaries given in Example 16.3 (and repeated here) and the random digits given in Example 20.2 (also repeated here) to answer parts (a) through (f) of this Exploration.

Table of States and Governors' Salaries

State	Salary	State	Salary	State	Salary
Alabama	112,895	Louisiana	130,000	Ohio	144,269
Alaska	125,000	Maine	70,000	Oklahoma	147,000
Arizona	95,000	Maryland	150,000	Oregon	93,600
Arkansas	87,352	Massachusetts	140,535	Pennsylvania	174,914
California	173,987	Michigan	177,000	Rhode Island	117,817
Colorado	90,000	Minnesota	120,303	South Carolina	106,078
Connecticut	150,000	Mississippi	122,160	South Dakota	115,348
Delaware	171,000	Missouri	133,821	Tennessee	170,340
Florida	130,273	Montana	100,121	Texas	150,000
Georgia	139,339	Nebraska	105,000	Utah	109,900
Hawaii	117,312	Nevada	141,000	Vermont	142,542
Idaho	115,348	New Hampshire	113,834	Virginia	175,000
Illinois	177,500	New Jersey	175,000	Washington	166,891
Indiana	95,000	New Mexico	110,000	West Virginia	95,000
Iowa	130,000	New York	179,000	Wisconsin	137,092
Kansas	110,707	North Carolina	139,590	Wyoming	105,000
Kentucky	145,885	North Dakota	105,036		

Random digits:

12975	13258	13048	45144	72321	81940	00360	02428	96767	35964	23822
96012	94591	65194	50842	53372	72829	50232	97892	63408	77919	44575
24870	04178	88565	42628	17797	49376	61762	16953	88604	12724	62964
99612	93465	64658	27402	56319	81103	46759	14520	19807	46845	30862

a. Starting in row 2 of the random digits, take a random sample of six states and find the average governor's salary of your sample.

b. Starting in row 3 of the random digits, take another random sample of six states and find the average governor's salary of this sample. How do the two samples compare?

c. What is the largest possible sample average salary that you could have obtained for a sample of size 6?

   d. What is the smallest possible sample average salary that you could have obtained for a sample of size 6?

   e. Would the largest possible sample average salary that you could have obtained increase or decrease if you took samples of size 10 instead of size 6?

   f. Would the smallest possible sample average salary that you could have obtained increase or decrease if you took samples of size 10 instead of size 6?

12. Here is a list of the states that are east of the Mississippi River: Alabama, Connecticut, Delaware, Florida, Georgia, Illinois, Indiana, Kentucky, Maine, Maryland, Massachusetts, Michigan, Mississippi, New Hampshire, New Jersey, New York, North Carolina, Ohio, Pennsylvania, Rhode Island, South Carolina, Tennessee, Vermont, Virginia, West Virginia, Wisconsin.

   a. Suppose you take a sample of six states, stratified by location relative to the Mississippi River. Explain how you would get such a sample.

   b. Starting with the first line of the random digits table shown in Exploration 11, carry out the stratified sampling procedure you described in part (a) of this Exploration.

   c. How is this sample different from those you collected in Exploration 11(a) and (b)?

13. Refer to Warner's randomized response model, described in Example 20.7, and find the probability that a respondent chosen at random answered yes, given that he or she answered Question 1. Assume there were 50 respondents in the survey and 23 of them answered yes.

14. Use Warner's randomized response model to carry out a survey in class, using the question "Have you ever used marijuana?" Set up a table to analyze the results. Find the probability that a student answered yes given that he or she answered the sensitive question.

Sampling and Surveys

In this activity, you will investigate why random sampling is important and also consider some issues involved in the design of surveys.

1. Here is a list of the 50 United States:

Alabama	Hawaii	Massachusetts	New Mexico	South Dakota
Alaska	Idaho	Michigan	New York	Tennessee
Arizona	Illinois	Minnesota	North Carolina	Texas
Arkansas	Indiana	Mississippi	North Dakota	Utah
California	Iowa	Missouri	Ohio	Vermont
Colorado	Kansas	Montana	Oklahoma	Virginia
Connecticut	Kentucky	Nebraska	Oregon	Washington
Delaware	Louisiana	Nevada	Pennsylvania	West Virginia
Florida	Maine	New Hampshire	Rhode Island	Wisconsin
Georgia	Maryland	New Jersey	South Carolina	Wyoming

   a. Using your sense of the land area of each state, choose what *you think* is a representative (that is, your *subjective*) sample of six states. List them in the following table. (It might help to try to visualize a map of the United States to get your sample of six "representative" states.)

State	Land Area
1.	
2.	
3.	
4.	
5.	
6.	

b. Refer to the table at the end of this activity and record the land area for each of your chosen states, in square miles given to the nearest integer, in the second column of the table (*Source:* Department of Commerce, Bureau of the Census, www.infoplease.com).

c. Compute the sample mean land area for your *subjective* sample of six states and record it here: _____.

d. Now use Excel's random number generator to generate six random integers between 00 and 49. (Recall how you generated random integers 0, 1, 2, and 3 in Activity 18.1 and adapt that technique to generate integers between 00 and 49. It is possible that you might get repeated numbers when you generate the random integers. If you do, discard the second occurrence of the repeated number and generate another integer so you have six distinct integers.) Describe how you generated the six integers and list them here.

e. Number the states in a systematic way, and use the numbers you generated in part (d) of this question to pick a *random* sample of six states. Explain your systematic method for numbering the states.

f. Enter the states you chose in the following table and record the land area for each of your states:

State	Land Area
1.	
2.	
3.	
4.	
5.	
6.	

g. Compute the sample mean land area for your *random* sample of six states and record it here: _____.

h. Recall that a **population** refers to the whole group about which you want to draw a conclusion and a **sample** refers to a subgroup of the population. The population mean land area for the population of all 50 states is approximately 70,748 square miles. Count how many of your classmates found *subjective* sample means (in part (c) of this question) greater than the population mean and how many found *subjective* sample means less than the population mean. Record those values here:

Number of subjective sample means greater than the population mean:

_____.

Number of subjective sample means less than the population mean:

_____.

i. Now count how many of your classmates found *random* sample means (in part (g) of this question) greater than the population mean and how many found *random* sample means less than the population mean. Record those values here:

Number of random sample means greater than the population mean:

_____.

Number of random sample means less than the population mean:

_____.

j. Explain how the class results using the two sampling methods (subjective and random) were different, if they were, or how they were similar. Were the class results unexpected?

2. Now you will generate a stratified random sample of states to estimate average water area.

a. Use Excel to generate a stratified random sample of six states, stratified by location relative to the Mississippi River. (Use the list of states east of the Mississippi given in Topic 20, Exploration 12.) Describe how you generated your stratified random sample of states.

b. List your six states in the following table; refer to the last table at the end of this activity and record the *water area* of each state in your sample.

State	Water Area
1.	
2.	
3.	
4.	
5.	
6.	

c. Compute the sample mean water area for your *stratified random* sample of six states and record it here: _____.

d. Describe how your sample mean water area compares to the population mean water area of approximately 5,133 square miles.

e. Explain why you might want to stratify as you did in part (b) of this question.

3. Work with one or two classmates on this question. Choose two of the following topics, and for each, discuss how you would design and implement a survey to obtain information about how college students feel about the issue. Be sure to consider what kinds of questions you would ask, how you would obtain truthful answers to sensitive questions, how you would get information from a representative sample, whether you would want to use a stratified sampling technique, how you would avoid bias, and any other factors involving sampling that you would need to bear in mind.

a. Students' attitudes about changing to an academic system in which no grades are given for college course work. (In this system, professors provide a written evaluation about each student's work in individual courses but no grades are given.)

b. Students' attitudes about the topic of health insurance coverage for abortions.

c. Students' attitudes about laws that establish a voting age of 18 and a drinking age of 21.

d. Students' attitudes about whether municipalities should contribute to the cost of professional sports stadiums.

e. Students' attitudes about restricting Internet access at public libraries to approved sites.

Summary

In this activity, you investigated why you would want to collect random samples rather than "subjective" samples. You also collected a stratified random sample. To collect these random samples, you used Excel's random number generator. Finally, you considered various aspects of how to design a survey to collect information about students' attitudes on a specific topic.

State	Land Area	Water Area	Total Area
Alabama	50744	1675	52419
Alaska	571951	91316	663267
Arizona	113635	364	113998
Arkansas	52068	1110	53179
California	155959	7736	163696
Colorado	103718	376	104094
Connecticut	4845	699	5543
Delaware	1954	536	2489
Florida	53927	11828	65755
Georgia	57906	1519	59425
Hawaii	6423	4508	10931
Idaho	82747	823	83570
Illinois	55584	2331	57914
Indiana	35867	551	36418
Iowa	55869	402	56272
Kansas	81815	462	82277
Kentucky	39728	681	40409
Louisiana	43562	8278	51840
Maine	30862	4523	35385
Maryland	9774	2633	12407
Massachusetts	7840	2715	10555
Michigan	56804	39912	96716
Minnesota	79610	7329	86939
Mississippi	46907	1523	48430
Missouri	68886	818	69704

State	Land Area	Water Area	Total Area
Montana	145552	1490	147042
Nebraska	76872	481	77354
Nevada	109826	735	110561
New Hampshire	8968	382	9350
New Jersey	7417	1304	8721
New Mexico	121356	234	121589
New York	47214	7342	54556
North Carolina	48711	5108	53819
North Dakota	68976	1724	70700
Ohio	40948	3877	44825
Oklahoma	68667	1231	69898
Oregon	95997	2384	98381
Pennsylvania	44817	1239	46055
Rhode Island	1045	500	1545
South Carolina	30109	1911	32020
South Dakota	75885	1232	77116
Tennessee	41217	926	42143
Texas	261797	6784	268581
Utah	82144	2755	84899
Vermont	9250	365	9614
Virginia	39594	3180	42774
Washington	66544	4756	71300
West Virginia	24078	152	24230
Wisconsin	54310	11188	65498
Wyoming	97100	713	97814
Total	**3,537,379**	**256,641**	

More on Decision Making

After completing this topic, you will be able to:

- Organize information for decisions that involve uncertainty.

- Calculate the expected value (also called mean or average value) of a probability distribution.

- Use the expected value to help make decisions that involve uncertainty.

- Use the maximin and maximax criteria (when "payoffs" are profits) or the minimax and minimin criteria (when "payoffs" are costs) to help make a decision involving uncertain information.

Suppose you have $1,000 that you want to invest for the next five years; you need to decide where to invest it. You could put your money in a standard savings account, in a mutual fund, or even buy stock with it. If you choose one kind of investment, your $1,000 might grow more than if you made a different choice. You can track the earnings records for each type of investment, which will enable you to attach probabilities to your potential earnings. In Topic 11, we considered decisions that involve information that we know for certain. But many decisions involve information that contains some element of uncertainty. In this topic we look at such decisions.

In situations like the "where to invest" one described above, there is uncertainty about what may happen with the economy, and we must choose among possible alternative actions before we know what happens. The weather can also be a source of uncertainty. For example, suppose we are planning a vacation in September, and we've narrowed our vacation choices to the areas along the outer banks of North Carolina or along the coast of Maine. Our two alternative actions are to choose a vacation in NC or in ME. The uncertainty involves the weather; there may be a hurricane in NC or an early snowstorm in ME. We need to reserve our vacation home several months before we know what the weather will be. This is typical of what is involved in a decision under uncertainty.

We use the terms **courses of action** for the choices from which the decision maker will choose, and **states of nature** for the things that will impact the decision, but over which the decision maker has no control. The weather and stock market conditions are examples of states of nature.

Example 21.1 ■ ■ ■ ■ ■ ■ ■ ■ ■

For each of the following situations, identify the possible courses of action and the unknown states of nature:

a. The manager of a college bookstore must decide how many copies of a particular monthly magazine to order. The manager does not know how many copies of that magazine will be purchased.

b. A college student is trying to decide on the amount of the deductible she is willing to pay when choosing collision insurance for her car.

Solution

a. In this scenario, the alternative courses of action are the number of copies of the magazine that the bookstore manager can choose to order. For example, it is possible that the manager can choose to order 5 copies, 6 copies, 7 copies, and so on up to 20 copies. The unknown states of nature are how many copies will be purchased during the month. For example, it is reasonable that customers can purchase 0 copies, 1 copy, 2 copies, 3 copies, and so on up to 20 copies. The manager must order the magazines before he or she knows how many copies will be purchased.

b. In this situation, the alternative courses of action are the possible deductible amounts that the student can choose to pay. The unknown states of nature are how much accident damage will be incurred by the student during the insurance period.

■ ■ ■ ■ ■ ■ ■ ■ ■ ■

The next example shows how, with detailed information about costs, we can set up a table to organize the information associated with a decision involving uncertainty.

Example 21.2 ■ ■ ■ ■ ■ ■ ■ ■ ■

You have been offered your first full-time job and one of the benefits is health insurance. You must decide which of three healthcare plans to choose. The first plan costs you $30 each month and there is a $500 deductible annually. (This means that the employee pays all expenses until expenses for the year are $500.) After expenses have reached $500, the employee pays 20% of the costs, with the rest paid by the insurance company. The second plan is the same as the first, except that it costs $10 per month and the deductible amount is $1,000. The third plan costs $25 per month and there is no deductible. The employee pays 30%, of all medical expenses, with the rest paid by the insurance company. (All plans cover the same services: office visits, hospital visits, surgery, and prescriptions.) After doing some research, you estimate that there are five basic annual healthcare cost levels: $200; $600; $1,200; $3,000; $15,000.

a. How is uncertainty involved in this decision?

b. Identify the courses of action and the states of nature.

c. Assess yearly costs under each of the plans and for the various healthcare expense levels. Make a table containing the results and be sure to include the cost of insurance.

Solution

a. You are uncertain what your healthcare costs will be for the year. You might need to make only a few visits to the doctor's office for colds and a serious case of poison ivy, or you might have considerably more healthcare expenses.

b. The courses of action are the three plans you can choose. The states of nature are the five estimated levels of healthcare costs for the year.

c. We need to determine the cost for each plan and each of the possible expense levels. We'll set up a table with the courses of action as the rows and the possible expense levels as the columns. For example, suppose we choose the first plan and our medical expenses are $200. The cost to us is $(\$30)(12) + 200 = \$560$. If we choose the second plan and our expenses are $200, the cost to us is $(\$10)(12) + 200 = \$320$. If we choose the third plan and our expenses are $200, the cost to you is $(\$25)(12) + (0.3)(\$200) = \$360$. We can verify that the costs for the other plans are as given in the following table:

	Medical Expense Level				
	$200	$600	$1,200	$3,000	$15,000
Plan 1	560	880	1,000	1,360	3,760
Plan 2	320	720	1,160	1,520	3,920
Plan 3	360	480	660	1,200	4,800

For some of these decisions, we will be able to estimate probabilities associated with the states of nature, but for others, we might not. In this topic, we'll develop tools to help us make decisions in both types of situations.

In the next two examples, we explore how we can interpret an average payout and the probabilities associated with states of nature. This will help us evaluate decisions for which we can attach probabilities to the states of nature.

Example 21.3

Suppose you are tracking a particular mutual fund and discover that during the previous 10-year period, the annual return was 7% for one year of that period, 8% for three of the years, 12% for two of the years, and 9% for four of the years. What was the average yearly return over the 10-year period?

Solution

We can answer the question by writing out the annual return for each of the 10 years: 7, 8, 8, 8, 12, 12, 9, 9, 9, 9; then we add these 10 numbers and divide by 10 to get the average: $\frac{7+8+8+8+12+12+9+9+9+9}{10} = 9.1\%$. We could carry out the computations more efficiently by writing the average, or mean, as $\frac{(7\cdot1)+(8\cdot3)+(12\cdot2)+(9\cdot4)}{10} = 7\frac{1}{10} + 8\frac{3}{10} + 12\frac{2}{10} + 9\frac{4}{10} = 9.1\%$.

We can use information like that given in the previous example to set up a probability table that presents each possible annual return and the probability associated with each possible return, as the following example shows.

Example 21.4

If we were to choose a year at random from the 10-year period given in Example 21.3, the probability is $\frac{1}{10}$ that we choose a year in which the annual return was 7% because out of the 10 years, only one year had an annual return of 7%. Set up a probability table showing each possible annual return and the probability of receiving that return for the mutual fund of Example 21.3.

Solution

In the probability table, we give the possible values of the annual return in the first row of the table. Then the probability associated with each annual return value is given in the corresponding cell in the second row of the table. We find the probability by counting the number of years out of the 10 in which the annual return was the particular value of interest. Here is the table:

Annual Return Values	7	8	9	12
Probabilities	$\frac{1}{10}$	$\frac{3}{10}$	$\frac{4}{10}$	$\frac{2}{10}$

Examples 21.3 and 21.4 suggest a definition for mean value, or average value, of a payoff if the probabilities of the various possible payoff values are known. Specifically, suppose we know possible values of a quantity and their associated probabilities, as shown in the following table:

Values v	v_1	v_2	v_3	$\cdots$	v_n
Probabilities	p_1	p_2	p_3	$\cdots$	p_n

The **mean value** or **expected value** is denoted by the Greek letter μ (mu) and is obtained by multiplying each possible value by its probability and then adding all these products. For the values and probabilities given in the previous table,

$$\mu = v_1 \cdot p_1 + v_2 \cdot p_2 + v_3 \cdot p_3 + \cdots + v_n \cdot p_n$$

Example 21.5

You get a special offer in the mail. A potentially winning ticket (along with special purchase price offers on a whole lot of magazines—no purchase necessary) arrives. One first prize of $1,000,000, 5 second prizes of $100,000, and 10 third prizes of $1,000 will be awarded. Your ticket contains the numbers 69-43-37-01-55-24. You read the fine print, which tells you that each individual was sent a ticket consisting of six distinct numbers between 1 and 70. To win, you must match all six numbers and the order of the numbers on the ticket is not important. The probabilities of winning first, second, and third prizes are, respectively, 7.63×10^{-9}, 3.81×10^{-8}, and 7.63×10^{-8}. To be entered into the drawing and eligible to win, you need to mail back the enclosed ticket containing your numbers to the company. Assuming you don't want to buy any magazines, is it worth the cost of a first-class postage stamp to send in your number?

Solution

We can summarize the prizes and probabilities in a table:

	1st Prize	2nd Prize	3rd Prize
Amount of Prize in $	$1,000,000 = 10^6$	$100,000 = 10^5$	$1,000 = 10^3$
Probability	7.63×10^{-9}	3.81×10^{-8}	7.63×10^{-8}

We'll compute the expected prize payout (in dollars) for each ticket as the sum of the amounts of the prizes times the probability: $\mu = 10^6 \times 7.63 \times 10^{-9} + 10^5 \times 3.81 \times 10^{-8} + 10^3 \times 7.63 \times 10^{-8} = 0.00763 + 0.00381 + 0.0000763 \approx 0.0115$ dollars. This means that, in the long run, if you sent in a large number of tickets of this type, your average payout per ticket would be $1.15¢$. It probably is not worth the cost of postage to return such a ticket.

We can use the expected value to help make decisions among several possible alternatives when there is uncertainty about what may happen. Expected value is appropriate to use when an action will be done repeatedly over an extended period of time and when we can reasonably estimate the probability associated with each state of nature. In that case, looking at an "average" payoff or cost makes sense.

Example 21.6 ■ ■ ■ ■ ■ ■ ■ ■ ■

Consider the scenario described in Example 21.2, where you must decide which of three healthcare options to choose. Suppose the following table gives reasonable probabilities of different healthcare expense levels for a healthy recent college graduate. (This table gives an approximation of the probabilities and a simplification of the healthcare expense levels, but it offers one way to obtain information to help make a decision.)

Total Medical Expenses	$200	$600	$1,200	$3,000	$15,000
Probability	0.50	0.30	0.10	0.07	0.03

a. Use these probabilities and the table from the solution to Example 21.2 (given here) to compute the expected yearly cost of healthcare under each of the plans:

	Medical Expense Level				
	$200	**$600**	**$1,200**	**$3,000**	**$15,000**
Plan 1	560	880	1,000	1,360	3,760
Plan 2	320	720	1,160	1,520	3,920
Plan 3	360	480	660	1,200	4,800

b. Recommend a healthcare plan.

Solution

a. We want to evaluate the expected or average cost under each of the plans. For Plan 1, expected cost $= (0.5)(\$560)+(0.3)(\$880)+(0.1)(\$1,000)+(0.07)(\$1,360)$ $+(0.03)(\$3,760) = \$852$. For Plan 2, expected cost $= (0.5)(\$320)+(0.3)(\$720)$ $+(0.1)(\$1,160) +(0.07)(\$1,520)+ (0.03)(\$3,920) = \$716$. Finally, for Plan 3, expected cost $= (0.5) (\$360)+(0.3)(\$480)+(0.1)(\$660)+(0.07)(\$1,200)$ $+(0.03)(\$4,800) = \$618$.

b. If we have confidence in the probability estimates from part (b), we would probably choose the plan that has the smallest expected cost: Plan 3.

■ ■ ■ ■ ■ ■ ■ ■ ■ ■

It is important to realize that in this situation we must choose a healthcare plan before we know what our healthcare expenses will be for the year. So, we make the decision first and commit to the plan and then experience one of the levels of healthcare expenses. But suppose we didn't have confidence in our ability to predict the probabilities associated with each of the healthcare levels? We'll look at two additional methods that don't require us to estimate probabilities, to help us make our decision. One could be described as an optimist's decision method and the other as a pessimist's decision method.

Suppose we have a problem such as the healthcare decision problem described in Example 21.6, where we want to choose the plan that will result in the least cost for healthcare for the year. Using the optimist's method, we want to consider each plan and record the **minimum cost** for each plan over all expense levels (since we don't know in advance what our healthcare expenses will be). That is, assuming that our medical expense levels will be either $200; $600; $1,200; $3,000; or $15,000, we will be optimistic that we will incur the least cost no matter what plan we choose. The minimum cost for each plan (that is, the minimum in each row) is shown in the last column of the following table:

	Medical Expense Level					Minimum Cost for Each Plan
	$200	$600	$1,200	$3,000	$15,000	
Plan 1	560	880	1,000	1,360	3,760	560
Plan 2	320	720	1,160	1,520	3,920	320
Plan 3	360	480	660	1,200	4,800	360

We will choose the plan for which the minimum cost is the smallest. This is called the **minimin** decision strategy because it is the minimum of all the minimums. It is used when the decision table represents costs. For this example, if we were using the optimistic or minimin strategy, we would select Plan 2.

The pessimist's method for making a decision would consider each plan and ask, "What is the worst that can happen?" that is, what is the **maximum cost** for each plan over all expense levels (since again we don't know in advance what our healthcare expenses will be—that is, we will be pessimistic that we will incur the most cost over each plan). The maximum cost for each plan (that is, the maximum in each row) is shown in the last column of the following table:

	Medical Expense Level					Minimum Cost for Each Plan
	$200	$600	$1,200	$3,000	$15,000	
Plan 1	560	880	1,000	1,360	3,760	3,760
Plan 2	320	720	1,160	1,520	3,920	3,920
Plan 3	360	480	660	1,200	4,800	4,800

We will choose the plan for which the maximum cost is the smallest. The pessimist thinks that the worst will happen but wants to make the smartest choice, given that the worst will happen. This is called the **minimax** decision strategy because it is the minimum of all the maximums. This strategy is used when the decision table represents costs. For this example, if we were using the pessimistic or minimax strategy, we would select Plan 1.

In the next example, we look at another decision involving costs and analyze it using several decision methods.

Example 21.7

We want to invest $5,000 for the next year. There are three possible investments we could make, and three possible states of the economy that would affect the return on our investment. The following table shows the loss, in dollars, that we would experience for each investment and each possible state of the economy:

	State 1	State 2	State 3
Investment 1	0	−300	−200
Investment 2	2,000	700	−600
Investment 3	−500	−1,000	1,000

a. Explain how to interpret a loss of "−500 dollars."

b. What investment would a pessimistic decision maker choose?

c. What investment would an optimistic decision maker choose?

Solution

a. A loss of −500 dollars is actually a profit of $500. A negative loss for the year represents a profit, so a loss of −500 dollars for the year means we end the year with $5,500.

b. The pessimistic decision maker would identify the maximum cost for each row and then choose the decision alternative for which the maximum cost is the smallest. The row maximums are shown in the next table. The pessimist would choose the minimax alternative, that is, Investment 1.

	Maximum
Investment 1	0
Investment 2	2,000
Investment 3	1,000

c. The optimistic decision maker identifies the minimum cost for each row.

	Minimum
Investment 1	−300
Investment 2	−600
Investment 3	−1000

Then he or she chooses the investment alternative for which the minimum is the smallest; that is, the optimist chooses the minimin alternative, which in this example is Investment 3.

In the previous example, we presented a table of the loss (or cost) associated with each investment and state of the economy pair, but we could have just as easily set up the table in terms of profits. If we had used a profit table, the sign of each table entry would change; negative entries would become positive and positive entries would become negative. In Exploration 8, you are asked to revisit this example using a profit table.

We now consider how an optimist and how a pessimist would make a decision if the decision table represents profits instead of costs. With a profit table, the strategy chosen by the optimist is the **maximax** decision strategy, while the pessimist would choose the **maximin** decision strategy.

Example 21.8

a. Describe how an optimistic decision maker would make a decision if the decision table contained profits instead of costs.

b. Describe how a pessimistic decision maker would make a decision if the decision table contained profits instead of costs.

Solution

a. If the table contained profits, an optimistic decision maker would look at the largest profit for each decision alternative. This optimistic decision maker would think that the best thing that could possibly happen (that is, maximum profits) will happen. Then this decision maker would choose the decision alternative for which the maximum is the greatest. This is the maximax decision alternative.

b. If the table contained profits, a pessimistic decision maker would think the worst. This decision maker would look at the minimum profit for each decision alternative. He or she is a pessimist, but is also smart, so the smart pessimist will choose the decision alternative for which the minimum profit is the greatest. This is the maximin decision alternative.

In the next example, we consider a profit table and find what option an optimistic decision maker would choose and what option a pessimistic decision maker would choose.

Example 21.9

The senior class is planning an end-of-the-year activity. The activity must be planned now, before the planners know what the weather will be. The amount of money the class takes in will depend on the weather. The following table gives the amount of money the class will make, in dollars, for each activity and weather condition:

	Sunny and Warm	Cloudy	Rainy	Cold
Indoor Concert	50	100	140	120
Beach Trip	250	150	50	−100
Outdoor Picnic	125	200	−50	−50

a. Suppose an optimistic decision maker prevails. What activity would he or she choose using the maximax decision method?

b. Suppose a pessimistic decision maker is in charge of deciding what activity to plan. What activity would the pessimist choose using the maximin decision method?

c. Suppose we have reliable information that each of the four weather possibilities is equally likely to occur. Find the expected profit for each activity and decide what activity to plan.

Solution

a. First, we find the maximum profit for each activity. The last column of the following table shows the maximum profit for each row:

	Sunny and Warm	Cloudy	Rainy	Cold	Max ($)
Indoor Concert	50	100	140	120	140
Beach Trip	250	150	50	−100	250
Outdoor Picnic	125	200	−50	−50	200

We choose the activity for which the maximum profit is the greatest; this is the beach trip, with a maximum profit, over all weather conditions, of $250.

b. To find what activity a pessimistic decision maker would choose, we find the minimum profit for each activity and then choose the activity for which the minimum profit is the greatest.

	Sunny and Warm	Cloudy	Rainy	Cold	Min
Indoor Concert	50	100	140	120	50
Beach Trip	250	150	50	−100	−100
Outdoor Picnic	125	200	−50	−50	−50

The pessimistic decision maker chooses the indoor concert, which is the maximin choice.

c. If each weather possibility is equally likely to occur, each has a probability of $\frac{1}{4}$. So, the expected profit if we choose the indoor concert is $\frac{1}{4}(50) + \frac{1}{4}(100) + \frac{1}{4}(140) + \frac{1}{4}(120) = 102.5$. Similarly, if we choose the beach trip, the expected profit is $\frac{1}{4}(250) + \frac{1}{4}(150) + \frac{1}{4}(50) + \frac{1}{4}(-100) = 87.5$, and if we choose the outdoor picnic, the expected profit is $\frac{1}{4}(125) + \frac{1}{4}(200) + \frac{1}{4}(-50) + \frac{1}{4}(-50) = 56.25$. Using this decision method, we would choose the indoor concert.

We've discussed several methods for informing decisions involving uncertain information. Ultimately, the decision still rests in the hands of the decision maker. He or she needs to decide which method, if any, to use to help make the decision.

Summary

In this topic, we looked at the concept of expected value, that is, expected profit and expected cost, in situations where uncertainty is involved. The expected value measures the average payoff or cost, in the long run if the activity is repeated over and over. We use estimates of the probability associated with uncertain events to calculate expected value. In situations where we cannot estimate probabilities or if a decision will be made just one time (rather than repeatedly), we can use the optimistic method (minimin if costs are given and maximax if profits are given) or the pessimistic method (minimax, if costs are given and maximin if profits are given). In either case, the decision still lies with the decision maker.

Explorations

1. For each of the following situations, identify the possible courses of action and the unknown states of nature:

   a. A homeowner wants to decide whether to refinance the mortgage on his home now or wait another month or two to apply for a new mortgage.

   b. A traveler who can travel approximately 50 more miles on the gas currently in his car's gas tank needs to decide whether to buy gas at the gas station at the upcoming exit that is 2 miles away or to wait until the following exit that is 30 miles away.

   c. You are driving home from college and need to decide which of two possible routes to take. There is often construction on one route and heavy truck traffic may exist on another of the potential routes and both routes have the potential to be slowed by accidents.

2. A typical roulette wheel has evenly sized "pockets" that are numbered from 1 to 36 and two additional pockets labeled 0 and 00. The 1 through 36 numbered pockets are alternately colored red and black, with the 0 and 00 colored green. You can place many different bets; for example, you can choose a single number, even numbers, or red numbers. Suppose you bet $2 on red; this means that if the ball lands on red after the wheel is spun, you double your money. Otherwise, you lose your $2.

   a. Find your expected payoff for this bet.

   b. Interpret your expected payoff from part (a) of this Exploration.

3. Refer to Example 21.5. How might you justify returning the ticket even though the expected winnings are quite a bit less than the cost of postage?

4. You need to decide how to invest a graduation gift of $1,000. The annual rate of return is given in the next table for each of three different types of investments and three different states of the economy:

	Recession	Stable Economy	Expansion
Investment A	2.5%	2.5%	2.5%
Investment B	2%	4%	5%
Investment C	−2%	4%	10%

   a. Create a table that gives the amount of money for each type of account and state of the economy after one year, if interest is compounded monthly.

   b. Compute the expected value of your account at the end of one year for each of the investment types, if the probability of a recession is 0.5 and the probability of a stable economy is 0.3.

   c. What investment would an optimistic decision maker choose?

   d. What investment would a pessimistic decision maker choose?

5. Consider again the healthcare table set up in the solution of Example 21.6.

   a. How would your decision change if you had assessed the following probabilities for the healthcare expense levels:

Total Medical Expenses	$200	$600	$1,200	$3,000	$15,000
Probability	0.10	0.20	0.20	0.20	0.30

   b. What decision would you make if each healthcare expense level is equally likely to occur?

6. You live near a river that is prone to flooding, so you are considering purchasing flood insurance at $150 per year. You look at storm patterns over the past 10 years and estimate the yearly chance of minor flood damage (approximately $1,000 damage) to be 0.01. The chance of moderate damage (approximately $5,000 damage) is estimated to be 0.005, and the chance of major damage ($20,000 damage) is estimated to be 0.001. The insurance will cover all damages.

   a. Find the expected yearly cost of flood damage if you purchase flood insurance.

   b. Find the expected yearly cost of flood damage if you do not purchase flood insurance.

   c. Do you think you should purchase flood insurance? Justify your answer.

7. A student group must order hot dogs for a concession stand at a sporting event. In the past, demand has been 7, 8, 9, or 10 dozen hot dogs for the event, depending on attendance and the weather. The group needs to order the hot dogs and buns the week before the event and will not be able to return any unused inventory. It purchases the hot dogs and buns for $2 per dozen and sells them for $6 per dozen.

   a. Fill in the following table to show the net profit (profit minus cost) for each of the possible order and demand levels—7, 8, 9, or 10 dozen:

	Demand 7 Dozen	Demand 8 Dozen	Demand 9 Dozen	Demand 10 Dozen
Order 7 Dozen				
Order 8 Dozen				
Order 9 Dozen				
Order 10 Dozen				

    b. Find what size order a pessimistic decision maker would place.

    c. Find what size order an optimistic decision maker would place.

    d. What size order would you place if you knew from past experience that each demand level is equally likely?

8. Consider again Example 21.7.

    a. Rewrite the loss table given in the example as a profit table.

    b. Use your profit table to determine what investment a pessimist would choose.

    c. Use your profit table to determine what investment an optimist would choose.

    d. Explain how your results and table compare with the results of Example 21.7.

9. A proud grandparent has promised to give you $10,000 for a college graduation gift. You need to decide what to do with your gift.

    a. List at least three alternative courses of action you can take.

    b. What states of nature should you consider?

    c. Set up a table of payoffs and describe how you decided what the payoffs should be.

10. A friend of yours falls while you are hiking together and receives a large gash on his leg. You need to help your friend decide what to do.

    a. List at least three alternative courses of action you can take.

    b. What states of nature should you consider?

    c. Set up a table of payoffs and describe how you decided what the payoffs should be.

To Purchase a Warranty
or Not: Making a Decision

In this activity, you will create and evaluate a decision table that will help you decide whether to purchase a limited warranty, an extended warranty, or no warranty at all for a major electronics purchase. You will consider various decision strategies to help you make your decision.

1. Suppose you just purchased a major (expensive) piece of electronic equipment and you are considering whether you want to purchase a two-year limited warranty, a full two-year warranty, or no warranty at all.

    a. What additional information do you need to collect to construct a decision table to help you make your decision?

    b. What are your three decision alternatives? (These alternatives will form the rows of the table.)

2. Suppose that a limited warranty costs $100, has a $50 deductible (you pay the first $50 of repairs), and covers repairs up to $500. (So, insurance pays a maximum of $450. If repairs cost more than $500, you are responsible for the first $50 and any amount greater than $500.) A full warranty costs $200 and covers all repair costs. Repair records have shown that the two-year repair costs can be grouped into three levels of repairs: minor repairs cost around $150; medium-level repairs cost approximately $500; major repairs cost $800.

    a. Fill in the following table, giving your cost for the warranty plus repairs for each decision alternative and each repair level:

	No Repairs Needed	Minor Repairs	Medium-level Repairs	Major Repairs

    b. Suppose you want to use the pessimistic (or minimax) decision strategy. What would you decide to do about purchasing a warranty? Explain how you made this decision.

    c. Suppose you want to use the optimistic (or minimin) decision strategy. What would you decide to do about purchasing a warranty? Explain how you made this decision.

3. You have a friend who has accessed the repair records for this type of electronic equipment. The following table shows the probability of each level of repairs. Use these probabilities and the table you filled in above to evaluate the expected cost for each decision alternative. What would you do about purchasing a warranty using these probabilities and the expected cost to decide?

Repair Level	No Repairs	Minor	Medium-level	Major
Probability	0.75	0.1	0.1	0.05

4. Suppose the probability of each level of repairs is as given in the following table. Use these probabilities to evaluate the expected cost for each decision alternative. What would you do about purchasing a warranty using these probabilities and the expected cost to decide?

Repair Level	No Repairs	Minor	Medium-level	Major
Probability	0.5	0.2	0.15	0.15

5. What would you do if you assumed that the four repair levels were equally likely to occur and you used expected cost to make a decision?

6. After considering several decision strategies, if you had to decide whether to purchase a warranty or not, what would you do? Explain how you made your decision.

7. In analyzing this decision, you made some simplifying assumptions. Explain how you could enhance your analysis.

Summary

In this activity, you looked at a decision that involved uncertain information. You analyzed it using a variety of decision strategies and obtained different recommendations about what to do. You then had to make and defend a decision, based on the analysis and your personal preferences about decision-making methods.

APPENDIX

A

Excel Commands by Activity

Activity 1.3: SATs and the Super Bowl: Creating and Interpreting Histograms

Topic 2

Activity 2.1: Estimating Dates: Scatterplots

Activity 2.2: State Governors' Salaries and Per Capita Income: More on Scatterplots

Topic 3

Activity 3.1: Temperature Patterns: Functions and Line Graphs

Topic 4

Activity 4.1: Blood Alcohol Levels and Credit Cards: Working with More Than Two Variables

Topic 5

Activity 5.1: Rates of Change and Linear Functions

Topic 6

Activity 6.1: The Genie's Offer: Exponential Growth and Linear Growth

Activity 6.2: Lines of Best Fit

Topic 7

Activity 7.1: Richter Scale and Logarithms

INDEX